WILDLIFE FEEDING
AND NUTRITION

ANIMAL FEEDING AND NUTRITION
A Series of Monographs and Treatises

Tony J. Cunha, Editor

Distinguished Service Professor Emeritus
University of Florida
Gainesville, Florida

and

Dean Emeritus, School of Agriculture
California State Polytechnic University
Pomona, California

Tony J. Cunha, SWINE FEEDING AND NUTRITION, 1977

W. J. Miller, DAIRY CATTLE FEEDING AND NUTRITION, 1979

Tilden Wayne Perry, BEEF CATTLE FEEDING AND NUTRITION, 1980

Tony J. Cunha, HORSE FEEDING AND NUTRITION, 1980

Charles T. Robbins, WILDLIFE FEEDING AND NUTRITION, 1983

WILDLIFE FEEDING AND NUTRITION

Charles T. Robbins
Wildlife Biology Program
Department of Zoology
Washington State University
Pullman, Washington

1983

ACADEMIC PRESS, INC.
Harcourt Brace Jovanovich, Publishers

Orlando   San Diego   New York
Austin   Boston   London   Sydney
Tokyo   Toronto

ACADEMIC PRESS, INC.
Orlando, Florida 32887

United Kingdom Edition published by
ACADEMIC PRESS, INC. (LONDON) LTD.
24/28 Oval Road, London NW1 7DX

Library of Congress Cataloging in Publication Data
Main entry under title:

Wildlife feeding and nutrition.

   (Animal feeding and nutrition)
   Includes index.
   1. Animals--Food habits. 2. Animal nutrition.
3. Game and game-birds--Feeding and feeds. 4. Wild
animals, Captive--Feeding and feeds. I. Robbins,
Charles T. II. Series.
QL756.5.W54   599'.013   82-13720
ISBN 0-12-589380-9

PRINTED IN THE UNITED STATES OF AMERICA

87 88 89   9 8 7 6 5 4 3 2

To William, Shorty, and Benjamin

"One thing I have learned in a long life: that all our science, measured against reality, is primitive and childlike— and yet it is the most precious thing we have."
                              —Albert Einstein (1879–1955)

Contents

15 Computer Models of the Nutritional Interaction 333

Foreword

This is the fifth in a series of books on animal feeding and nutrition. The first four were *Swine Feeding and Nutrition, Dairy Cattle Feeding and Nutrition, Beef Cattle Feeding and Nutrition,* and *Horse Feeding and Nutrition.* All five books, and others to follow in this series, are designed to keep the reader abreast of the rapid developments in feeding and nutrition that have occurred in recent years. The volume of scientific literature is expanding rapidly and becomes increasingly larger. As this occurs, interpretation becomes more complex and requires a continuing need for summarization in up-to-date books. This necessitates that top scientists and authorities in the field collate all available information in one volume for each group of animals.

More attention is being paid to the seriousness of the world's food problem. A recent World Bank report indicates that over 1 billion people suffer from chronic malnutrition. Half of them are children under 5 years of age. Every 2½ years, the world's population increases by over 200 million people, which is almost the equivalent of another United States to feed. Animals provide excellent quality protein, plus many important minerals and vitamins in the diet. Wild animals have consistently been used as a food source throughout man's evolutionary history. In many areas of the world today, wildlife continue to be an important source of food. Wildlife rarely compete with domestic animals since they usually rely on food sources not utilized by them. They also cause little, if any, environmental damage when harvested properly to control population densities in the absence of natural controls.

This fifth book in the series, *Wildlife Feeding and Nutrition,* is written by Dr. Charles T. Robbins, who has done an excellent job in assembling a wealth of information pertinent to wildlife nutrition. The text will be invaluable to wildlife biologists, to those who are interested in captive animal nutrition and management, and to those who are interested in improving the feed supply and nutrition of free-ranging wildlife. It will also be valuable to those who are interested in wildlife as creatures of beauty which enrich their joy of nature. The book is unique in that it is the first to bring the subject of wildlife nutrition into one volume. It should be very helpful to undergraduate and graduate students as well as teachers of biology and wildlife management. The book will be a useful reference for all who are interested and concerned with wildlife throughout the world.

Tony J. Cunha

Preface

Wildlife Feeding and Nutrition fills a serious gap in the wildlife and animal nutrition literature by providing a discussion of the basic principles of nutrition and their application to the broader field of wildlife ecology. This book is based on lectures presented in an upper-level wildlife nutrition course taught at Washington State University. Consequently, the needs and interests of the students in that course are reflected in the text.

For example, although my own research has been on the nutrition and ecology of ungulates, wildlife undergraduates typically are interested in the entire natural world and all of its life forms. Thus, in order to support and expand the undergraduate's appetite for knowledge, this book is written from a very broad comparative perspective as it focuses on both wild mammals and birds. The wildlife nutritionist often does have much more of a comparative perspective than does the domestic animal nutritionist. Unfortunately, although many of the principles of nutrition could be learned by wildlife students taking the standard animal nutrition course taught in the animal science department, most of these departments have chosen virtually to ignore the nondomesticated and nonlaboratory species. Therefore, because wildlife students usually lack any nutritional background, the various chapters start at an introductory level and progress to a more advanced level that will challenge the advanced wildlife student and professional biologist.

ACKNOWLEDGMENTS

Many people have contributed to my studies and to the writing of this text. Julius Nagy of Colorado State University first stimulated my interest in wildlife nutrition while I was an undergraduate. Aaron Moen of Cornell University supported and encouraged my further progress as a graduate student. He also introduced me to the all-encompassing topic of carrying capacity. While at Cornell, I thoroughly enjoyed and benefitted from my association with two particularly outstanding animal scientists, Peter J. Van Soest and J. Tom Reid. Since coming to Washington State, James R. King and his students have been helpful in providing a stimulating exchange of ideas. Since all university research programs are dependent upon students of the highest caliber, I am indebted to those of mine who have investigated several topics contained within this book. They include Yosef Cohen, Eric Mould, Katherine Parker, Mark Wickstrom, and Don Spalinger. Their research has been supported by the National Science Foundation, Washington Department of Game, and U.S. Forest Service. Very thorough and helpful reviews of the manuscripts were provided by Aaron Moen, Tom Hanley, D. E. Ullrey, D. C. Church, Bruce Watkins, Mary E. Allen, James R. King, and Frederick F. Gilbert. Finally, I wish to thank my parents and grandparents for introducing me to the natural world and my wife, Barb, for providing me the time necessary to complete this book and sharing with me the fun and frustrations of research.

Charles T. Robbins

1

Introduction

The physiology of wild animals is almost entirely unknown. . . . Our
understanding of food and water is limited at the outset by our deficient
understanding of game physiology.
                                        —Leopold, 1933, pp. 253, 302

Knowledge of wildlife nutrition, as a component of both wildlife ecology and
management, is central to understanding the survival and productivity of all
wildlife populations, whether free-ranging or captive. Although it is difficult to
identify the earliest interest in wildlife nutrition, the science of wildlife nutrition
is an extremely young area of investigation. The historical roots in North America largely began during the late 1870s and early 1880s when biologists, primarily ornithologists and entomologists, started investigating the food habits of
wildlife in relation to the welfare of human beings (McAtee, 1933). This new
area of investigation was entitled *economic ornithology* because of the effort to
relate the ingestion of agricultural crops or insects to the economic benefit or
detriment of the farmer.

Economic ornithology officially began in 1885 when Congress instructed the
Department of Agriculture to initiate "the study of [the] interrelation of birds and
agriculture, an investigation of the food, habits, and migration of birds in relation
to both insects and plants [McAtee, 1933, p. 114]." Because of this early
legislative and economic focus, studies of food habits were the major efforts of
most early nutritionists (Fig. 1.1). Although techniques and emphasis on food
habits research have changed over the years, food habit studies have continued to
be a major percentage of all wildlife nutrition investigations. However, classical
food habit studies usually tell us only what has been eaten and rarely how much,
for what reason, or the physiological role or importance of the different ingested
foods (Leopold, 1933). Thus, the use of only food habits information to develop
management schemes is all too often destined to failure because of the absolute
need to understand the much broader nutritional interaction from an ecological
perspective. Unfortunately, preoccupation with food habits has reduced our investigations of other equally important areas of nutrition (Bartholomew and

1

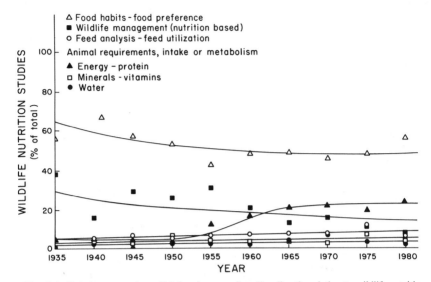

Fig. 1.1. Relative proportion of bird and mammal studies directly relating to wildlife nutrition published in 5-year intervals since 1935. Statistics were compiled by reviewing all titles (and abstracts if available) published in *Wildlife Reviews* for the appropriate years and assigning each study to one of six areas on the basis of its major contribution. The compilation is based on the assumptions that the content of an article can be adequately judged from the title, that the abilities of authors to correctly title their articles have not changed, and that the biases by *Wildlife Reviews* in surveying the available literature have been constant.

Cade, 1963). Similarly, early attempts to understand wildlife productivity based on correlations to soil fertility (Albrecht, 1944; Denney, 1944; Crawford, 1949; Williams, 1964) were not adequate to develop an understanding of nutritional interactions and, therefore, provide a base for meaningfully altering the animal–environment interaction (Jones *et al.*, 1968). Plants and the subsequent nutritional interactions are not mere passive reflectors of soil quality. Wildlife researchers were warned as early as 1937 (McAtee, 1937) that they would ''do well to obtain the cooperation in their experimental work of experts in animal nutrition [p. 37].''

The number of wildlife nutrition articles has increased markedly since 1937 (Fig. 1.2). Their increasing number should be indicative of a growing body of knowledge relative to all aspects of wildlife nutrition. Food habits and wildlife management studies based on nutritional perspectives have averaged 73% (ranging from 94 to 59%) of all such inquiries. The absolute and relative number of studies in these two areas is probably underestimated because the content of papers beginning with ''The ecology . . .'' or ''The biology . . .'' cannot always be identified. Nutritional investigations, especially of food habits, may be included in many of these articles. Water requirement, intake, and metabolism investigations have averaged 3% (0–7%) of the nutrition articles in any one year, minerals and vitamins 4% (0–7%), and feed analysis or utilization studies 7% (0–12%). Research dealing with energy or protein requirements, intake, or metabolism averaged only 5% of all studies between 1935 and 1950, but increased markedly after 1950 and averaged 19% between 1955 and 1980. The increase in energy and protein studies is primarily due to the proliferation of bioenergetic investigations dating back to Brody (1945) and Scholander *et al.* (1950).

The relative decline in management programs or studies with a nutritional basis partially reflects the decline in the popularity of food plantings for wildlife. While it is certainly saddening to see such a decline, the increasing numbers of inquiries in other equally important areas will strengthen the entire field of wildlife nutrition. The pyramid of the science of wildlife nutrition must rest firmly on a base of studies of animal physiology and ecology, with management programs at the apex. Only then will efforts to maximize management returns by altering wildlife productivity through nutrition be rewarded.

Nutrition research and management of wildlife species offer many challenges not always encountered by the scientist working with domestic animals, primarily because of the need to maintain an ecological perspective in designing and implementing any wildlife nutrition research (Watson, 1973). The free-ranging animal is in a constantly changing environment, and human beings have little control over many of the animal–environment interactions. For example, a survey of southern and eastern wildlife management agencies indicated that 90% or more of the total bobwhite quail production and harvest occurred without deliber-

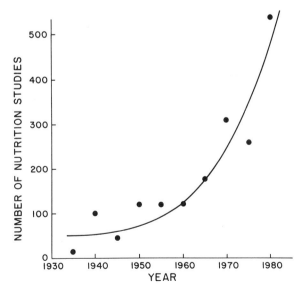

Fig. 1.2. Total number of wildlife nutrition studies per year that were cited in *Wildlife Reviews* in 5-year intervals beginning in 1935.

ate management efforts (Frye, 1961). The lack of control over wild free-ranging animals often requires that the wildlife ecologist have a greater base of knowledge to understand interactions and implement meaningful management programs than the scientist working with domestic animals. The wildlife ecologist often has a minimal incentive relative to the domestic-animal scientist to maximize meat production or economic return, but may attempt simply to understand and appreciate an ecological system. However, many nutrition results generated in such basic investigations are often essential to understanding other living systems and, as such, are the grist of applied research. Although one might assume that our nutritional knowledge of captive animals is complete because of our long history of captive animal efforts, 60 to 70% of the animals dying in captivity die because of poor management and husbandry, with nearly 25% dying from nutritional problems (Ratcliffe, 1966; Wallach, 1970).

The use of captive wild animals in nutrition research often requires special facilities and great perseverence in handling or training the animals. Although captive and instrumented wild animals may provide the only means for answering many questions, one must always be concerned about the effects of captivity and handling upon the results. The nutrition of captive primates, particularly the rhesus monkey (*Macaca mulatta*) (Harris, 1970), and common laboratory rodents, such as the guinea pig, rat, and mouse, has been studied the most, with the emphasis on these species as models for understanding human nutrition. While these studies are very applicable to understanding captive primate or rodent nutrition, their lack of an ecological perspective suitable for increasing our knowledge of free-ranging animals is unfortunate.

Wildlife nutrition overlaps with many other disciplines. Environmental physiology, wildlife ecology and management, and range and forest management often produce or implement wildlife nutrition results. Wildlife nutrition is indeed a science, since the nutritional interactions between the animal and its environment are not random events but highly predictable interactions forming the basis for the science of wildlife nutrition. However, the application of much of the wildlife nutrition data to field management is both an art and a science because of the lack of adequate knowledge of many control mechanisms determining the outcome of any manipulation. Many of the problems encountered by the ecologist or animal and land manager may involve basic nutritional questions about starvation, competition, winter feeding of wildlife, diet formulation, habitat manipulations such as clearcuts, fertilization, and reseeding, predator–prey interactions, and carrying capacity estimations. Consequently, wildlife nutrition is a basic and yet broad field of investigation with many challenges to be met.

REFERENCES

Albrecht, W. A. (1944). Soil fertility and wildlife–cause and effect. *Trans. North Am. Wildl. Conf.* **9,** 19–28.

Bartholomew, G. A., and Cade, T. J. (1963). The water economy of land birds. *Auk* **80,** 504–539.

Brody, S. (1945). "Bioenergetics and Growth." Hafner, New York.

Crawford, B. (1949). Relationships of soils and wildlife. *Missouri Cons. Comm., Circ. No.* **134,** 10–18.

Denney, A. H. (1944). Wildlife relationships to soil types. *Trans. North Am. Wildl. Conf.* **9,** 316–322.

Frye, O. E., Jr. (1961). A review of bobwhite quail management in eastern North America. *Trans. North Am. Wildl. Nat. Resour. Conf.* **26,** 273–281.

Harris, R. S. (Ed.). (1970). "Feeding and Nutrition of Nonhuman Primates." Academic Press, New York.

Jones, R. L., Labisky, R. F., and Anderson, W. L. (1968). Selected minerals in soils, plants, and pheasants: An ecosystem approach to understanding pheasant distribution in Illinois. *Ill. Nat. Hist. Surv., Biol. Notes No.* **63,** 8 pp.

Leopold, A. (1933). "Game Management." Charles Scribner's Sons, New York.

McAtee, W. L. (1933). Economic ornithology. *In* "Fifty Years' Progress of American Ornithology 1883–1933," pp. 111–129. American Ornithologists' Union 50th Anniversary, New York, N.Y. November 13–16, 1933.

McAtee, W. L. (1937). Nutritive value of maize. *Wildl. Rev.* **9,** 36–38.

Ratcliffe, H. L. (1966). Diets for zoological gardens: Aids to conservation and disease control. *Int. Zoo Ybk.* **6,** 4–23.

Scholander, P. F., Hock, R., Walters, V., and Irving, L. (1950). Adaptation to cold in arctic and tropical mammals and birds in relation to body temperature, insulation, and basal metabolic rate. *Biol. Bull.* **99,** 259–271.

Wallach, J. D. (1970). Nutritional diseases of exotic animals. *J. Am. Vet. Med. Assoc.* **157,** 583–599.

Watson, A. (1973). Discussion: Nutrition in reproduction–Direct effects and predictive functions. *In* "Breeding Biology of Birds" (D. S. Farner, ed.), pp. 59–68. Nat. Acad. Sci., Washington, D.C.

Williams, C. E. (1964). Soil fertility and cottontail body weight: A reexamination. *J. Wildl. Manage.* **28,** 329–338.

2

General Nutrient and Energy Requirements

The number of species whose nutritional requirements are known with any
precision is relatively few. Of the mammals only about a dozen species have been
studied out of a total of over 5000; the situation with birds is worse.
—Evans and Miller, 1968, p. 121

Wildlife nutrition provides an understanding of specific biochemical and biophysical interactions critical to the survival and productivity of individuals and
populations. Nutrition is the process whereby the animal processes portions of its
external chemical environment for the functioning of internal metabolism. All
animals are located somewhere on an internal tissue metabolism gradient (Fig.
2.1). The wildlife nutritionist need not consider catabolism and weight loss as
undesirable but rather as essential components of life strategies of many wild
animals (Le Maho, 1977; Sherry *et al.*, 1980). However, the position of an
animal along the gradient represents a dynamic balancing between cellular and
organismal requirements and the rate and efficiency at which specific components of the external environment can be acquired.

Wildlife nutritionists are primarily interested in the basic biochemical– biophysical interactions between the animal and its environment. These interactions
may be grouped into five major nutritional categories of constituents that animals
must acquire from their external environments. These are energy, which is
discussed here, and protein (either amino acids or, in some cases, nonprotein
nitrogen), water, minerals, and vitamins which are discussed in Chapters 3–6.
Other items are also necessary for proper metabolic functioning, such as essential
fatty acids, particularly linoleic acid (Maynard and Loosli, 1969; McDonald *et
al.*, 1973; Goodwin, 1976). Thus, the nutritionist dealing with extremely purified diets should be aware that there are requirements in addition to the five
broad nutrient categories.

Energy, the capacity to do work or produce motion against a resisting force, is
of interest to the wildlife nutritionist since biochemical transformations, muscle
contractions, nerve impulse transmissions, excretion processes, and all other
active body functions require energy. Although physicists make no ultimate
distinction between energy and matter, nutritionists treat energy, particularly
chemical energy, as a property of matter. Energy transformations may be de-

6

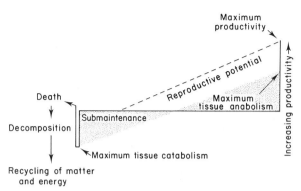

Fig. 2.1. The productivity gradient experienced by all wildlife. (From *Wildlife Ecology: An Analytical Approach* by Aaron N. Moen. W. H. Freeman and Company. Copyright © 1973.)

scribed by the laws of thermodynamics. The first law is that energy can be neither created nor destroyed but merely changed in form. Thus, if a specific amount of one form of energy disappears (such as chemical energy), an equal amount of another form of energy (such as heat) must appear.

The second law of thermodynamics is that energy transformations produce heat and, therefore, increase entropy within the system. Simply stated, chemical energy transformations are not 100% efficient in the production of chemical work. Although many machines can use heat to perform work, the living cell cannot. Heat is useful only in maintaining the relatively high, constant body temperature of endotherms. Fortunately, a significant portion of the chemical energy liberated during catabolic processes can be used to synthesize other useful forms of chemical energy, particularly adenosine triphosphate (ATP). The total amount of useful free energy and heat released when organic compounds are oxidized is totally independent of the speed of the reaction, but dependent only on initial and final states. Thus, a gram of glucose will yield the same amount of energy irrespective of whether it is burned in a flame or in the controlled reactions of the animal body as long as the end products, carbon dioxide and water, are the same. This is often referred to as Hess's law of constant heat summation.

Because of the constant summation of heat in biological systems, the measurement of chemical energy as the heat of combustion is most frequently used to evaluate animal energetics. The most commonly used measures of heat in nutrition have been the calorie (1000 cal = 1 kcal) and the joule (1 cal = 4.184 joules). The calorie is the amount of heat needed to raise the temperature of 1 g of water from 14.5° to 15.5°C. The bomb calorimeter (Fig. 2.2) is used to measure the amount of heat released when a sample of plant or animal tissue is completely oxidized. The sample is placed in a combustion chamber containing excess oxygen, which is immersed in an insulated water jacket and ignited. The temperature rise in the surrounding water is measured and is proportional to the sample's chemical energy content. For instance, when a 1 g sample is ignited and produces a temperature rise of 4°C in 1000 g of water, the chemical energy content is 4 kcal. The chemical energy content of the various organic compounds in Table 2.1 varies inversely with the carbon to hydrogen ratio because carbon when oxidized to carbon dioxide yields approximately one-quarter the energy of hydrogen when oxidized to water. Similarly, molecular oxygen and nitrogen reduce the energy content because oxygen in itself is not energy yielding and is readily available via respiration, and nitrogen, while oxidized in the calorimeter, is not oxidized in the body.

Differences in the chemical or gross energies of either plants or animals are due to the energy contents of the specific chemical compounds and their relative proportions. For example, the energy content of dried animal tissue increases as the fat content increases and decreases as the carbohydrate or mineral content increases. Plants contain many different compounds, but the gross energies of plant tissue are often very uniform (Table 2.2). The higher energy content of

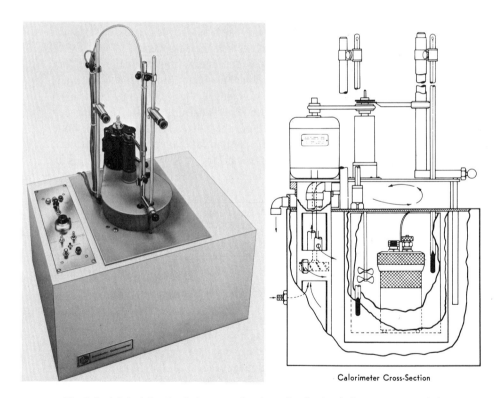

Calorimeter Cross-Section

Fig. 2.2. Adiabatic bomb calorimeter used to determine the chemical energy content of plant or animal samples. (Courtesy of Parr Instrument Company, Moline, Illinois.)

roots, seeds, and evergreen and alpine communities is due to their higher content of oils or fats, waxes, resins, and other high-energy compounds. Since the degree of dietary energy use may vary from 0 to 100% depending on the completeness of digestion and oxidation, gross energies must be further evaluated to understand animal energetics.

Although energy is required during all chemical transformations within the animal body, estimations of the animal's energy requirements would be an

TABLE 2.1
Chemical Energy Content of Dietary Substances and Metabolic or Excretion Products[a]

Element or substance	Energy content (kcal/g)	Composition (% weight)			
		Carbon	Hydrogen	Oxygen	Nitrogen
Carbon	8.0	100			
Hydrogen	34.5		100		
Methane	13.3	75	25		
Glucose	3.75	40	7	53	
Sucrose	3.96	42	6	52	
Starch and glycogen	4.23	45	6	49	
Cellulose	4.18	45	6	49	
Acetic acid	3.49	40	7	53	
Propionic acid	4.96	49	8	43	
Butyric acid	5.95	55	9	36	
Palmitic acid	9.35	75	13	12	
Stearic acid	9.53	76	13	11	
Oleic acid	9.50	76	12	11	
Average triglyceride[b]	9.45	75	13	12	
Glycine	3.11	32	7	42	19
Alanine	4.35	40	8	36	16
Tyrosine	5.92	60	6	26	8
Average protein[b]	5.65	52	7	23	16
Urea	2.53	20	7	26	47
Uric acid	2.74	36	2	29	33
Creatine	4.24	37	7	24	32
Creatinine	4.60	43	6	14	37
Oxalic acid	0.67	27	2	71	
Citric acid	2.48	38	4	58	
Succinic acid	3.00	41	5	54	
Acetaldehyde	6.12	55	9	36	

[a]From Brody, 1945; Maynard and Loosli, 1969.

[b]The average caloric content of complex mixtures of triglycerides, proteins, or their constituents is dependent on the concentration and caloric content of each component. Consequently, these values should not be used as constants throughout the animal and plant kingdoms.

TABLE 2.2
Average Energy Content of Plant Parts or Communities[a]

Plant part or community	Number of samples	Average energy content (kcal/g)	Coefficient of variation
Leaves	260	4.229	0.116
Stems and branches	51	4.267	0.081
Roots	52	4.720	0.092
Litter	82	4.298	0.104
Seeds	22	5.065	0.219
Tropical rain forest	15	3.897	0.060
Herb old field	35	4.177	0.096
Poa old field	115	4.075	0.064
Pinus sylvestris stand	14	4.787	0.078
Alpine meadow	3	4.711	0.005
Alpine *Juncus* dwarf heath	2	4.790	0.003

[a]From Golley (1961).

extremely tedious process if each biochemical reaction were measured. Since the oxidation of most organic compounds requires oxygen and produces carbon dioxide and heat, measurements of these parameters in the living animal indicate the energy required for all ongoing metabolic transformations. Similarly, net energy retention as growth or other productive processes is equal to the gross energies of the tissues or products. Thus, one can speak of energy requirements of the whole animal for basal metabolism, activity, thermoregulation, growth, reproduction, lactation or brooding, pelage or plumage growth, the support of parasites, and other energy-demanding processes even though each of these includes many individual energy transformations.

REFERENCES

Brody, S. (1945). "Bioenergetics and Growth." Hafner, New York.
Evans, E., and Miller, D. S. (1968). Comparative nutrition, growth and longevity. *Proc. Nutr. Soc.* **27**, 121–129.
Golley, F. B. (1961). Energy values of ecological materials. *Ecology* **42**, 581–584.
Goodwin, L. G. (1976). Research. *Symp. Zool. Soc. London* **40**, 215–222.
Le Maho, Y. (1977). The emperor penguin: A strategy to live and breed in the cold. *Am. Sci.* **65**, 680–693.
McDonald, P., Edwards, R. A., and Greenhalgh, J. F. D. (1973). "Animal Nutrition." Longman, London.
Maynard, L. A., and Loosli, J. K. (1969). "Animal Nutrition." McGraw-Hill, New York.
Moen, A. N. (1973). "Wildlife Ecology: An Analytical Approach." Freeman, San Francisco.
Sherry, D. F., Mrosovsky, N., and Hogan, J. A. (1980). Weight loss and anorexia during incubation in birds. *J. Comp. Physiol. Psychol.* **94**, 89–98.

3

Protein

Proteins are major constituents of the animal's body. They are important constituents of animal cell walls (whether in soft tissues or in feathers, hair, and bone) and active as enzymes, hormones, and lipoproteins in fat transport; as antibodies and clotting factors in blood; and as carriers in active transport systems. Thus, *protein* is a very general term encompassing a heterogeneous grouping of compounds having many different functions in the body. A continuous supply of dietary protein must be available to the animal for these functions.

The unifying principle of protein composition is that they are all composed of amino acids. Classically, proteins are mixtures of 20–25 different amino acids (Table 3.1), although over 200 nonprotein amino acids also exist (Bell, 1972; Fowden, 1974). Amino acids are linked via peptide bonds

$$-\text{N}-\text{C}-$$
$$\begin{array}{cc} | & \| \\ \text{H} & \text{O} \end{array}$$

to form plant and animal proteins. Since animal and plant proteins are usually degraded during digestion to their constituent amino acids or very small peptide groups by enzymatic hydrolysis of the peptide bond, the specific kind of protein is not as important as the amino acid content and availability relative to animal requirements. The relative concentrations and degradation rates of amino acids in the animal's body should be related to the amino acids required in the diet (Dunn *et al.*, 1949; Williams *et al.*, 1954). Although differences in the amino acid content of animals do exist (Table 3.2) and could reflect differing amino acid requirements, many of the differences are relatively small when compared to the major anatomical differences between the species.

Some amino acids can be produced within the animal, and some must be ingested. Those that cannot be produced by the animal in sufficient amounts to meet requirements and, therefore, must come from the diet are called *essential* amino acids. Those that can be produced adequately from nonspecific precursors within the animal, are called *nonessential* amino acids. Animals with simple

TABLE 3.1
Names and Structures of the Most Important Amino Acids

Amino acid	Formula	Structure
Aliphatic		
Glycine	$C_2H_5O_2N$	$\begin{array}{c} NH_2 \\ \mid \\ H-C-COOH \\ \mid \\ H \end{array}$
Alanine	$C_3H_7O_2N$	$\begin{array}{c} NH_2 \\ \mid \\ CH_3-C-COOH \\ \mid \\ H \end{array}$
Serine	$C_3H_7O_3N$	$\begin{array}{c} NH_2 \\ \mid \\ HO-CH_2-C-COOH \\ \mid \\ H \end{array}$
Threonine	$C_4H_9O_3N$	$\begin{array}{c} H\ \ NH_2 \\ \mid\ \ \ \mid \\ CH_3-C-C-COOH \\ \mid\ \ \ \mid \\ OH\ H \end{array}$
Valine	$C_5H_{11}O_2N$	$\begin{array}{c} CH_3 \diagdown \quad NH_2 \\ \quad CH-C-COOH \\ CH_3 \diagup \quad\ \ H \end{array}$
Leucine	$C_6H_{13}O_2N$	$\begin{array}{c} CH_3 \diagdown \qquad\quad NH_2 \\ \quad CH-CH_2-C-COOH \\ CH_3 \diagup \qquad\quad H \end{array}$
Isoleucine	$C_6H_{13}O_2N$	$\begin{array}{c} CH_3-CH_2 \diagdown \quad NH_2 \\ \qquad\quad CH-C-COOH \\ \qquad CH_3 \quad\ \ H \end{array}$
Basic		
Histidine	$C_6H_9O_2N_3$	$\begin{array}{c} NH_2 \\ \mid \\ CH=C-CH_2-C-COOH \\ \mid\quad \mid \qquad\quad \mid \\ NH\ \ N \qquad\quad H \\ \diagdown\!\!\diagup\!\!/ \\ CH \end{array}$
Arginine	$C_6H_{14}O_2N_4$	$\begin{array}{c} NH_2 \\ \mid \\ NH_2-C-NH-CH_2-CH_2-CH_2-C-COOH \\ \parallel \qquad\qquad\qquad\qquad\qquad \mid \\ NH \qquad\qquad\qquad\qquad\qquad\ H \end{array}$
Lysine	$C_6H_{14}O_2N_2$	$\begin{array}{c} NH_2 \\ \mid \\ NH_2-CH_2-CH_2-CH_2-CH_2-C-COOH \\ \mid \\ H \end{array}$
Aromatic		
Phenylalanine	$C_9H_{11}O_2N$	$\begin{array}{c} NH_2 \\ \mid \\ \bigcirc\!-CH_2-C-COOH \\ \mid \\ H \end{array}$

TABLE 3.1 *Continued*

Amino acid	Formula	Structure
Tyrosine	$C_9H_{11}O_2N$	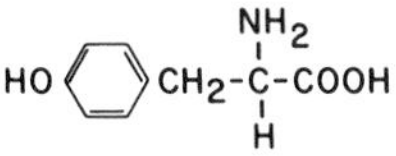 HO⟨⟩CH$_2$-C-COOH with NH$_2$ and H
Sulfur-containing		
Cysteine	$C_3H_7O_2NS$	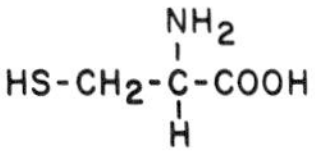 HS-CH$_2$-C-COOH with NH$_2$ and H
Cystine	$C_6H_{12}O_4N_2S_2$	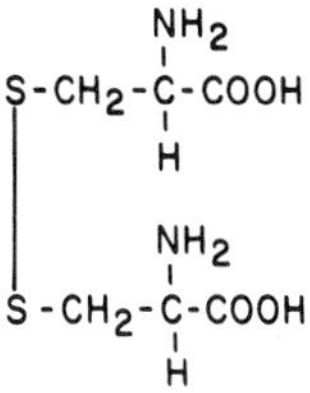 S-CH$_2$-C-COOH (NH$_2$, H) / S-CH$_2$-C-COOH (NH$_2$, H)
Methionine	$C_5H_{11}O_2NS$	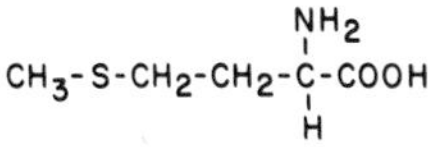 CH$_3$-S-CH$_2$-CH$_2$-C-COOH with NH$_2$ and H
Heterocyclic		
Tryptophan	$C_{11}H_{12}O_2N_2$	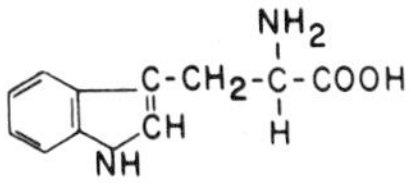 C-CH$_2$-C-COOH, CH, NH (indole ring) with NH$_2$, H
Proline	$C_5H_9O_2N$	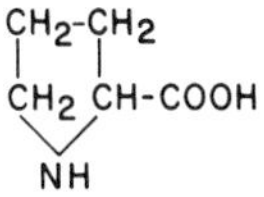 CH$_2$-CH$_2$ / CH$_2$ CH-COOH / NH
Hydroxyproline	$C_5H_9O_3N$	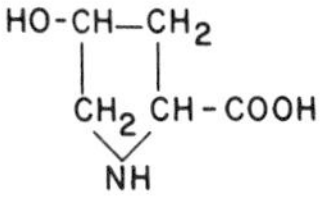 HO-CH—CH$_2$ / CH$_2$ CH-COOH / NH
Acidic		
Aspartic acid	$C_4H_7O_4N$	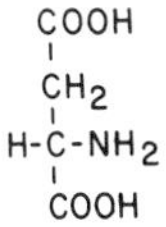 COOH / CH$_2$ / H-C-NH$_2$ / COOH
Asparagine	$C_4H_8O_3N_2$	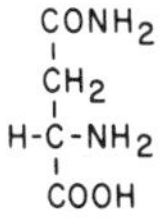 CONH$_2$ / CH$_2$ / H-C-NH$_2$ / COOH

(*continued*)

TABLE 3.1 *Continued*

Amino acid	Formula	Structure
Glutamic acid	$C_5H_9O_4N$	$COOH$ CH_2 CH_2 $H-C-NH_2$ $COOH$
Glutamine	$C_5H_{10}O_3N_2$	$CONH_2$ CH_2 CH_2 $H-C-NH_2$ $COOH$

stomachs usually require 10 essential amino acids: arginine, histidine, iso-leucine, leucine, threonine, lysine, methionine, phenylalanine, tryptophan, and valine. These are essential because of the inability or limited rate of the animal's metabolic pathways to produce certain ring structures and molecular linkages.

TABLE 3.2
Amino Acid Composition (%) of the Body Protein of Several Species[a]

Amino acid	Cat	Chicken	Duck	Mink	Mouse	Pig	Rabbit	Rat	White-tailed deer
Arginine	6.3	5.8	5.9	6.4	5.9	7.1	6.3	5.9	—
Aspartic acid	8.7	8.2	9.8	7.3	8.8	—	8.9	8.1	10.3
Glutamic acid	11.5	11.5	11.2	12.7	12.9	—	12.7	13.9	15.6
Glycine	10.0	10.2	8.2	8.4	10.4	—	8.9	10.0	10.0
Histidine	1.7	1.6	1.8	1.9	1.8	2.6	1.8	2.2	3.6
Isoleucine	3.3	4.2	4.7	3.0	3.7	3.8	3.6	3.5	3.6
Leucine	6.1	7.2	7.5	6.7	7.3	7.1	6.6	6.5	8.5
Lysine	5.6	6.3	6.0	5.7	7.9	8.5	6.6	7.6	8.0
Methionine	1.8	1.6	1.9	1.7	2.0	1.8	1.9	1.7	1.5
Phenylalanine	3.2	4.2	4.0	3.5	3.6	3.8	3.4	3.7	4.4
Threonine	3.9	3.7	4.3	3.7	3.7	3.8	4.1	3.9	4.5
Valine	4.4	5.4	5.7	4.4	4.5	6.0	4.7	5.5	4.5
Serine	—	—	—	3.8	—	—	—	—	4.1
Proline	—	—	—	5.9	—	—	—	—	5.1
Alanine	—	—	—	5.5	—	—	—	—	8.7
Cystine	—	1.7	—	2.0	—	1.0	—	1.5	0.1
Tyrosine	—	2.5	—	2.9	—	2.6	—	2.9	2.7

[a]Data from Dunn *et al.*, 1949; Williams *et al.*, 1954; Levin *et al.*, 1973; Glem-Hansen, 1979. Note that differing methods of protein hydrolysis and amino acid analyses between various authors may affect the results.

The requirements for essential amino acids relative to necessary dietary intake are reduced or eliminated in ruminants, animals with active cecal fermentation (such as grouse and equines), and animals practicing coprophagy (such as many rodents and rabbits). Bacterial modification of dietary protein within the gastrointestinal tract or the production of amino acids from other nitrogen sources, such as urea, reduces their essential amino acid requirements.

Although amino acids are essential for body functioning, very few wildlife studies have examined amino acid composition or requirements (Crawford *et al.*, 1968; Krapu and Swanson, 1975; Parrish and Martin, 1977). The lack of amino acid studies may become increasingly troublesome as wildlife management intensifies and as zoological gardens increase their use of processed diets. Wildlife nutritionists have typically determined the nitrogen content of a food in order to provide an index of its protein content. Since proteins characteristically have 16% nitrogen (ranging from 18.9% for nuts to 15.7% for milk protein), the nitrogen content multiplied by 6.25 (or 100/16) is termed the crude protein content. The estimate is crude since not all plant or animal nitrogen is in the form of protein (other examples are nucleic acids and nitrates) and the exact nitrogen to protein ratio may not be known. Estimates of protein requirements of animals for maintenance and production are in many ways more difficult to make than estimates of energy requirements because dietary protein use may depend on its amino acid composition, the proportion of protein in relation to usable energy within the diet, and the total amount of food consumed (Payne, 1967; Parrish and Martin, 1977). Further, amino acids can be deaminated and used as an energy source when dietary energy is insufficient. Thus, estimations of protein requirements without consideration of energy and other dietary nutrients may be of little value.

Scientists who work with domestic animals often express protein requirements as a percentage of the diet. However, free-ranging wildlife are often on rather varied diets because of changes in range condition, and the ratios of usable energy to protein may vary widely. Similarly, the animal may not be able to consume an unlimited food source. Thus, the estimation and understanding of protein requirements and efficiencies of dietary utilization by free-ranging animals are important areas of investigation.

REFERENCES

Bell, E. A. (1972). Toxic amino acids in the Leguminosae. *In* "Phytochemical Ecology" (J. B. Harborne, ed.), pp. 163–177. Academic Press, New York.

Crawford, M. A., Patterson, J. M., and Yardley, L. (1968). Nitrogen utilization by the cape buffalo (*Syncerus caffer*) and other large mammals. *Symp. Zool. Soc. London* **21**, 367–379.

Dunn, M. S., Camien, M. N., Malin, R. B., Murphy, E. A., and Reiner, P. J. (1949). Percentages of twelve amino acids in blood, carcass, heart, kidney, liver, muscle, and skin of eight animals. *Univ. Calif. Publ. Physiol.* **8**, 293–325.

Fowden, L. (1974). Nonprotein amino acids from plants: Distribution, biosynthesis, and analog functions. *In* "Metabolism and Regulation of Secondary Plant Products. Recent Advances in Phytochemistry, Vol. 8" (V. C. Runeckles and E. E. Conn, eds.), pp. 95–122. Academic Press, New York.

Glem-Hansen, N. (1979). Protein requirement for mink in the lactation period. *Acta Agric. Scandinavica* **27,** 129–138.

Krapu, G. L., and Swanson, G. A. (1975). Some nutritional aspects of reproduction in prairie nesting pintails. *J. Wildl. Manage.* **39,** 156–162.

Levin, E., Collins, V., Varner, D. S., Williams, G., and Hardenbrook, H. J. (1973). Dietary protein for man and animal. *Illinois Veterinarian* **16,** 10–14.

Parrish, J. W., Jr., and Martin, E. W. (1977). The effect of dietary lysine level on the energy and nitrogen balance of the dark-eyed junco. *Condor* **79,** 24–30.

Payne, P. R. (1967). The relationship between protein and calorie requirements of laboratory animals. *In* "Husbandry of Laboratory Animals" (M. L. Conalty, ed.), pp. 77–95. Academic Press, New York.

Williams, H. H., Curtin, L. V., Abraham, J., Loosli, J. K., and Maynard, L. A. (1954). Estimation of growth requirements for amino acids by assay of the carcass. *J. Biol. Chem.* **208,** 277–286.

4

Water

Water is one of the most important essential nutrients because of the variety of its functions and magnitude of its requirements (Maynard and Loosli, 1969). Water is essential within the animal body as a solvent and is involved in hydrolytic reactions, temperature control, transport of metabolic products, excretion, lubrication of skeletal joints, and sound and light transport within the ear and eye (Robinson, 1957). Water comprises 99% of all molecules within the animal's body (MacFarlane and Howard, 1972). The very high molecular concentration of water within the body is, in part, due to the very small size of water molecules relative to protein, fat, or carbohydrate molecules. Although the actual gravimetric concentration of water within the body is lower than the molecular concentration, the neonatal bird and mammal often has a water concentration between 71 and 85% of its body weight (Sugden and Harris, 1972; Fedyk, 1974; Sawicka-Kapusta, 1970, 1974; Robbins *et al.*, 1974; Kaufman and Kaufman, 1975). The very high water content of the young animal generally decreases as the animal grows and matures (Moulton, 1923). Water concentrations in very fat, adult animals can be as low as 40%.

The body water content of both mammals and birds is quite similar and is a linear function of increasing species weight (Fig. 4.1). However, interspecific comparisons of body water content mask the dynamic relationships occurring within each species. Generally, body fat and water content are inversely proportional. Weight gain, weight loss, and dehydration can alter the total body water content. The actual increase in body water content within a given species is usually less than suggested by the 0.98 exponent of the interspecific relationship because water tends to be accumulated at a decreasing rate as body weight increases. For example, the average exponent of the equations relating water content to body weight for six mammalian species was 0.90 ± 0.04 (Table 4.1). In addition to the water incorporated during growth, water requirements are dependent on many other animal–environment interactions. Water requirements are affected by ambient air temperatures, solar and thermal radiation, vapor pressure deficits, metabolic rates, feed intake, productive processes, amount and

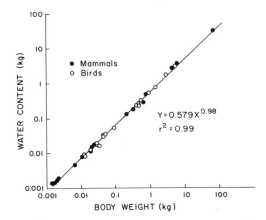

Fig. 4.1. Total body water content determined by desiccation of whole-body homogenates in birds and mammals ranging in weight from 1.59 g to 57.65 kg. For those species having multiple values available in the literature, only two points representing the range of values were used. (Data from Gorecki, 1965; Yarbrough and Johnston, 1965; Myrcha, 1968; Sawicka-Kapusta, 1968, 1970, 1974; Helms and Smythe, 1969; Szwykowska, 1969; Moreau and Dolp, 1970; Penney and Bailey, 1970; Yarbrough, 1970; Sugden and Harris, 1972; Fedyk, 1974; Morton *et al.*, 1974; Robbins *et al.*, 1974; Kaufman and Kaufman, 1975; Cain, 1976; Galster and Morrison, 1976; Husband, 1976; Williams *et al.*, 1977.)

TABLE 4.1
The Relationships between Body Weight (*W*) and the Weight of Body Water (*Y*) during Growth in Several Wildlife Species

Species	Weight range	Predictive equation	R	Reference
Mammals				
Old-field mouse	1.6–11.4 g	$Y = 0.876W^{0.88}$	0.99	Kaufman and Kaufman (1977)
Common vole	1.9–16.0 g	$Y = 0.892W^{0.88}$	0.99	Sawicka-Kapusta (1970)
Bank vole				
Laboratory	2.1–17.4 g	$Y = 0.940W^{0.86}$	0.99	Sawicka-Kapusta (1974)
Captive outdoors	2.2–18.4 g	$Y = 0.910W^{0.88}$	0.99	
Free-ranging	14.0–21.3 g	$Y = 0.790W^{0.97}$	0.98	
Brown lemming	3.5–63.1 g	$Y = 1.082W^{0.82}$	0.99	Holleman and Dieterich (1978)
Mink	274.0–750.8 g	$Y = 1.418W^{0.88}$	0.99	Harper *et al.* (1978)
White-tailed deer	5.4–58.4 kg	$Y = 0.921W^{0.90}$	0.99	Robbins *et al.* (1974)
Birds				
Lesser scaup	29.4–537.9 g	$Y = 0.869W^{0.94}$	0.99	Sugden and Harris (1972)
Lesser snow geese	0.1–3.3 kg	$Y = 0.620W^{0.98}$	0.99	Campbell and Leatherland (1980)

temporal distribution of activity, and physiological, behavioral, and anatomical water conservation adaptations.

Water to meet the various requirements comes from three sources: (a) *free water,* such as in streams, lakes, puddles, rain, snow, or dew; (b) *preformed water* contained in food; and (c) *oxidative* or *metabolic water* produced as a product of the oxidation of organic compounds containing hydrogen (Bartholomew and Cade, 1963). Preformed water may be less than 10% of the weight of air-dried feed (Taylor, 1972) to 70% or more of the fresh weight of such feeds as animal tissue or succulent plant parts. In relation to oxidative water, anhydrous carbohydrates typically yield 56% of their weight as water; proteins yield 40%; and fats yield approximately 107% when completely oxidized (Bartholomew and Cade, 1963). However, catabolism of protein tissue by the starving, water-deprived animal potentially yields two to three times more water per unit of metabolized energy than does body fat because of the much greater water content, for instance, of skeletal muscle (72%) than adipose tissue (3–7%) (Bintz *et al.,* 1979, Bintz and Mackin, 1980).

Measurements of free water intake underestimate total water requirements because of the omission of preformed and metabolic water. Free-ranging kangaroo rats (Richmond *et al.,* 1962; Yousef *et al.,* 1974), ground squirrels (Byman, 1978), peccaries (Zervanos and Day, 1977), pronghorn antelope (Beale and Smith, 1970), rock wallabies and brush-tailed opossums (Kennedy and Heinsohn, 1974), fennec (*Fennecus zerda,* a small foxlike carnivore of the Sahara desert; Noll-Banholzer, 1976), several African antelopes (Taylor, 1969, 1972), and numerous carnivorous and insectivorous birds and mammals (Bartholomew and Cade, 1963; Morton, 1980) have been able to meet their water requirements from preformed and metabolic water. Estimates of preformed and metabolic water can be made from fairly detailed knowledge of food intake, although such measurements are often confined to experiments on captive animals where both feed and water intake can be measured accurately.

Isotopes of water, such as tritium or deuterium oxide, have provided a simpler means than direct intake measurements of determining water turnover or requirements in either the captive or free-ranging animal. This method involves injecting the experimental animal with a known amount of isotopic water that mixes with and is diluted by the total water pool in the animal. The time necessary for complete mixing has ranged from 2–3 hr in small animals with simple stomachs to 7–8 hr in ruminants (MacFarlane *et al.,* 1969; Mullen 1970). Dehydration also increases the time necessary for equilibration (Denny and Dawson, 1975).

Once complete mixing has occurred, a blood or urine sample is taken and the isotopic water concentration determined. Since water is constantly lost in urine, feces, and evaporation and the isotopic water is further diluted by ingested water, succeeding samples will have less and less isotopic water. The loss of isotopic water following its injection, assuming that the body metabolizes the different molecular forms equally, provides an estimate of total water turnover without distinguishing between free, preformed, and metabolic water. The loss and dilution of isotopic water follows an exponential pattern in which the intercept of the curve at the Y axis (time 0) is an estimate of the dilution of the injected dose by the total body water pool. Total body water is estimated by dividing the total amount of isotope injected by its concentration at time 0 (Fig. 4.2).

The slope of the regression (b) multiplied by the body water content is an estimate of the turnover or flux rate of water and, thus, an estimate of the water requirement. The use of turnover rates as an estimate of a requirement should not be considered synonymous with the minimum necessary intake. Turnover rates of any nutrient are directly related to intake and excretion of the nutrient (Cameron *et al.*, 1976; Green *et al.*, 1978). Thus, if the animal has an ad libitum access of free water, turnover rates may reflect an optimum level of water metabolism that is above the absolute, minimum need. Furthermore, insectivores, carnivores, and some frugivores ingesting relatively moist, low energy density diets will have a water turnover rate far in excess of any requirement. Although several authors have conducted validation experiments of the isotopic dilution method and found it an accurate indicator of water kinetics (Holleman and Dieterich, 1975; Cameron *et al.*, 1976), others have suggested that the body will differentiate in its metabolism of the differing water forms (Rubsamen *et al.*, 1979; Grubbs, 1980). The method usually slightly overestimates the water pool and corresponding turnover rate (King *et al.*, 1978). Further discussions of sources of error in using isotopic water may be found in Mullen (1970, 1971) and Grubbs (1980).

Water turnover in both captive and free-ranging wild animals increases as a curvilinear function of body weight (Fig. 4.3; Table 4.2). Although the average water turnover in free-ranging wild mammals is often higher than in captives (Green *et al.*, 1978), differences are not statistically significant. The variation is

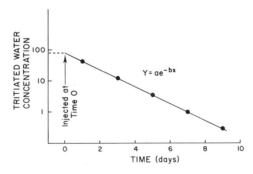

Fig. 4.2. The change in the concentration of isotopic water in the blood, urine, or other body water sample as a function of time following its injection into a test animal.

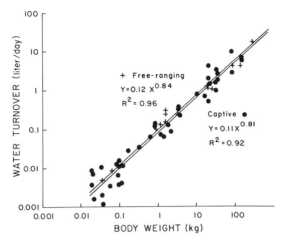

Fig. 4.3. Water turnover rates of wild mammals (captive and free-ranging) as a function of body weight. (Data from Richmond *et al.*, 1962; Taylor and Lyman, 1967; Farrell and Wood, 1968; Knox *et al.*, 1969; Ghobrial, 1970; Longhurst *et al.*, 1970; MacFarlane and Howard, 1970; Wesley *et al.*, 1970; MacFarlane *et al.*, 1971; Maloiy and Hopcraft, 1971; Mullen, 1971; Hulbert and Gordon, 1972; Denny and Dawson, 1973; Holleman and Dieterich, 1973; Kennedy and Heinsohn, 1974; Yousef *et al.*, 1974; Soholt, 1975; Cameron *et al.*, 1976; Noll-Banholzer, 1976; Reaka and Armitage, 1976; Zervanos and Day, 1977; Green and Dunsmore, 1978; Green *et al.*, 1978; King *et al.*, 1978; Reese and Haines, 1978; Deavers and Hudson, 1979; Green and Eberhard, 1979; Noll-Banholzer, 1979; Grubbs, 1980; Ward and Armitage, 1981.) Similar equations were generated by Adolph (1949) ($Y = 0.10X^{0.88}$) and Richmond *et al.* (1962) ($Y = 0.12X^{0.80}$) for several domestic and captive wild mammals and by Nicol (1978) ($Y = 0.10X^{0.82}$) for captive and free-ranging wild mammals.

in part due to the comparison of animals in differing thermal environments and productive states with varying access to free water. For example, water lost in milk will increase the water requirements of lactating females if compensatory adjustments do not occur (Hulbert and Gordon, 1972). Similarly, the kangaroo rat (*Dipodomys* sp.) has been used as one of the classical examples of a species having an extremely low water requirement and excellent physiological water conservation capabilities. Its water requirements have often been determined on caged animals in the laboratory environment, with only the preformed water of an air-dried ration available. Mullen (1970, 1971) has criticized the ecological relevancy of such observations because body water turnover rates were 2.4–4.0 times faster for free-ranging kangaroo rats than for those observed in captivity. Fossoriality, nocturnality, and more extensive consumption of preformed water reduced the need for extensive physiological water conservation. Paradoxically, desert rodents that do not normally consume free water may develop an addiction to water in captivity, consume far more than needed, and die of dehydration

TABLE 4.2
Water Turnover Rates (Y in liter/day) in Captive Birds as a Function of Body Weight (W in kg)

Conditions of measurement	Predictive equation	Reference
Nonthermally stressed, water ad libitum, at or above maintenance	$Y = 0.111W^{0.69}$ $Y = 0.203W^{0.81}$ $Y = 0.119W^{0.75}$	Ohmart et al. (1970) Thomas and Phillips (1975) Walter and Hughes (1978)
Restricted water intake, weight stasis	$Y = 0.052W^{0.64}$	Thomas and Phillips (1975)

when the free water is removed (Boice, 1972). Thus, estimates of water requirements should be judged in the context of the experimental conditions.

The curvilinear relationship between water turnover and body weight may be due to weight-dependent metabolic rates and surface area to volume ratios relative to evaporation and kidney and respiratory function (Adolph, 1949). The specific exponent (generally between 0.80 and 0.88 when comparing a wide range of mammals) and its meaning have been debated by several authors, particularly MacFarlane and Howard (1970), who suggested 0.82 as the standard. In general, points below the relationship of body weight and water turnover often represent species, such as the kangaroo rat or fennec, that had no access to free water, animals whose water requirements were determined without thermal stress and perhaps marginal dehydration, or species that are normally desert dwelling. Similarly, observations above the regression line often represent animals experiencing thermal stress, having additional requirements such as lactation, or having an ad libitum access to free water requiring minimal physiological or behavioral water conservation. Although macropod marsupials have a generally lower water turnover rate ($Y = 0.09W^{0.80}$, where Y is in liters per day and body weight (W) is in kilograms; Denny and Dawson, 1975) than eutherian mammals, Nicol (1978) has suggested that water turnover in marsupials is not significantly different from that in eutherians and that the effect of "normal" habitat is more significant than phylogenetic differences.

Water turnover of nonthermally stressed captive birds having an ad libitum access to free water is similar to the values for mammals, according to the more recent equation of Walter and Hughes (1978) (Table 4.2). However, the slopes of all equations for water turnover in birds may be exaggerated relative to the nonproductive animal, since egg-laying chickens are usually included in the observations for each equation. Egg laying increased the water turnover rate 69% when compared to nonproducing female chickens, and this increase is far greater than can be accounted for by simply the water content of the eggs (Chapman and

Black, 1967; Chapman and Mihai, 1972). Water restriction in captive birds reduced the water turnover rate 50–75% when compared to nonstressed birds (Table 4.2). Because water turnover is a power function of body weight, water ingestion when expressed as a percentage of body weight in captive birds decreased curvilinearly as body weight increased (Fig. 4.4). The smaller bird has a much higher water requirement relative to its body weight than does the larger bird when both are under similar conditions.

Several major physiological differences between birds and mammals exist that affect water requirements. Because of their intense metabolism and correspondingly high rates of pulmocutaneous water loss, small passerines generally dehydrate during flight at air temperatures above 7°C (Torre-Bueno, 1978). Water loss during prolonged, nonstop migrations by birds may actually limit their flight range or require flight at moderate temperatures or high altitudes (Hart and Berger, 1972). High evaporative water loss in birds is partially balanced by their reduced urinary water excretion relative to mammals. Urea excretion in mam-

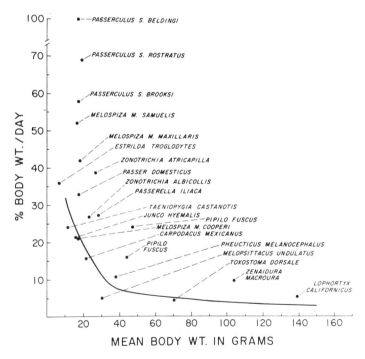

Fig. 4.4. Ad libitum water consumption by birds of various weights in the absence of thermal stress. The curve is the average evaporative water loss. (From Bartholomew and Cade, 1963, courtesy of *Auk*.)

mals requires 20–40 times more water than is required to excrete a similar amount of uric acid in birds (Bartholomew and Cade, 1963). However, because of their almost universal diurnal habits, birds living within environmental conditions at or above their thermoneutral zone evaporate much more water than they produce metabolically (Bartholomew, 1972; Dawson *et al.*, 1979). Thus, unlike many fossorial and nocturnal mammals, few seed-eating birds (only several small desert or salt marsh birds) are able to meet their water requirements solely from metabolic water and very minimal preformed water (Williams and Koenig, 1980).

Wildlife management has been concerned with water requirements because of the opportunities to increase wildlife populations when water is limited by improving natural watering facilities or providing artificial watering facilities, such as "gallinaceous guzzlers." Many species, including mule deer (Elder, 1954), bighorn sheep (Blong and Pollard, 1965; Jones *et al.*, 1957), and pronghorn antelope (Sundstrom, 1968) living in relatively arid conditions are quite dependent on free water. However, the larger ruminants are often able to withstand even more extensive water deprivation than are simple-stomached animals, because water within the ruminant stomach is one of the first used and most enduring sources during deprivation (Turner, 1973; Zervanos and Day, 1977).

Smaller mammals and birds may also respond to increased water supplies. Free-ranging rabbit populations without access to free water during droughts have declined when the water content of the available vegetation dropped below approximately 60% (Richards, 1979). The reduced water supplies apparently impeded milk production and therefore total reproductive effort, as well as directly killed rapidly growing juveniles. Artificial water supplies apparently stimulated reproduction in several small desert rodents (*Peromyscus truei, Gerbillurus paeba,* and *Rhabdomys pumilio*), but did not increase long-term population densities in the latter two species because of either increased mortality or emigration (Bradford, 1975; Christian, 1979).

Even though artificial water sources often attract large numbers of birds, Snyder (1967) has cautioned against their installation without very careful reviews of water requirements relative to existing supplies from all sources. Although the provision of artificial water sources in Utah altered Gambel's quail distribution and possibly the survival of young quail, water did not limit the population during drought conditions (Nish, 1964). Similar results occurred when Gambel's quail in Nevada were provided with artificial water sources; the populations were limited by food, rather than water (Gullion, 1966). When the improvement of water supplies is contemplated, water budget approximations should be compared to minimal field evaluations of water supplies. Such analyses are necessarily species and area specific but should quickly indicate whether water might be limiting. Although building water facilities in arid areas may be worthwhile public relations efforts for conservation agencies, the rewards on a strictly biological basis may not be as apparent in all cases.

REFERENCES

Adolph, E. F. (1949). Quantitative relations in the physiological constitutions of mammals. *Science* **109,** 579–585.

Bartholomew, G. A. (1972). The water economy of seed-eating birds that survive without drinking. *Proc. Int. Ornith. Congr.* **15,** 237–254.

Bartholomew, G. A., and Cade, T. J. (1963). The water economy of land birds. *Auk* **80,** 504–539.

Beale, D. M., and Smith, A. D. (1970). Forage use, water consumption, and productivity of pronghorn antelope in western Utah. *J. Wildl. Manage.* **34,** 570–582.

Bintz, G. L., and Mackin, W. W. (1980). The effect of water availability on tissue catabolism during starvation in Richardson's ground squirrels. *Comp. Biochem. Physiol.* **65A,** 181–186.

Bintz, G. L., Palmer, D. L., Mackin, W. W., and Blanton, F. Y. (1979). Selective tissue catabolism and water balance during starvation in Richardson's ground squirrels. *Comp. Biochem. Physiol.* **64A,** 399–403.

Blong, B., and Pollard, W. (1965). Summer water requirements of desert bighorn in the Santa Rosa Mountains, California, in 1965. *Calif. Fish Game* **54,** 289–296.

Boice, R. (1972). Water addiction in captive desert rodents. *J. Mammal.* **53,** 395–398.

Bradford, D. F. (1975). The effects of an artificial water supply on free-living *Peromyscus truei. J. Mammal.* **56,** 705–707.

Byman, D. (1978). Energy exchange relations and water economics of diurnal small mammals. *Am. Zool.* **18,** 632.

Cain, B. W. (1976). Energetics of growth for black-bellied tree ducks. *Condor* **78,** 124–128.

Cameron, R. D., White, R. G., and Luick, J. R. (1976). Accuracy of the tritium water dilution method for determining water flux in reindeer (*Rangifer tarandus*). *Can. J. Zool.* **54,** 857–862.

Campbell, R. R., and Leatherland, J. F. (1980). Estimating body protein and fat from water content in lesser snow geese. *J. Wildl. Manage.* **44,** 438–446.

Chapman, T. E., and Black, A. L. (1967). Water turnover in chickens. *Poult. Sci.* **46,** 761–765.

Chapman, T. E., and Mihai, D. (1972). Influences of sex and egg production on water turnover in chickens. *Poult. Sci.* **51,** 1252–1256.

Christian, D. P. (1979). Comparative demography of three Namib Desert rodents: Responses to the provision of supplementary water. *J. Mammal.* **60,** 679–690.

Dawson, W. R., Carey, C., Adkisson, C. S., and Ohmart, R. D. (1979). Responses of Brewer's and chipping sparrows to water restriction. *Physiol. Zool.* **52,** 529–541.

Deavers, D. R., and Hudson, J. W. (1979). Water metabolism and estimated field water budgets in two rodents (*Clethrionomys gapperi* and *Peromyscus leucopus*) and an insectivore (*Blarina brevicauda*) inhabiting the same mesic environment. *Physiol. Zool.* **52,** 137–152.

Denny, M. J. S., and Dawson, T. J. (1973). A field technique for studying water metabolism of large marsupials. *J. Wildl. Manage.* **37,** 574–578.

Denny, M. J. S., and Dawson, T. J. (1975). Comparative metabolism of tritiated water by macropod marsupials. *Am. J. Physiol.* **228,** 1794–1799.

Elder, J. B. (1954). Notes on summer water consumption by desert mule deer. *J. Wildl. Manage.* **18,** 540–541.

Farrell, D. J., and Wood, A. J. (1968). The nutrition of the female mink (*Mustela vison*). III. The water requirements for maintenance. *Can. J. Zool.* **46,** 53–56.

Fedyk, A. (1974). Gross body composition in postnatal development of the bank vole. I. Growth under laboratory conditions. *Acta Theriol.* **19**, 381–401.

Galster, W., and Morrison, P. (1976). Seasonal changes in body composition of the arctic ground squirrel, *Citellus undulatus*. *Can. J. Zool.* **54**, 74–78.

Ghobrial, L. I. (1970). The water relations of the desert antelope *Gazella dorcas dorcas*. *Physiol. Zool.* **43**, 249–256.

Gorecki, A. (1965). Energy values of body in small mammals. *Acta Theriol.* **10**, 333–352.

Green, B., and Dunsmore, J. D. (1978). Turnover of tritiated water and ^{22}sodium in captive rabbits (*Oryctolagus cuniculus*). *J. Mammal.* **59**, 12–17.

Green, B., and Eberhard, I. (1979). Energy requirements and sodium and water turnovers in two captive marsupial carnivores: The Tasmanian devil, *Sarcophilus harrisii*, and the native cat, *Dasyurus viverrinus*. *Aust. J. Zool.* **27**, 1–8.

Green, B., Dunsmore, J., Bults, H., and Newgrain, K. (1978). Turnover of sodium and water by free-living rabbits, *Oryctolagus cuniculus*. *Aust. Wildl. Res.* **5**, 93–99.

Grubbs, D. E. (1980). Tritiated water turnover in free-living desert rodents. *Comp. Biochem. Physiol.* **66A**, 89–98.

Gullion, G. W. (1966). A viewpoint concerning the significance of studies of game bird food habits. *Condor* **68**, 372–376.

Harper, R. B., Travis, H. F., and Glinsky, M. S. (1978). Metabolizable energy requirement for maintenance and body composition of growing farm-raised male pastel mink (*Mustela vison*). *J. Nutr.* **108**, 1937–1943.

Hart, J. S., and Berger, M. (1972). Energetics, water economy and temperature regulation during flight. *Proc. Int. Ornithol. Congr.* **15**, 189–199.

Helms, C. W., and Smythe, R. B. (1969). Variation in major body components of the tree sparrow (*Spizella arborea*) sampled within the winter range. *Wilson Bull.* **81**, 280–292.

Holleman, D. F., and Dieterich, R. A. (1973). Body water content and turnover in several species of rodents evaluated by the tritiated water method. *J. Mammal.* **54**, 456–465.

Holleman, D. F., and Dieterich, R. A. (1975). An evaluation of the tritiated water method for estimating body water in small rodents. *Can. J. Zool.* **53**, 1376–1378.

Holleman, D. F., and Dieterich, R. A. (1978). Postnatal changes in body composition of laboratory maintained brown lemmings, *Lemmus sibiricus*. *Lab. Anim. Sci.* **28**, 529–535.

Hulbert, A. J., and Gordon, G. (1972). Water metabolism of the bandicoot, *Isoodon macrourus* Gould, in the wild. *Comp. Biochem. Physiol.* **41A**, 27–34.

Husband, T. P. (1976). Energy metabolism and body composition of the fox squirrel. *J. Wildl. Manage.* **40**, 255–263.

Jones, F. L., Flittner, G., and Gard, R. (1957). Report on a survey of bighorn sheep in the Santa Rosa Mountains, Riverside County. *Calif. Fish Game* **43**, 103–111.

Kaufman, D. W., and Kaufman, G. A. (1975). Caloric density of the old-field mouse during postnatal growth. *Acta Theriol.* **20**, 83–95.

Kaufman, G. A., and Kaufman, D. W. (1977). Body composition of the old-field mouse (*Peromyscus polionotus*). *J. Mammal.* **58**, 429–434.

Kennedy, P. M., and Heinsohn, G. E. (1974). Water metabolism of two marsupials—the brush-tailed possum, *Trichosurus vulpecula* and the rock-wallaby, *Petrogale inornata* in the wild. *Comp. Biochem. Physiol.* **47A**, 829–834.

King, J. M., Nyamora, M. R., Stanley-Price, M. R., and Heath, B. R. (1978). Game domestication for animal production in Kenya: Prediction of water intake from tritiated water turnover. *J. Agric. Sci.* **91**, 513–522.

Knox, K. L., Nagy, J. G., and Brown, R. D. (1969). Water turnover in mule deer. *J. Wildl. Manage.* **33**, 389–393.

Longhurst, W. M., Baker, N. F., Connolly, G. E., and Fisk, R. A. (1970). Total body water and water turnover in sheep and deer. *Am. J. Vet. Res.* **31,** 673–677.

MacFarlane, W. V., and Howard, B. (1970). Water in the physiological ecology of ruminants. *In* "Physiology of Digestion and Metabolism in the Ruminant" (A. T. Phillipson, ed.), pp. 362–374. Proc. Third Int. Symp., Oriel Press, Newcastle.

MacFarlane, W. V., and Howard, B. (1972). Comparative water and energy economy of wild and domestic mammals. *Symp. Zool. Soc. London* **31,** 261–296.

MacFarlane, W. V., Howard, B., and Siebert, B. D. (1969). Tritiated water in the measurement of milk intake and tissue growth of ruminants in the field. *Nature* **221,** 578–579.

MacFarlane, W. V., Howard, B., Haines, H., Kennedy, P. M., and Sharpe, C. M. (1971). Hierarchy of water and energy turnover of desert mammals. *Nature* **234,** 483–484.

Maloiy, G., and Hopcraft, D. (1971). Thermoregulation and water relations of two East African antelope: The hartebeest and impala. *Comp. Biochem. Physiol.* **38A,** 525–534.

Maynard, L. A., and Loosli, J. K. (1969). "Animal nutrition." McGraw-Hill, New York.

Moreau, R. E., and Dolp, R. M. (1970). Fat, water, weights and winglengths of autumn migrants in transit on the northwest coast of Egypt. *Ibis* **112,** 209–228.

Morton, M. L., Maxwell, C. S., and Wade, C. E. (1974). Body size, body composition, and behavior of juvenile Belding ground squirrels (*Spermophilus beldingi beldingi*). *Great-Basin Nat.* **34,** 121–134.

Morton, S. R. (1980). Field and laboratory studies of water metabolism in *Sminthopsis crassicaudata* (Marsupialia: Dasyuridae). *Aust. J. Zool.* **28,** 213–227.

Moulton, C. R. (1923). Age and chemical development in mammals. *J. Biol. Chem.* **57,** 79–97.

Mullen, R. K. (1970). Respiratory metabolism and body water turnover rates of *Perognathus formosus* in its natural environment. *Comp. Biochem. Physiol.* **32A,** 259–265.

Mullen, R. K. (1971). Energy metabolism and body water turnover rates of two species of free-living kangaroo rats, *Dipodomys merriami* and *Dipodomys microps*. *Comp. Biochem. Physiol.* **39A,** 379–380.

Myrcha, A. (1968). Caloric value and chemical composition of the body of the European hare. *Acta Theriol.* **13,** 65–71.

Nicol, S. C. (1978). Rates of water turnover in marsupials and eutherians: A comparative review with new data on the Tasmanian devil. *Aust. J. Zool.* **26,** 465–473.

Nish, D. (1964). An evaluation of artificial water catchment basins on Gambel's quail populations in southern Utah. Master's thesis, Utah St. Univ., Logan.

Noll-Banholzer, U. (1976). Water balance, metabolism, and heart rate in the Fennec. *Naturwissenschaften* **63,** 202–203.

Noll-Banholzer, U. (1979). Water balance and kidney structure in the fennec. *Comp. Biochem. Physiol.* **62A,** 593–597.

Ohmart, R. D., Chapman, T. E., and McFarland, L. Z. (1970). Water turnover in roadrunners under different environmental conditions. *Auk* **87,** 787–793.

Penney, J. G., and Bailey, E. D. (1970). Comparison of the energy requirements of fledging black ducks and American coots. *J. Wildl. Manage.* **34,** 105–114.

Reaka, M. L., and Armitage, K. B. (1976). The water economy of harvest mice from zeric and mesic environments. *Physiol. Zool.* **49,** 307–327.

Reese, J. B., and Haines, H. (1978). Effects of dehydration on metabolic rate and fluid distribution in the jackrabbit, *Lepus californicus*. *Physiol. Zool.* **51,** 155–165.

Richards, G. C. (1979). Variation in water turnover by wild rabbits, *Oryctolagus cuniculus*, in an arid environment, due to season, age group and reproductive condition. *Aust. Wildl. Res.* **6,** 289–296.

Richmond, C. R., Langham, W. H., and Trujillo, I. I. (1962). Comparative metabolism of tritiated water by mammals. *J. Cell. Comp. Physiol.* **59**, 45–53.

Robbins, C. T., Moen, A. N., and Reid, J. T. (1974). Body composition of white-tailed deer. *J. Anim. Sci.* **38**, 871–876.

Robinson, J. R. (1957). Functions of water in the body. *Proc. Nutr. Soc.* **16**, 108–112.

Rubsamen, K., Nolda, U., and Von Engelhardt, W. (1979). Difference in the specific activity of tritium labelled water in blood, urine and evaporative water in rabbits. *Comp. Biochem. Physiol.* **62A**, 279–282.

Sawicka-Kapusta, K. (1968). Annual fat cycle of field mice, *Apodemus flavicollis* (Melchior, 1834). *Acta Theriol.* **13**, 329–339.

Sawicka-Kapusta, K. (1970). Changes in the gross body composition and the caloric value of the common voles during their postnatal development. *Acta Theriol.* **15**, 67–79.

Sawicka-Kapusta, K. (1974). Changes in the gross body composition and energy value of the bank voles during their postnatal development. *Acta Theriol.* **19**, 27–54.

Snyder, W. D. (1967). Experimental habitat improvement for scaled quail. *Colo. Game Fish Parks Dept., Tech. Publ. No. 19.*

Soholt, L. F. (1975). Water balance of Merriam's kangaroo rat, *Dipodomys merriami,* during cold exposure. *Comp. Biochem. Physiol.* **51A**, 369–372.

Sugden, L. G., and Harris, L. E. (1972). Energy requirements and growth of captive lesser scaup. *Poult. Sci.* **51**, 625–633.

Sundstrom, C. (1968). Water consumption by pronghorn antelope and distribution related to water in Wyoming's red desert. *Proc. Third Biennial Antelope States Workshop,* 39–46.

Szwykowska, M. M. (1969). Seasonal changes of the caloric value and chemical composition of the body of the partridge (*Perdix perdix* L.). *Ekologia Polska-Ser. A* **17**, 795–809.

Taylor, C. R. (1969). The eland and the oryx. *Sci. Am.* **220**, 88–95.

Taylor, C. R. (1972). The desert gazelle: A paradox resolved. *Symp. Zool. Soc. London* **31**, 215-227.

Taylor, C. R., and Lyman, C. P. (1967). A comparative study of the environmental physiology of an East African antelope, the eland, and the hereford steer. *Physiol. Zool.* **40**, 280–295.

Thomas, D. H., and Phillips, J. G. (1975). Studies in avian adrenal steroid function II: Chronic adrenalectomy and the turnover of $(^3H)_2O$ in domestic ducks (*Anas platyrhynchos* L.). *Gen. Comp. Endocrinol.* **26**, 404–411.

Torre-Bueno, J. R. (1978). Evaporative cooling and water balance during flight in birds. *J. Exp. Biol.* **75**, 231–236.

Turner, J. C., Jr. (1973). Water, energy, and electrolyte balance in the desert bighorn sheep, *Ovis canadensis.* Doctoral dissertation, Univ. of Calif., Riverside.

Walter, A., and Hughes, M. R. (1978). Total body water volume and turnover rate in fresh water and sea water adapted glaucous-winged gulls, *Larus glaucescens. Comp. Biochem. Physiol.* **61A**, 233–237.

Ward, J. M., Jr., and Armitage, K. B. (1981). Water budgets of montane-mesic and lowland-zeric populations of yellow-bellied marmots. *Comp. Biochem. Physiol.* **69A**, 627–630.

Wesley, D. E., Knox, K. L., and Nagy, J. G. (1970). Energy flux and water kinetics in young pronghorn antelope. *J. Wildl. Manage.* **34**, 908–912.

Williams, A. J., Siegfried, W. R., Burger, A. E., and Berruti, A. (1977). Body composition and energy metabolism of moulting eudyptid penguins. *Comp. Biochem. Physiol.* **56A**, 27–30.

Williams, P. L., and Koenig, W. D. (1980). Water dependence of birds in a temperate oak woodland. *Auk* **97**, 339–350.

Yarbrough, C. G. (1970). Summer lipid levels of some subarctic birds. *Auk* **87**, 100–110.

Yarbrough, C. G., and Johnston, D. W. (1965). Lipid deposition in wintering and premigratory myrtle warblers. *Wilson Bull.* **77,** 175–191.

Yousef, M. K., Johnson, H. D., Bradley, W. G., and Seif, S. M. (1974). Tritiated water-turnover rate in rodents: Desert and mountain. *Physiol. Zool.* **47,** 153–162.

Zervanos, S. M., and Day, G. I. (1977). Water and energy requirements of captive and free-living collared peccaries. *J. Wildl. Manage.* **41,** 527–532.

5

Minerals

Minerals are an extremely diverse group of nutrients that have many essential functions. However, each mineral element represents a very small fraction of the total body (Table 5.1). The total ash or mineral content of the animal's body does vary but is usually less than 5%. The body's mineral content and requirements vary with age, sex, species, season, maturity, and reproductive and productive condition.

Carbon, hydrogen, oxygen, and nitrogen are usually grouped as organic constituents and are therefore not discussed as required minerals. Of the remainder, those required or represented in the animal body by relatively large amounts (milligrams per gram) are called *macroelements* (calcium, phosphorus, sodium, potassium, magnesium, chlorine, and sulfur), whereas those required in relatively small amounts are called *trace elements* (iron, zinc, manganese, copper, molybdenum, iodine, selenium, cobalt, fluoride, and chromium). When animals are consuming purified diets in sterile, dust-free environments, silicon, tin, vanadium, nickel, and arsenic must be added to the diet for proper growth, hair or feather development, and bone formation (Frieden, 1972). However, the purity of the diet and the environment required to produce these deficiencies is so great that they do not cause practical problems for wildlife (Underwood, 1977).

Deficiencies and imbalances of minerals are well recognized as important determinants of animal condition, fertility, productivity, and mortality (Underwood, 1977). Pathological symptoms of mineral deficiencies or imbalances have been reported by many investigators for both captive and free-ranging wildlife (Encke, 1962; Christian, 1964; Jarrett *et al.*, 1964; Karstad, 1967; Mahon, 1969; McCullough, 1969; Wallach, 1970; Wallach and Flieg, 1969, 1970; Hebert and Cowan, 1971b; Flieg, 1973; Van Pelt and Caley, 1974; Henderson and Winterfield, 1975; Mundy and Ledger, 1976; Newman and Yu, 1976; Fleming *et al.*, 1977; Flynn *et al.*, 1977; Hyvarinen *et al.*, 1977a; Wilson and Hirst, 1977), although chronic deficiencies or imbalances that only marginally reduce growth, reproductive rate or success, and general fitness may be more important (Franzmann *et al.*, 1975). Marginal deficiencies are easily overlooked as population fluctuations are attributed to more obvious factors, such as severe weather, food

TABLE 5.1
Mineral Content of the Entire Body of Several Wildlife Species[a,b]

Mineral	Mammals						Birds			
	White-tailed deer	Short-tail shrew	Cotton mouse	Golden mouse	Old-field[c] mouse	Fox[d] squirrel	Blue tit	Coal tit	Goldcrest	Meadow[e] pipit
Calcium (mg/g)	30.9 ± 2.6	34.40 ± 1.53	40.50 ± 5.13	37.40 ± 4.33	5.69–26.16	16.6–34.6	32.84	33.10	28.43	13.0–27.7
Phosphorus (mg/g)	22.6 ± 1.6	17.20 ± 0.51	16.70 ± 1.32	19.20 ± 0.77	17.07–19.96	16.7–19.3	20.42	20.84	18.78	14.0–19.0
Potassium (mg/g)	9.53 ± 0.50	—	—	—	8.31–15.71	8.7–12.7	5.83	6.29	5.79	8.5–17.0
Sodium (mg/g)	3.88 ± 0.19	4.22 ± 0.15	2.38 ± 0.59	3.57 ± 0.25	2.74–5.84	4.0–12.7	3.74	3.93	3.97	3.5–13.0
Magnesium (mg/g)	0.91 ± 0.10	1.44 ± 0.07	1.23 ± 0.14	1.38 ± 0.12	0.36–0.84	1.2–1.4	1.04	1.11	1.11	1.0–1.5
Iron (µg/g)	164.5 ± 10.8	500.0 ± 38.0	200.0 ± 13.0	240.0 ± 15.0	221.0–396.0	—	—	—	—	300.0–500.0
Zinc (µg/g)	68.4 ± 4.7	120.0 ± 8.0	98.0 ± 8.0	110.0 ± 10.0	108.0–142.0	—	—	—	—	94.0–120.0
Manganese (µg/g)	28.5 ± 4.0	—	—	—	3.5–17.6	—	—	—	—	10.0–17.0
Copper (µg/g)	26.1 ± 1.8	—	—	—	—	—	—	—	—	—
Molybdenum (µg/g)	3.0 ± 0.3	—	—	—	—	—	—	—	—	—
Iodine (µg/g)	—	—	—	—	1.9 – 2.4	—	—	—	—	—

[a]Grimshaw et al., 1958; Beyers et al., 1971; G. A. Kaufman and Kaufman, 1975; D. A. Kaufman and Kaufman, 1975; Skar et al., 1975; Wiener et al., 1975; Havera, 1978.

[b]Data are not available on chlorine and selenium.

[c]0–42 days of age.

[d]Range of values for embryos and adults.

[e]Range of values for juveniles and adults.

shortages, parasites, and infectious agents (Underwood, 1977). As the practice of wildlife management becomes more intense, the need for field management and research personnel to recognize and suggest remedies for mineral deficiencies or imbalances becomes more important.

Mineral requirements of wildlife have traditionally been evaluated in relation to deficiency symptoms and the establishment of maximum growth or reproductive rates. Wildlife ecologists have also delineated the characteristics of mineral pools (i.e., standing crops) within various communities. However, much of the very detailed information necessary for understanding mineral metabolism and requirements in relation to maintenance and production is lacking. Most of the information currently available on minerals in nutrition is on domestic and laboratory animals. Studies on virtually all minerals deemed important to wildlife are needed relative to understanding efficiencies of absorption and retention, turnover rates, interactions with other minerals or dietary ingredients, effects of chronic or marginal deficiencies, requirements for various body processes, feeding strategies that may optimize mineral intake, deficiency signs, and possible routes or methods of dietary supplementation.

I. MACROELEMENTS

A. Calcium–Phosphorus

1. Functions

Calcium and phosphorus are major mineral constituents of the animal's body and are largely associated with skeletal formation. Because of their obligatory interactions in bone, calcium and phosphorus metabolism in the animal body are very closely related and are therefore discussed together. In the mature animal, 99% of its calcium and 80% of its total body phosphorus are contained in bone (Bronner, 1964; Irving, 1964). The composition of bone varies depending particularly on the age of the animal. Calcium and phosphorus densities also vary within each bone, although the calcium to phosphorus concentration in bone ash is usually in an approximate 2:1 ratio (Morgulis, 1931; Widdowson and Dickerson, 1964). Calcium is also associated with blood clotting, excitability of nerves and muscles, acid-base balance, egg shell formation, and muscle contraction, whereas phosphorus is involved in almost every aspect of animal metabolism, such as energy metabolism, muscle contractions, nerve tissue metabolism, transport of metabolites, nucleic acid structure, and carbohydrate, fat, and amino acid metabolism.

2. Requirements

The requirements for calcium, and most other nutrients for captive wildlife, are often expressed as a percentage of the diet because of the opportunity in captivity to provide food ad libitum. Thus, excluding economic or other constraints limiting the amounts of food offered, nutrient requirements for captive animals can be determined by feeding a diet in which consistent variation occurs in only one component. The dietary concentration of the specific ingredient is then compared to the requirement criteria, such as weight gain, reproductive success, survival, or mature weight. The technique is based on the observation that the various requirement criteria measurements increase to an asymptotic level as the concentration of the limiting nutrient increases as long as toxicities or imbalances do not occur. If a second nutrient becomes limiting prior to that of the specific nutrient being studied, the regression curve will normally be shifted downward, with simply lower asymptotic growth rates (as an example) occurring at a lower nutrient concentration.

Several processes, such as early growth when bone mineralization is being maximized, antler growth, and egg laying, are associated with elevated calcium–phosphorus requirements (Table 5.2). Antlers are composed of about 22% calcium and 11% phosphorus, with much of the remainder being protein (Rush, 1932; Chaddock, 1940; Bernhard, 1963; McCullough, 1969; Hyvarinen et al., 1977b). Egg shells are over 98% calcium carbonate and less than 1% phosphorus (Romanoff and Romanoff, 1949). The calcium and phosphorus requirements for

TABLE 5.2
Dietary Calcium and Phosphorus Requirements or Allowances for Several Species of Wild Birds and Mammals

Species	Productive status	Requirement (% of dry diet)		Reference
		Calcium	Phosphorus	
Birds				
Bobwhite quail	Growth	1.00	0.75	Nestler et al. (1948)
	Growth (0–6 weeks)	—	0.60	M. L. Scott (1963a, 1963b)
	Growth (6–12 weeks)	—	0.48	
	Growth	1.00	0.75	
	Growth (0–6 weeks)	0.65	0.65	H. R. Wilson et al. (1972)
	Growth	0.9–0.7	0.45–0.35	M. L. Scott (1977–1978)
	Growth	0.65	0.65	NRC (1977a)
	Egg production	2.1	1.0	Nestler (1949)
	Egg production	2.3	1.0	NRC (1977a)
	Egg production	2.7	0.5	M. L. Scott (1977–1978)
Ducks and geese	Growth	1.2	0.8	Holm and Scott (1954)
(domestic)	Growth	1.1–0.9	0.55–0.45	M. L. Scott (1977–1978)
	Growth	0.6–0.8	0.4–0.6	NRC (1977a)
	Egg production	2.25	0.75	Scott and Holm (1964)
	Egg production	2.7	0.5	M. L. Scott (1977–1978)
	Egg production	2.25–2.75	0.6	NRC (1977a)

(continued)

TABLE 5.2 *Continued*

Species	Productive status	Requirement (% of dry diet)		Reference
		Calcium	Phosphorus	
Pheasant	Growth	1.2	0.6	Reynells and Flegal (1979b)
	Growth (0–6 weeks)	1.0	0.8	NRC (1977a)
	Growth (0–6 weeks)	1.33	0.70	M. L. Scott (1963 a, 1963b)
	Growth (6–12 weeks)	0.50	0.48	
	Growth (3–5 weeks)	0.90–1.06	0.8	Hinkson *et al.* (1971)
	Growth	1.0–0.7	0.5–0.35	M. L. Scott (1977–1978)
	Growth (6–20 weeks)	0.7	0.6	NRC (1977a)
	Egg production	1.50–3.50		Chambers *et al.* (1966)
	Egg production	2.5	0.6	Hinkson *et al.* (1970)
	Egg production		1.0	NRC (1971b)
	Egg production	2.7	0.5	M. L. Scott (1977–1978)
	Egg production	2.1–2.7		Reynells and Flegal (1979a)
Turkey	Growth (0–8 weeks)	1.2	0.8	NRC (1977a)
(domestic)	Growth (8–16 weeks)	0.8	0.7	NRC (1977a)
	Growth	1.2–0.5	0.6–0.45	M. L. Scott (1977–1978)
	Egg production	2.25	0.75	NRC (1977a)
	Egg production	2.7	0.5	M. L. Scott (1977–1978)
Mammals				
Fox	Growth (7–37 weeks)	0.5–0.6	0.6–0.4	NRC (1968)
Guinea pig	Growth	0.8–1.0	0.4–0.7	NRC (1978a)
Hamster	Growth	0.59	0.30	NRC (1978a)
Laboratory mouse	Growth and reproduction	0.4	0.4	NRC (1978a)
Laboratory rat	Growth and reproduction	0.5	0.4	NRC (1978a)
Mink	Growth	0.4–1.0	0.4–0.8	NRC (1968)
Nonhuman primate	Growth	0.5	0.3	NRC (1978b)
Rabbit (domestic)	Growth	0.4	0.22	NRC (1977b)
	Gestation	0.45	0.37	
	Lactation	0.75	0.5	
White-tailed deer	Growth and/or antler formation	0.59	0.54	Magruder *et al.* (1957)
		0.64	0.56	McEwen *et al.* (1957)
		0.40	0.26	Ullrey *et al.* (1973, 1975)

captive, indeterminate laying birds (species such as the pheasant or quail that continue to lay large numbers of eggs if earlier ones are removed) during maximum egg production are approximately 2.0–2.5% dietary calcium and 0.6–1.0% phosphorus. Ecologically, calcium and phosphorus requirements may be lower for determinate layers (such as many passerines, shorebirds, and large raptors laying a constant number of eggs even when eggs are removed or added to a clutch) and indeterminate layers that successfully raise the first clutch. Ring-necked pheasants consuming 1.5 and 3.2% dietary calcium produced first clutches of equal sizes, but if the eggs were destroyed and renesting occurred, the birds consuming 3.2% dietary calcium produced significantly more eggs in the second clutch (Chambers *et al.*, 1966).

The calcium requirements of bobwhite quail when expressed as a percentage of the diet were dependent on ambient air temperature (Case and Robel, 1974). Although such a relationship might seem incongruous, the relationship between ambient air temperature and calcium required may be used to illustrate one of the fundamental problems of expressing a mineral requirement as a percentage of the diet. The absolute requirement for calcium was probably constant and indepen-dent of air temperature, but total feed intake was dependent on air temperature. Consequently, less food was ingested at 25° and 35°C, with inadequate amounts of a 2.3% calcium diet being consumed to meet the calcium requirement for egg laying, even though the dietary calcium concentration and intake were adequate at 5° and 15°C.

Requirements for maintenance of calcium balance in the nonproducing animal can be determined by comparing calcium intake to excretion. The minimal loss of calcium via the urine and feces of jackrabbits, fox squirrels, and rock hyraxes (Nagy *et al.*, 1976; Havera, 1978; Leon and Belonje, 1979) consuming nuts, seeds, or herbage at an extrapolated zero calcium intake is 27.1 $mg/W_{kg}/day$, with calcium balance being achieved only when intake exceeded 230 $mg/W_{kg}/day$ (Fig. 5.1).

However, several experimental and theoretical problems exist in using such an approach for estimating a mineral requirement. Experimentally, the regression technique assumes that the availability of the mineral is independent of the source. This assumption is not correct for calcium. For example, some of the

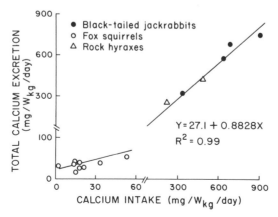

Fig. 5.1. Calcium excretion as a function of intake in black-tailed jackrabbits, fox squirrels, and rock hyraxes. (Data from Nagy *et al.*, 1976; Havera, 1978; Leon and Belonje, 1979.)

calcium in herbages—particularly alfalfa, which was fed to both the jackrabbits and hyraxes—occurs as calcium oxalate crystals, which are poorly available in comparison to other calcium salts (Harbers *et al.*, 1980). Similarly, calcium absorption in the squirrels may have been reduced by the very low calcium to phosphorus ratios in several of the seed and nut diets.

Theoretically, one should expect the relationship between intake and total excretion of any required element to be curvilinear rather than linear. A linear relationship suggests that excretion is merely a passive reflector of intake. However, very active homeostatic mechanisms are involved in calcium metabolism. Consequently, animals consuming calcium at levels below their necessary requirement retain the absorbed element with maximum efficiency. The efficiency of absorption and retention should decline as calcium is consumed at increasing levels above the requirement. Such a relationship becomes apparent if the two basic sets of data (the squirrels at the very low end and the jackrabbits and hyraxes at the upper end of the scale) are plotted separately. The efficiency of retention (i.e., $1 -$ slope of the excretion curve) decreases from 49% for the squirrels to 22% for the jackrabbits and hyraxes. Because of the higher retention in the squirrels, calcium equilibrium occurs at 58 mg/W_{kg}/day. The two estimated maintenance requirement levels (58–230 mg/W_{kg}/day) essentially cover the range of requirement estimates available for other species (150 for growing primates, 144–354 for growing mink and foxes, and 22–76 for the maintenance of cattle and sheep) (NRC 1968, 1970, 1975, 1978b; Nagy and Milton, 1979). Productive processes, such as milk or egg production, growth, or gestation, must increase the requirement in proportion to the amount used in the particular process.

3. Deficiencies and Dietary Sources

Calcium deficiencies are probably the major mineral problem encountered in captive wildlife (Wallach, 1970; Fowler, 1978b). Calcium deficiencies result in retarded growth, decreased food consumption, high basal metabolic rate, reduced activity and sensitivity, osteoporosis and rickets, abnormal posture and gait, susceptibility to internal hemorrhage, egg shell thinning, transient paralysis and tetany, retarded feather growth, reduced antler growth, and elevated serum alkaline phosphatase (Rings *et al.*, 1969; Wallach and Flieg, 1969). Phosphorus deficiencies result in loss of appetite, abnormal appetite (pica), reduced antler growth and strength, rickets, reduced body growth, and weakness and death (French *et al.*, 1956; McCullough, 1969; Mahon, 1969; Krausman and Bissonette, 1977).

Excesses of other dietary components can also precipitate a calcium or phosphorus deficiency. Excesses of phosphorus or manganese, for example, form insoluble complexes with calcium, thereby reducing calcium absorption. Diets containing high levels of oils and fats may form insoluble calcium soaps in the digestive tract (Wallach and Flieg, 1967, 1969). Excesses of iron, aluminum, and magnesium form insoluble phosphorus complexes. Dietary calcium to phosphorus ratios ranging from 1:1 to 2:1 are best for proper absorption and metabolism, although higher ratios can be handled as indicated by the extensive use of alfalfa (Table 5.3) as a single feed for many captive herbivores. However, since a significant part of the calcium in alfalfa can exist as insoluble calcium oxalate, the effective calcium to phosphorus ratio may be much lower (Harbers *et al.*, 1980). Similarly, excesses of calcium have a far lesser effect on phosphorus absorption than do excesses of phosphorus on calcium absorption.

Excesses of dietary phosphorus associated with low or marginal dietary calcium levels produce a pathological osteomalacia termed *nutritional secondary hyperparathyroidism*, or NSH. Since blood calcium levels are controlled by the

TABLE 5.3
Calcium and Phosphorus Content and Their Relative Proportions in Several Food Sources[a]

Feed	Content (% as fed)		Ratio
	Calcium	Phosphorus	
Animal products			
Beef (meat, fresh)	0.01	0.19	1:17
Beef liver	0.01	0.35	1:44
Chicken	0.01	0.20	1:17
Horsemeat	0.01	0.15	1:15
Mackerel	0.01	0.27	1:34
Sardines	0.15	0.97	1:6
Fruits and seeds			
Apples	0.01	0.01	1:1
Barley	0.08	0.72	1:9
Corn (air dry)	0.02	0.30	1:15
Nuts (oak, hickory, and walnuts)	0.07	0.24	1:3.4
Oats	0.10	0.35	1:3.5
Orange pulp	0.03	0.02	1.4:1
Soybeans	0.25	0.60	1:2.4
Sunflower seeds (hulled)	0.15	1.01	1:7
Wheat	0.06	0.42	1:7
Forages			
Alfalfa (dry)	1.60	0.31	5.2:1
Buffalograss	0.57	0.14	4.1:1
Cheatgrass	0.64	0.28	2.3:1
Reed canarygrass	0.36	0.33	1.1:1
Smooth brome	0.36	0.19	1.9:1

[a]From Rings *et al.*, 1969; Wallach and Flieg, 1970; NRC, 1971a; Fowler, 1978a; Havera, 1978.

interaction of calcitonin (a parathyroid hormone that promotes bone accretion) and parathyroid hormone that stimulates bone resorption, inadequate dietary calcium absorption produces a hyperplasia of the parathyroid, calcium resorption from the bones, and, if continued, eventual exhaustion of bone mineral stores even though renal excretion is decreased (Arnaud, 1978). NSH has been observed in captive psittacines (parrots, cockatoos, and parakeets) fed a diet of sunflower seeds, peanuts, and oats (Ca:P ratio of 1:10, Table 5.3) (Wallach and Flieg, 1967), golden-mantled ground squirrels fed orange pulp and sunflower seeds (Rings *et al.*, 1969), and carnivorous birds and mammals fed pure meat diets (Slusher *et al.*, 1965; P. P. Scott, 1968; Wallach and Flieg, 1969, 1970; Gorham *et al.*, 1970; Dieterich and Van Pelt, 1972; Brambell and Mathews, 1976) (Table 5.3). Many seeds contain calcium to phosphorus ratios ranging from 1:2 to 1:15 and meat and fish contain ratios from 1:6 to 1:44 (Wallach and Flieg, 1969). Dominance hierarchies and overfeeding of heterogeneous diets to captive wildlife can also produce calcium deficiencies via food preferences in which items inadequate in calcium are preferentially consumed even though the total diet offered may be adequate (Tomson and Lotshaw, 1978) (Fig. 5.2).

Free-ranging wildlife can also suffer from NSH. Osteoporosis has been reported in wild-caught red fox and arctic fox kits in Alaska (Van Pelt and Caley, 1974; Conlogue *et al.*, 1979). The red fox kits had apparently been consuming snowshoe hare meat retrieved by the parents, and the arctic foxes had consumed seal meat without an adequate ingestion of bones. The kits were thin and unthrifty, with extensive bone porosity. Small rodents or birds, which could have been consumed whole and thereby have provided adequate calcium, were apparently an insignificant part of the diet. The red fox kits recovered when fed a balanced commercial dog food. Serum calcium concentrations of reindeer consuming lichens during late winter and early spring were below levels associated with tetany in domestic livestock, even though classical tetany signs were absent (Hyvarinen *et al.*, 1977a).

Calcium and phosphorus deficiencies in captive wildlife can easily be corrected by the addition of bone meal, calcium phosphates, or calcium carbonates (limestone or oyster shells) to the diet, depending on the existing dietary calcium to phosphorus ratio and assuming adequate vitamin D availability. Limestone is readily eroded in the gastrointestinal tract of birds and either absorbed or excreted (Korschgen *et al.*, 1965; Trost, 1981). Dissolved calcium in drinking water is an inadequate source of calcium (Fowler, 1978b). For example, drinking "extremely" hard water (800 mg $CaCO_3$/liter) would supply only about 5% of the daily calcium requirement of captive birds.

Free-ranging wildlife consuming meat or seed diets or having very high requirements must actively seek and consume calcium supplements (Fig. 5.3). Many birds ingest snails and their associated calciferous shell to help meet the calcium requirement of egg production (Krapu and Swanson, 1975; Beasom and

Pattee, 1978; Ankney and Scott, 1980). Lemming bones and egg shell fragments are essential sources of calcium to arctic sandpipers (MacLean, 1974) and Lapland longspurs (Seastedt and MacLean, 1977). The consumption of fungi, insects, bones, and other rich calcium sources is essential for free-ranging fox squirrels, since nuts and seeds commonly consumed are inadequate in calcium (Havera, 1978). Osteophagia or bone chewing is often observed in free-ranging

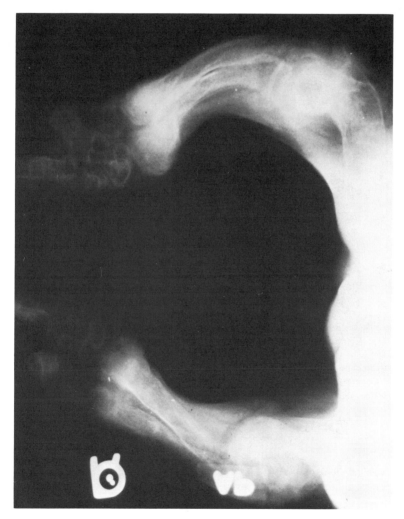

Fig. 5.2. Radiograph of the front legs of a ricketic lion fed only red meat. Note the bone curvature and poor mineralization of the bones. (Courtesy of D. E. Ullrey, Michigan State University, East Lansing.)

Fig. 5.3. A free-ranging Philippine eagle meeting its calcium requirement by ingesting a very large bone. While such a bone can be ingested by the adult, one can appreciate the need for small bones and bone fragments in meeting the chick's requirement. The chick (top of downy head just barely visible) died later from asphyxiation as it attempted to ingest too large a bone, which became lodged. (Courtesy of Films and Research for an Endangered Environment, Ltd., Suite 1735, 201 N. Wells Street, Chicago, Illinois.)

mammals and is usually ascribed to a lack of calcium or phosphorus (Krausman and Bissonette, 1977; Sekulic and Estes, 1977; Langman, 1978).

The occasional inability of free-ranging wildlife to ingest adequate calcium has been suggested as an important determinant of their productivity (Dale, 1954; Harper and Labisky, 1964; Kopischke and Nelson, 1966; Cowles, 1967; Anderson and Stewart, 1969, 1973; Kreulen, 1975). One of the better examples of the interaction of ecosystem components in affecting the intake of calcium and subsequent productivity by specific organisms was given by Mundy and Ledger

TABLE 5.4
Relative Proportions (%) of Main Items Found at Vulture Nests

Item	Cape vulture (no large carnivores)	White-backed vulture (carnivores)
Bone fragments	37	92
Artifacts (glass, china, plastic)	31	0
Whole bones (ribs and phalanges)	18	3
Other (teeth, stones)	14	5
	100	100

[a]From Mundy and Ledger (1976), courtesy of the *South African Journal of Science.*

(1976). Breeding white-backed vultures were studied in two wildlife preserves in southern Africa where large mammalian carnivores (such as lions and hyenas) existed, and cape vultures were studied in a predominant ranching area where large carnivores had been eliminated. Since meat is very deficient in calcium, the altricial nestlings were dependent on the adults retrieving appropriate-sized bone fragments to the nest and the subsequent ingestion of the bone fragments. Of the 231 cape vulture chicks examined, 4.8% had broken, mended, or deformed wing bones and 11% had rickets. However, all 35 white-backed vulture chicks examined had completely normal bone formation. The two species were apparently very dependent on large carnivores breaking the bones of prey into suitable-sized fragments. White-backed vultures coexisting with large carnivores brought many more bone fragments to the nests than did the cape vultures nesting in the ranching area. The latter brought larger bones and pieces of glass, plastic, china, stones, and teeth in an apparent effort to provide calcium for the growing chicks (Table 5.4). Calcium-deficient birds often exhibit a greater interest in bizarre objects of edible size (Joshua and Mueller, 1979). Thus, the interaction between vultures and large carnivores is apparently essential in meeting calcium requirements of young vultures.

B. Sodium

1. Functions

The terrestrial animal maintains an internal, marine-like environment by bathing all body cells in a saline solution. The chief cation of the marine and the extracellular environment is sodium. Physiologically, sodium is important in the regulation of body fluid volume and osmolarity, acid-base balance and tissue pH, muscle contraction, and nerve impulse transmission, and is therefore necessary for growth and reproduction. Although the sodium ion is quite soluble, approximately 25% of the total body sodium content is in a nonionizable, poorly available form in bone (Forbes, 1962; Weeks and Kirkpatrick, 1976; Green, 1978).

2. Requirements

Sodium requirements for growth and reproduction of birds and mammals range from 0.05 to 0.15% of the diet (Table 5.5) (Morris, 1980). However, sodium requirements, particularly in mammals, are not constant but can be increased by behavioral stress, reproduction, and excessive potassium or water intake. Sodium balances and therefore requirements are controlled primarily by aldosterone, a steroid hormone secreted by the adrenal cortex that stimulates sodium reabsorption and retention by the kidney, sweat glands, salivary glands, and gastrointestinal mucosa. Glucocorticoids, also secreted by the adrenal cortex during stress with pronounced effects on carbohydrate metabolism, may antagonize the sodium-retaining effects of aldosterone (Uete and Venning, 1962). Such an interaction is supported by the observation that the ad libitum consumption of sodium chloride solutions by captive meadow voles increased per animal as population densities increased (Aumann and Emlen, 1965). The effect of stress on sodium requirements may provide a major problem in relating the sodium kinetics of captive wildlife to free-ranging animals.

Reproduction increases the requirement, since fetal growth and milk production require sodium. Sodium appetite in free-ranging animals has often been highest during gestation and lactation. The desire for sodium during reproduction in herbivores may also be due to the consumption of excessive water and potassium in very succulent forage, both of which may reduce sodium retention (Bintz, 1969; Scoggins et al., 1970; Weeks and Kirkpatrick, 1976, 1978; H. C. Smith et al., 1978).

Just as water turnover measurements were helpful in estimating water requirements, mineral turnover studies might also indicate an optimum intake level. Sodium intake and turnover measured with isotopic sodium are linearly related (Fig. 5.4). However, the two variables are not direct reflections of each other, since turnover averages 84% of the observed intake. Green (1978) and Green and Dunsmore (1978) have suggested that the difference between intake and turnover may be a digestibility or absorption estimate, an error due to the very slow interchange between the readily exchangeable and poorly exchangeable pools, or an error due to the preferential retention of isotopic sodium. The regression of sodium intake versus fecal excretion in jackrabbits consuming herbage supports the contention that ingested sodium is readily available, since the net gastrointestinal retention was 95.7% (data of Nagy et al., 1976). However, most of the absorbed sodium was excreted in the urine. Jackrabbits, fox squirrels, howler monkeys, and wallabies at an extrapolated zero sodium intake continued to lose 2.77 mg sodium/W_{kg}/day via the urine and feces (Fig. 5.4).

TABLE 5.5
Additional Mineral Requirements or Allowances for Several Species of Wildlife

Species	Mineral										Reference
	Na	K	Mg	Cl	Fe	Zn	Mn	Cu	I	Se	
	% of diet				ppm in diet						
Birds											
Bobwhite quail											
Growth (0–12 weeks)	0.085	—	—	0.11	—	—	—	—	0.3	—	M. L. Scott *et al.* (1960), Scott (1963a), Scott (1963b)
Growth	0.085	—	0.06	0.11	—	50	90	—	0.3	—	NRC (1977a)
Growth	0.1	0.6	0.05–0.04	0.15	130–90	62–60	95–70	13–9	0.3	0.1	M. L. Scott (1977–1978)
Egg production	0.1	0.6	0.035	0.15	130	60	70	13	0.3	0.1	M. L. Scott (1977–1978)
Egg production	0.15	—	0.04	0.15	—	50	70	—	0.3	—	NRC (1977a)
Chukar partridge											
Growth	0.095	—	—	—	—	—	—	—	—	—	Anthony *et al.* (1978)
Ducks and geese (domestic)											
Growth	0.15	—	0.05	—	—	—	40	—	—	—	NRC (1977a)
Growth	0.1	0.6	0.05–0.04	0.15	130–90	65–60	75–60	13–9	0.3	0.1	M. L. Scott (1977–1978)
Egg production	0.1	0.6	0.035	0.15	130	60	60	13	0.3	0.1	M. L. Scott (1977–1978)
Egg production	0.15	—	0.05	—	—	—	25	—	—	—	NRC (1977a)
Pheasant											
Growth (0–12 weeks)	0.085	—	—	0.11	—	62	95	—	0.3	—	M. L. Scott *et al.* (1960), Scott (1963a), Scott (1963b)
Growth	0.1	—	0.06	0.11	—	60	90	—	0.3	—	NRC (1977a)
Growth	0.1	0.6	0.05–0.04	0.15	130–90	62–60	95–70	13–9	0.3	0.1	M. L. Scott (1977–1978)
Egg production	0.1	0.6	0.035	0.15	130	60	70	13	0.3	0.1	M. L. Scott (1977–1978)
Egg production	0.1	—	0.04	0.11	—	50	70	—	0.3	—	NRC (1977a)

(continued)

43

TABLE 5.5 *Continued*

Species	Mineral										Reference
	Na	K	Mg	Cl	Fe	Zn	Mn	Cu	I	Se	
	% of diet				ppm in diet						
Turkey (domestic)											
Growth (0–8 weeks)	0.15	0.4	0.05	0.08	60	75	55	6	0.4	0.2	NRC (1977a)
Growth (8–16 weeks)	0.15	0.4	0.05	0.08	40	40	25	4	0.4	0.2	NRC (1977a)
Growth	0.1	0.6	0.05–0.04	0.15	130–90	70–60	95–70	13–9	0.3	0.2	M. L. Scott (1977–1978)
Egg production	0.15	0.4	0.05	0.08	60	65	35	6	0.4	0.2	NRC (1977a)
Egg production	0.1	0.6	0.035	0.15	130	60	70	13	0.3	0.2	M. L. Scott (1977–1978)
Mammals											
Guinea pig											
Growth	—	0.5–1.4	0.1–0.3	—	50	20	40	6	1.0	0.1	NRC (1978a)
Hamster											
Growth	0.15	0.6	0.06	—	140	9.2	3.7	1.6	1.6	0.1	NRC (1978a)
Laboratory mouse											
Growth and reproduction	—	0.2	0.05	—	25	—	45	4.5	0.25	—	NRC (1978a)
Laboratory rat											
Growth and reproduction	0.05	0.36	0.04	0.05	35	12	50	5	0.15	0.1	NRC (1978a)
Nonhuman primate											
Growth	—	—	0.26	—	—	—	—	—	—	—	NRC (1978b)
Rabbit (domestic)											
Growth, maintenance, and reproduction	0.2	0.6	0.03–0.04	—	—	—	8.5	3	0.2	—	NRC (1977b)

The slope of the relationship between sodium intake and excretion (0.99) also illustrates one of the fundamental principles of sodium homeostasis. Excretion in the nonproductive animal virtually always equals intake. The regression covering the entire range of values indicates that sodium balance is not achieved until intake equals 250.0 mg/W_{kg}/day. However, as will be discussed later, the body has a tremendous ability to conserve sodium at low intake. Consequently, the relationship between intake and excretion is undoubtedly slightly curvilinear at

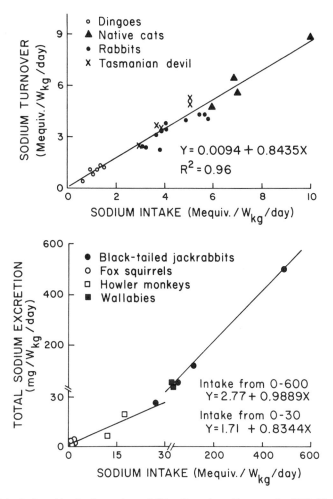

Fig. 5.4. Sodium kinetics in captive wildlife. (Data from Nagy *et al.*, 1976; Havera, 1978; Green, 1978; Green and Dunsmore, 1978; Green *et al.*, 1978; Green and Eberhard, 1979; Nagy and Milton, 1979; Hume and Dunning, 1979.)

the very low end of the scale, with retention being maximized until the minimal urine and fecal losses are balanced by intake. If one examines only the excretion values at intakes below 30 mg/W_{kg}/day, sodium balance occurs at 10.3 mg/W_{kg}/ day.

3. Deficiencies and Dietary Sources

Few terrestrial plants require sodium, and only halophytes actively concentrate it, even though sodium is the sixth most abundant mineral in the earth's crust (Botkin et al., 1973). Therefore, the potential exists, particularly for herbivores, to incur sodium deficiencies in areas low in available soil or water sodium. Symptoms of sodium deficiencies are reduced growth, softening of bones, corneal keratinization, gonadal inactivity, loss of appetite, weakness and incoordination, decreased plasma sodium concentration, adrenal hypertrophy, impaired dietary energy and protein utilization, and decreased plasma and fluid volumes producing shock and death.

Carnivores normally would not incur sodium deficiencies, since the tissues of even sodium-stressed herbivores have a relatively high sodium content (Blair-West et al., 1968). However, captive pinnipeds (and possibly other marine mammals) maintained in freshwater pools can develop sodium deficiencies when fed fish diets without mineral supplementation (Geraci, 1972). The sodium concentration of marine fishes ranges from 0.0003 to 0.0015% and is often further reduced when frozen fish are thawed by immersion in distilled water (an average 25% reduction after 3 hr).

Sodium supplements are easily provided for captive animals by either mixing directly into the diet or providing free choice. Dose rates for marine mammals in freshwater pools of 3 g NaCl/kg food (i.e., 0.12% dietary sodium) or higher if necessary are suggested (Geraci, 1972). The practice is not recommended and may be detrimental for marine mammals or birds maintained in sea water. Captive animals having free access to sodium salts may consume more than their minimum daily requirements, but excesses are readily excreted when adequate drinking water is provided. Salt (NaCl) toxicity occurred in pheasants only when the sodium content of the diet reached 2.95%, or about 30 times the estimated daily requirement (M. L. Scott et al., 1960).

Wildlife ecologists have long been interested in understanding how free-ranging herbivores and granivores meet their sodium requirement and whether low sodium intake might regulate populations. Moose on Isle Royale, caribou in Alaska, deer in Indiana, mountain goats in British Columbia, and fox squirrels in Illinois consumed feeds having from 0.0003 to 0.0251% sodium (Hebert and Cowan, 1971a; Botkin et al., 1973; Weeks and Kirkpatrick, 1976; Havera, 1978; Staaland et al., 1980). These dietary levels are well below estimated requirements of captive animals and are not simply the extreme examples. Soils and vegetation of alpine, mountain, and continental areas having moderate to high snow or rainfall are usually sodium depleted, whereas coastal and desert areas are sodium repleted. For example, vegetation of coastal and desert areas of Australia had sodium concentrations from 110 to 130 times higher than in sodium-depleted alpine regions (Scoggins *et al.*, 1970). However, coastal alpine areas with very high rainfall, such as the Olympic Mountains of Washington, can also be quite sodium depleted.

Sodium is easily leached from ecosystems receiving moderate precipitation since it is not accumulated by plants. Ecologically, coastal rainfall and marine aerosols can be very significant sources of sodium. The sodium in rainwater collected 8 km inland from the ocean in New Zealand deposited approximately 1.3 kg of sodium chloride per hectare per centimeter of rainfall (R. B. Miller, 1961). Lesser though still very significant amounts of sodium (0.08 kg/ha/cm of rainfall or 4–5 kg/ha/year) were deposited in the Santa Yenz Mountains of southern California 5–10 km inland (Schlesinger and Hasey, 1980).

Animals ingesting foods of minimal sodium content must either conserve ingested sodium or augment sodium intake by actively consuming natural sodium supplements. Again, the primary physiological means for sodium retention is the increased aldosterone secretion and corresponding sodium retention. Sodium-stressed mountain rabbits, kangaroos, and wombats excreted urine almost devoid of sodium (Blair-West *et al.*, 1968), whereas rabbits in sodium-replete grassland and desert areas had urinary sodium concentrations 280 times higher (Scoggins *et al.*, 1970). Sodium-stressed animals had adrenal cortex aldosterone secreting areas up to 5 times larger than sodium-replete animals (Myers, 1967; Scoggins *et al.*, 1970; M. C. Smith *et al.*, 1978).

Sodium is also actively resorbed from the intestinal tract to minimize fecal sodium concentrations in herbivores existing in sodium-poor environments. Sodium resorption in the intestinal tract occurs concurrent to water resorption and is more efficient when drier feces are produced. The soft, more moist feces of wild rabbits on sodium-poor diets had 2.5 times the sodium concentration of hard feces (Scoggins *et al.*, 1970). Environmental pollutants, such as crude oil, can reduce intestinal sodium absorption and thereby contribute to mortality of contaminated wildlife (Crocker *et al.*, 1974).

Many observations of herbivores consuming soil or water from mineral licks have been reported (Blair-West *et al.*, 1968; Hebert and Cowan, 1971a; Weir, 1972, 1973; Botkin *et al.*, 1973; Weeks and Kirkpatrick, 1976; Calef and Lortie, 1975). Use of mineral licks has been attributed to the fact that they often have higher sodium content than do surrounding vegetation, soil, or water (Fraser *et al.*, 1980). The preference of animals for sodium in sodium-poor environments was eloquently displayed by Blair-West *et al.* (1968), Weeks and Kirkpatrick (1978), and Fraser and Reardon (1980) by observing rabbits, fox squirrels, woodchucks, white-tailed deer, and moose either directly consuming solutions of sodium salts or chewing and licking wooden stakes soaked in either NaCl, NaI, or $NaHCO_3$, while largely ignoring stakes soaked in distilled water, $MgCl_2$, $CaCl_2$, $KHCO_3$, Kl, and KCl. Similarly, elephants have been observed consuming burned wood and ash in an apparent sodium drive (Weir 1972, 1973), and other animals, such as porcupines, have damaged buildings where wood was impregnated with salt from evaporated urine (Blair-West et al., 1968).

Mountain goats in the Olympic National Park of Washington avidly consume urine or urine-soaked ground and table salt provided by park visitors. Sodium chloride was as effective a bait in spring and summer as food was during winter for trapping white-tailed deer in the Adirondack Mountains of New York (Mattfeld et al., 1972). Moose on Isle Royale actively seek aquatic plants having from 50 to 500 times higher sodium concentrations than terrestrial plants (Botkin et al., 1973). Although many of the field observations of food preferences and mineral lick use are difficult to interpret as unequivocally indicating that sodium is the element being sought, the accumulated information collectively indicates that animals in sodium-depleted areas are actively seeking this element (Weeks, 1978).

Thus, although sodium is a ubiquitous element, its acquisition and retention may require extensive effort, particularly in herbivores. Limited availability of sodium in some areas may restrict animal distribution and productivity (Aumann, 1965; Aumann and Emlen, 1965; Weir, 1972, 1973; Weeks and Kirkpatrick, 1978; Belovsky, 1981; Belovsky and Jordan, 1981). Efficient sodium acquisition and retention may require several generations to develop and therefore may limit newly introduced populations from sodium-replete environments. For example, expanding populations of eastern cottontails invading sodium-depleted areas from sodium-repleted areas had higher mortality during cold stress when glucocorticoids also would be released than did the more adapted resident New England cottontails (Chapman, personal communication).

C. Potassium

Potassium occurs primarily within the cells and functions in nerve and muscle excitability, carbohydrate metabolism and enzyme activation, tissue pH, and osmotic regulation. The very high potassium content of growing plants and animals, usually in excess of animal requirements, greatly reduces the chances of potassium deficiencies occurring in free-ranging or captive wildlife. The accumulation of potassium by both aquatic and terrestrial plants has created many terrestrial environments via biological and geological cycles that are potassium replete (Wilde, 1962).

Symptoms of potassium deficiencies, produced experimentally in many animals, include muscle weakness, poor intestinal tone with intestinal distension, cardiac and respiratory weakness and failure, retarded growth, and tubular degeneration of the kidneys (McDonald et al., 1973). Potassium may have been involved in a die-off of sika deer on James Island, Maryland (Christian *et al.*, 1960; Christian, 1964). The deer population decreased from 300 to 100 without apparent, classical starvation. Potassium deficiencies were suggested from renal pathology at necropsy, even though potassium intake would normally have been adequate. Stress-induced adrenal hyperplasia and excessive potassium excretion may have been produced by excessive population levels rather than by an initial mineral insufficiency (Christian, 1964). However, further studies of the potassium balance of wildlife consuming cured forages in winter is certainly warranted.

The diffusion of cellular potassium into the circulatory system, its subsequent excretion, and, if severe, cardiac and respiratory failure (Wilde, 1962) are also associated with prolonged stress or trauma in laboratory animals. Potassium deficiencies may be rare in wildlife, but deficiencies could occur because of a complex of precipitating factors and should always be suspected in captive wildlife subjected to stress and prolonged diarrhea (Newberne, 1970).

D. Magnesium

The major portion of body magnesium, approximately 70%, is in bone, with bone ash being 0.31–0.75% magnesium (Morgulis, 1931; Wacker and Vallee, 1964; Hyvarinen *et al.*, 1977b). While magnesium is an essential constituent of bone and tooth formation, it is of paramount importance in enzyme activation relative to energy metabolism (McDonald *et al.*, 1973). Like potassium, magnesium is very abundant in both plant and animal dietary sources, since magnesium is the chelated metal in chlorophyll.

Deficiency symptoms include vasodilation, hyperirritability, convulsions, reluctance to stand and loss of equilibrium, tetany, increased heat production due to tonic muscular activity, reduced appetite and weight loss, impaired blood clotting, liver damage, soft tissue calcification, defective bones and teeth, and death (W. J. Miller *et al.*, 1972; McDonald *et al.*, 1973; Church, 1979). Reports of magnesium deficiencies have been extremely rare in either captive or free-ranging wildlife. Free-ranging pheasants apparently even avoided consuming grit high in magnesium during egg production (Kopischke, 1966).

One major pathological magnesium deficiency is grass tetany or grass staggers, which occasionally occurs in herbivores consuming very early spring forage. The forage may be relatively high in magnesium, but the deficiency is apparently produced by dietary factors that affect the absorption or use of magnesium, such as high nitrogen, potassium, long-chain fatty acids, or organic acids (Fontenot, 1972; W. J. Miller *et al.*, 1972; Wilkinson *et al.*, 1972). The onset of grass tetany is so rapid that the resorption of bone magnesium is insufficient to maintain serum and extracellular magnesium levels (W. J. Miller *et al.*, 1972).

Such a deficiency apparently occurred in captive deer grazing a heavily fertilized pasture at Washington State University (K. Farrell, personal communication) and may have been a contributing factor to mortality in emaciated reindeer consuming early spring vegetation (Hyvarinen *et al.*, 1977a). While magnesium deficiencies in herbivores have been difficult to understand because of the mineral's preponderance in the diet, the deficiency may be an important mortality factor when animals consume very lush, spring forage.

D. Chloride and Sulfur

Chloride, the principle anion of body fluids, is involved in acid-base relations, gastric acidity (hydrochloric acid), and digestion. Deficiency symptoms may include reduced growth and feed intake, hemoconcentration, dehydration, nervous disorders, and reduced blood chloride levels. Dietary chloride concentrations can be extremely variable, ranging from approximately 0.1 to 1.0% of the dry matter (NRC, 1971a).

Sulfur is primarily a constituent of the sulfur-containing amino acids (cystine, cysteine, and methionine), with much smaller amounts in biotin, thiamin (two B-complex vitamins), and the hormone insulin (McDonald *et al.*, 1973). Thus, a deficiency of sulfur is synonymous with a deficiency of the sulfur-containing metabolites, particularly the amino acids, and is usually not viewed as a simple deficiency of inorganic sulfur.

II. TRACE ELEMENTS

Trace element research with wildlife has largely been confined to studies of major deficiencies resulting from either improper feeding of captive animals or the inability of free-ranging animals to acquire adequate dietary sources. Trace elements act primarily as catalysts in cellular enzyme systems, although many other functions exist (Table 5.6). Persons having a major interest in trace elements should consult Underwood (1977).

A. Iron

Iron deficiencies are relatively common in zoological gardens, particularly in young, bottle-raised animals (Wallach, 1970; Flieg, 1973). Cow's milk is usually very low in iron relative to levels occurring in milk of other species, although there is considerable variation among sampling times because the iron content of milk usually decreases throughout lactation. For example, while human, cow, and domestic goat milk are all fairly similar in iron content (about 0.5 μg/ml), the iron content of rat and rabbit milk decrease from 15 to 5 μg/ml and from 10 to 4 μg/ml, respectively, during the first 25 days of lactation. The milk iron content of the Australian macropod marsupial *Setonix brachyurus* averages 25 μg/ml while the young is confined to the pouch but rapidly falls to 5 μg/ml as weaning begins (Loh and Kaldor, 1973). Significant differences also occur even among ungulates. For example, moose, Dall sheep, and muskox milk during mid-lactation contain 80, 21, and 12 μg iron per gram of milk, respectively (Baker *et al.*, 1970; Cook *et al.*, 1970a, 1970b). Thus, the anemia in bottle-raised wildlife when fed cow's milk results simply from a deficient diet. Unfortunately, iron supplements for lactating females are generally ineffective in altering the iron content of the milk produced. Similarly, one must be very careful in adding iron to milk, since a phosphorus deficiency can be produced by its precipitation into a nonabsorbable iron complex.

Animal tissue is generally very high in iron relative to carnivore requirements, and adequate amounts are available in most forages with the exception of cereal grains and grasses grown on very sandy soil (Underwood, 1977). Deficiencies in adult animals, therefore, are probably confined to animals fed very purified or

TABLE 5.6
Functions and Deficiency Signs of Trace Elements[a]

Mineral	Major functions	Deficiency signs
Iron (Fe)	Metal chelate of hemoglobin, myoglobin, and oxidizing enzymes	Anemia, listlessness, weight loss
Zinc (Zn)	Essential for synthesis of DNA, RNA, and proteins, component or cofactor of many enzyme systems	Enlarged hocks of birds, poor feathering, retarded growth, rough hair coat and hair loss, weight loss, parakeratosis, reproductive difficulties, loss of appetite, impaired wound healing
Manganese (Mn)	Bone formation, energy metabolism (cofactor in oxidative phosphorylation)	Slipped tendon (perosis), weight loss or reduced growth, impaired reproduction, weakness, nervous disorders, loss of equilibrium, bone malformations
Copper (Cu)	Necessary for hemoglobin and melanin formation, component of several blood proteins and enzyme systems	Anemia, diarrhea, loss of appetite, nervous disorders, loss of hair color, reduced hair growth, defective keratinization of hair and hooves, bone deformities, impaired reproduction
Iodine (I)	Hormones thyroxine and triiodothyronine from thyroid gland	Reduced growth, goiter, bilateral, posterior alopecia in large carnivores, reduced production and energy metabolism

(continued)

TABLE 5.6 *Continued*

Mineral	Major functions	Deficiency signs
Selenium (Se)	Interacts with vitamin E to maintain tissue integrity	Nutritional muscular dystrophy, pale areas in muscles and degeneration of muscle fibers, labored breathing, difficulty in feeding, stiffness and disinclination to move far or fast, diarrhea, liver necrosis, reduced fertility, lung edema
Cobalt (Co)	Vitamin B_{12}	Anemia, wasting away, listlessness, loss of appetite, weakness, fatty degeneration of the liver, reduced hair growth
Molybdenum (Mo)	Component of the enzymes xanthine oxidase, aldehyde oxidase, and sulfite oxidase	Renal calculi, reduced growth
Fluoride (F)	Specific biochemical roles are still uncertain	Reduced rate of growth, infertility, increased susceptibility to dental problems
Chromium (Cr)	Synergism with insulin to promote glucose uptake and metabolism	Reduced rate of growth and longevity, hyperglycemia

[a]Depending on the severity of the deficiency, time span, and species, specific signs may or may not be present.

simplified diets, or to those with injuries, diseases, or parasites that alter iron excretion, since the healthy animal also tends to conserve iron very efficiently. The incorporation of high levels of several species of marine fish, such as Atlantic whiting and Pacific hake, in the diets has induced an iron deficiency called cotton-fur in captive mink. The signs of deficiency include anemia, reduced growth, and a whitish discoloration of the underfur. The deficiency is induced by other dietary factors that apparently reduce iron absorption, since cotton-fur-producing diets have essentially the same iron content as adequate diets (Stout *et al.,* 1960).

Geophagia, or soil consumption not necessarily associated with mineral licks, is commonly observed in wildlife and has been suggested as an important means of meeting trace element requirements, including iron (Arthur and Alldredge, 1979). Although the minerals of ingested soil may be nutritionally important, clay particles can also effectively chelate metal ions and prevent their absorption. Thus, mineral analyses of soils can be misleading relative to their nutritional interpretation if not combined with availability measurements. Geophagia can be either a useful source of iron or a contributing factor to anemia in wild animals, depending on soil iron content and chelating capacity of soil clay (Underwood, 1977).

B. Iodine

The functions of iodine in the animal body are entirely related to the many metabolic controlling mechanisms of the thyroid hormones, thyroxine and tri-iodothyronine. If iodine intake is inadequate for necessary thyroxine production, enlargement of the thyroid (goiter) occurs. The uptake of iodine by the thyroid is inversely proportional to its dietary concentration and ranges from 20% in mule deer consuming 0.28 mg iodine/$W_{kg}^{0.75}$ to 32% in those consuming 0.19 mg/$W_{kg}^{0.75}$ (Gist and Whicker, 1971).

Iodine deficiencies producing both simple and congenital goiters occur with "astonishing frequency" in captive wildlife, since many natural foods, particularly red meats, are iodine deficient (Wallach, 1970). Iodine deficiencies commonly occur in captive felines fed unsupplemented, fresh beef. Compared to the estimated requirement of 1000 μg/day for a lion, fresh beef at 33 μg/kg provides only a fraction of the needed intake (P. P. Scott, 1968; Brambell and Mathews, 1976). Liver may contain as much as 1100 μg/kg. Marine plants and animals and iodized salts can be very useful iodine supplements in the formulation of diets for captive wildlife.

Deficiencies of iodine in free-ranging animals are usually characteristic of areas in which soil iodine and therefore vegetation concentrations are inadequate. Iodine is not required by plants. Thus, the concentration in vegetation can be quite varied and influenced by soil content, chemical form, and soil pH. Although a belt of soils that are deficient in iodine relative to the needs of humans and domestic livestock traverses the northern half of the United States from the Great Lakes region to the Pacific Northwest (Kubota and Allaway, 1972), only recently has goiter been observed in free-ranging white-tailed deer in the same area (Seal and Jones, 1979). Average thyroid weights of the goitrous deer were approximately three times normal, whereas their serum thyroxine levels were one-half to one-third control levels. Iodine supplementation of the goitrous deer increased both thyroxine levels and testicular weights.

Although the iodine requirement of white-tailed deer has not been determined, a diet containing 0.26 ppm was adequate for maintenance and reproduction of captive deer (Watkins, 1980). Since reproduction normally increases the requirement, the maintenance requirement is probably far less and comparable to the 0.10–0.25 ppm in the diet for domestic livestock. Iodine concentrations in deer foods in Michigan ranged from a minimum of 0.008 ppm in fruits, nuts, and grains to 3.10 ppm in aquatic plants (Watkins, 1980). Thus, while the iodine content of many forage species is below the estimated requirement, selective foraging and efficient retention should enable deer to meet their requirements or at least minimize pathological problems.

C. Copper

Copper is widely distributed in foods, with forage concentrations generally reflecting soil content. Milk is usually very low in copper, and liver, the primary storage site of absorbed copper, is very high in those animals receiving adequate dietary copper. Because the neonate consuming only milk must rely on copper acquired prior to parturition, the fetal liver copper content is often many times higher than that occurring in the adult. For example, the copper content of red deer livers fell from 5.9 mg/kg in the fetus and neonate to 2.4 mg/kg in the adult (Reid *et al.*, 1980). Copper absorption tends to be quite low and is affected by chemical form, copper status of the ingesting animal, and the levels of other metal ions, notably calcium, cadmium, zinc, iron, lead, silver, molybdenum, and sulfur, which may interfere with its absorption (McCullough, 1969; Kubota, 1975; Erdman *et al.*, 1978).

Thus, deficiencies can easily occur in captive carnivores fed unsupplemented red meat diets and in herbivores receiving fodder grown on soils either deficient in copper or high in molybdenum. Anemias and faded coat color are characteristic of copper deficiencies in captive felines fed red meat diets supplemented with excessive calcium (Wallach, 1971). Induced copper deficiency in 3-month or older, nursing rhesus monkeys was characterized by achromotrichia (loss of hair color), alopecia (loss of hair), decreased vigor and activity, anemia with little or no evidence of erythrocyte production, and eventually either cessation of nursing and death or spontaneous recoverery (Fig. 5.5) (Obeck, 1978). Deficiencies occurred in monkeys in galvanized cages but not in those kept in stainless steel cages. Apparently, since the infants were continually licking and mouthing portions of their caging, they were ingesting enough zinc from the galvanized coating, which may have been solubilized by urine, to interfere with dietary copper absorption and produce a copper deficiency from a normally adequate diet. Spontaneous recoveries probably occurred in those animals who simply stopped mouthing their cages.

Induced copper deficiencies in moose on the Alaska Kenai Peninsula apparently reduced hair and hoof keratinization, since 1–3% of the population had an overgrowth of hoof in which the hoof doubled in length and curled upward (Flynn *et al.*, 1976, 1977). Mobility of the animals was affected, possibly predisposing many to predation. Although a sex differential in copper metabolism may occur since estrogens apparently can increase the copper-binding capacity of plasma proteins (Hanson and Jones, 1974), the Kenai moose population of adult females had only a 53.3% pregnancy rate as compared to 91.6% for moose in another area of Alaska in which copper intake was adequate.

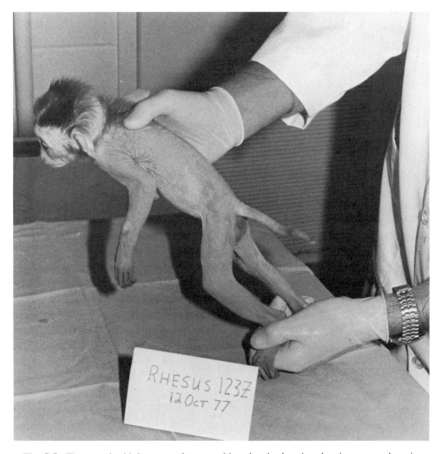

Fig. 5.5. Five-month-old rhesus monkey reared in galvanized caging showing severe alopecia as the result of a zinc-induced copper deficiency. (From Obeck, 1978, courtesy of the American Association for Laboratory Animal Science.)

Copper deficiencies have also been reported in the bontebok (*Damaliscus dorcas dorcas*) grazing on a nature reserve in South Africa known for being in an area of low soil copper (Zumpt and Heine, 1978). Copper contents of the livers of free-ranging bontebok were approximately 20% of supplemented control animals. Their hair coats were less intense in color and looked dirty, faded, and rough, and osteoporosis and healed fractures were common in the deficient animals. When chased, spontaneous fractures of limb bones occurred in several animals, and the animals were generally more susceptible to capture myopathy or overstraining disease that terminated in paralysis, heart failure, and death. Reproduction, however, was apparently not affected in these animals.

Copper toxicities can also occur. Excess copper, primarily associated with copper sulfate treatment of ponds to control algae, produced necrosis and slough-ing of the proventriculus and gizzard, small hemorrhages in the liver, greenish discoloration of the lungs and the ingesta of the digestive tract, and death in Canada geese (Henderson and Winterfield, 1975).

D. Zinc

Because of zinc's role in protein synthesis and enzyme systems, reduced growth and feed intake are the first signs of a zinc deficiency. Other signs of zinc deficiency, such as parakeratosis (thickening and keratinization of the tongue epithelium), a rough, unkempt pelage, alopecia, and reduced play and explorato-ry activity, were observed in squirrel (*Saimisi sciureus*) and rhesus monkeys as well as in numerous domestic species (Barney *et al.*, 1967; Macapinlae *et al.*, 1967; Sandstead *et al.*, 1978). The estimated zinc requirements for wild and domestic animals typically range from 10 ppm to 70 ppm in the diet (Table 5.5) (Henkin *et al.*, 1979). Higher requirement estimates do occur, such as 190 ppm for moustached marmosets (*Saguinus mystax*) (Chadwick *et al.*, 1979), but prob-ably represent either errors or a reduced availability of dietary zinc.

Zinc is generally quite available when consumed in animal tissue. Likewise, because the zinc content of animal tissue is similar to estimated requirements, such as 10–80 ppm in poultry, red meat, and eggs, zinc deficiencies in car-nivores are unlikely. Milk is the major exception for animal products, in that cow's milk averages 3 ppm zinc, giraffe milk 5.4 ppm, and reindeer milk 10.4 ppm (Luick *et al.*, 1974; Hall-Martin *et al.*, 1977; Henkin *et al.*, 1979). The availability of zinc in plants is often less than that in animal tissues because of its complexing with phytate (inositol hexaphosphoric acid), which is most prevalent in seeds. Similarly, high levels of dietary calcium can further reduce zinc avail-ability. Although the zinc content of liver and bone samples varies seasonally in free-ranging granivores and herbivores (Ojanen *et al.*, 1975; Albl *et al.*, 1977; Anke *et al.*, 1980), pathological zinc deficiencies have not been reported in these groups. Zinc requirements of captive wildlife can usually be met by feeding and watering from galvanized pails, pipes, and troughs (Wallach, 1970).

E. Selenium

Selenium has been of interest because plants in many areas of the world contain concentrations that are either deficient or toxic relative to animal metabo-lism (Fig. 5.6). Selenium deficiencies become more probable as dietary con-centration drops below 0.05 ppm (Hebert and Cowan, 1971b), although white-tailed deer consuming a diet of 0.04 ppm selenium were not pathologically deficient (Brady *et al.*, 1978). Chlorinated hydrocarbons may raise the require-

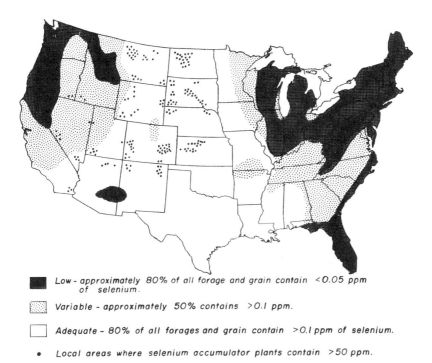

Fig. 5.6. Selenium concentration in plants relative to domestic animal needs and areas of toxicity in the United States. Note the correspondence of selenium deficiency in American wildlife to the map. (Courtesy of J. Kubota, U.S. Plant, Soil and Nutrition Laboratory, Ithaca, New York.)

ment, since they apparently interfere in selenium metabolism (Combs and Scott, 1977). High dietary sulfur can further reduce selenium's availability by forming sulfur–selenium complexes. The various ionic forms of selenium have differing availabilities. For example, selenites, selenates, and selenoamino acids are readily available, whereas metallic selenium and selenides are not. A major portion of ingested selenium in ruminants is apparently incorporated by microorganisms into the selenoanalogues of methionine and cystine.

One of the major functions of selenium is as a component of the enzyme glutathione peroxidase, which prevents oxidation of cell membranes (Combs and Scott, 1977). When dietary selenium is inadequate, membrane oxidation produces characteristic pathological lesions of white striations and degeneration of muscle fibers, or white muscle disease. White muscle disease should not be confused with the similar symptoms of capture myopathy, in which severe exertion or stress of chased or captured animals produces acidosis and death (Harthoorn and Young, 1974; Colgrove 1978). Selenium deficiencies have been observed in free-ranging mountain goats (British Columbia), pronghorn antelope (central Idaho), woodchucks (New York), Hunter's antelope, Grant's and Thomson's gazelles, wildebeest and topi (Kenya and Tanzania), and Nyala antelope (New York Zoo) (Jarret *et al.*, 1964; Gandal, 1966; Murray, 1967; Hebert and Cowan, 1971b; Fleming *et al.*, 1977; Stoszek *et al.*, 1980). The pathological manifestations of muscular degeneration, paralysis, and death have usually been precipitated by stress associated with chasing, trapping, drugging, or handling. However, harassment by predators, hunters, or high population densities could theoretically produce the acute symptoms. While some animals displayed symptoms at trapping, others took several days to develop paralysis and die (Hebert and Cowan, 1971b). Selenium injections are not effective under field conditions in combating the acute pathological signs because of the severity of muscle damage and the lengthy recovery once they appear.

Selenium was first investigated by nutritionists because of its toxicity when ingested at high levels. Selenium toxicity, also called alkali disease or blind staggers in domestic animals, produces an abundance of symptoms ranging in severity from reduced rates of gain to sloughing of the hoofs, atrophy of the heart, cirrhosis of the liver, blindness, partial paralysis, and death due to respiratory failure or starvation (Underwood, 1977). Toxicity is apparently due to the extensive substitution of selenium for sulfur in methionine and cysteine to produce malfunctioning selenoenzymes (Shrift, 1972). Several plants, including milk vetch (*Astragalus bisulcatus* and *A. grayi*), poison vetch (*A. pectinatus*), and prince's plume (*Stanleya pinnata*), are notorious for accumulating toxic selenium concentrations when grown on high-selenium soils. These plants are normally unpalatable to herbivores, but they will be consumed under starvation conditions. More importantly, when these plants drop their leaves or die and decay, readily absorbable selenium is increased in the root area of all adjacent plants. Consequently, plants that are highly palatable but not normally selenium-accumulator species, such as western wheatgrass (*Agropyron smithii*), blue grama (*Bouteloua gracilis*), and winter fat (*Eurotia lanata*), become toxic (Leininger *et al.*, 1977).

F. Cobalt and Fluoride

Although cobalt deficiencies occur in domestic livestock in many areas of the Midwest and eastern United States (Fig. 5.7), white-tailed deer (New York) and ring-necked pheasants (Minnesota) examined in marginal cobalt areas were not deficient in or limited by cobalt (S. E. Smith *et al.*, 1963; Nelson *et al.*, 1966). Ingested fluoride is primarily concentrated in bones and teeth. Although well recognized as an essential trace element, it has been studied in wildlife primarily as a toxic element when ingested in excess of the requirement. Fluorosis (excess flouride) has been observed in white-tailed and black-tailed deer ingesting industrial wastewater (Karstad, 1967; Newman and Yu, 1976). Excess fluoride can

COBALT

Fig. 5.7. Geographic distribution of cobalt-deficient areas in the eastern United States relative to domestic livestock and, therefore, possibly wildlife. (Courtesy of J. Kubota, U.S. Plant, Soil and Nutrition Laboratory, Ithaca, New York.)

▓ Areas where legumes usually contain less than 0.07 ppm of cobalt.

▒ Areas where legumes usually contain from 0.05 to 0.1 ppm of cobalt.

Grasses generally contain less than 0.10 ppm of cobalt throughout most of the U.S.

produce broken, pitted, blackened teeth; excessive or abnormal wear due to softening of the teeth; fractures of the teeth and jaw bones; anorexia; and possibly impaired reproduction.

G. Manganese

The largest store of body manganese is in bone, which ranges from 3.5 ppm in rabbits to 9.2 ppm in red deer antlers (Fore and Morton, 1952; Hyvarinen *et al.*, 1977b). The consumption of deficient diets will lower bone manganese content (Hidiroglou, 1980). Because manganese is essential for the proper formation of the cartilaginous matrix of bone, many of the deficiency signs represent improper bone formation and growth.

The estimated requirements for game birds (Table 5.5) are probably higher than the absolute minimum, but they represent the widespread practice of supplementing poultry diets because manganese salts are relatively inexpensive, manganese has a relatively wide safety margin, and other minerals, such as calcium and phosphorus, can reduce its absorption. Manganese has been largely ignored in studies of free-ranging wildlife (Anke *et al.*, 1979). Pathological manganese deficiencies in free-ranging wildlife have not been observed and probably will seldom occur because manganese is widely distributed in most seeds, forages, and animal tissue in concentrations that are apparently adequate relative to the requirement estimates for domesticated species.

H. Molybdenum

While molybdenum is an essential trace element, it has been studied largely because of its toxicity and complexing potential to induce copper deficiencies. Molybdenum deficiencies are unlikely for both free-ranging and captive wildlife. Toxicity signs, independent of inducing copper deficiencies, include reduced feed intake, weight loss, and diarrhea. Deer have been able to withstand higher levels of dietary molybdenum than have domestic livestock and have also been able to avoid diets high in molybdenum (Ward and Nagy, 1976). The ability of wildlife to purposefully select low-molybdenum feeds would reduce the chances of toxicity.

III. CONCLUSIONS

The various aspects of mineral requirements and metabolism in wildlife have been studied mainly when acute deficiencies have been observed. The chance observation of numerous mineral deficiencies and the obvious alterations in the ecology of many free-ranging animals to cope with potential insufficiencies of such minerals as sodium or iodine suggest a very fruitful area for far more intensive studies. Mineral deficiencies in free-ranging wildlife may become even more prevalent as human beings interrupt the various mineral cycles by cropping and removing domestic livestock, wildlife, trees, and forage (Stoszek *et al.*, 1980). One must surely conclude that the entire field of wildlife-related mineral studies will be one of the more interesting and exciting areas of future development in wildlife nutrition and management.

REFERENCES

Albl, P., Boyazoglu, P. A., and Bezuidenhout, J. P. (1977). Observations on the mineral status of springbok (*Antidorcas marsupialis* Zimmerman) in South West Africa. *Madoqua* 10, 79–83.

Anderson, W. L., and Stewart, P. L. (1969). Relationships between inorganic ions and the distribution of pheasants in Illinois. *J. Wildl. Manage.* 33, 254–270.

Anderson, W. L., and Stewart, P. L. (1973). Chemical elements and the distribution of pheasants in Illinois. *J. Wildl. Manage.* 37, 142–153.

Anke, M., Kronemann, H., Dittrich, G., and Neumann, A. (1979). The supply of wild ruminants with major and trace elements. II. The manganese content of winter grazing and the manganese status of red deer, fallow deer, roes and mouflons. *Arch. Tierernahr.* **29,** 845–858.

Anke, M., Riedel, E., Bruckner, E., and Dittrich, G. (1980). The supply of wild ruminants with major elements and trace elements. III. The zinc content of winter grazing and the zinc status of red deer, fallow deer, roes and mouflons. *Arch. Tierernahr.* **30,** 479–490.

Ankney, C. D., and Scott, D. M. (1980). Changes in nutrient reserves and diet of breeding brown-headed cowbirds. *Auk* **97,** 684–696.

Anthony, D. L., Lumijarvi, D. H., and Vohra, P. (1978). Estimation of sodium requirement of growing chukar partridge (*Alectoris graeca*). *Poult. Sci.* **57,** 307–308.

Arnaud, C. D. (1978). Calcium homeostasis: Regulatory elements and their integration. *Fed. Proc.* **37,** 2557–2560.

Arthur, W. J., III, and Alldredge, A. W. (1979). Soil ingestion by mule deer in North central Colorado. *J. Range Manage.* **32,** 67–71.

Aumann, G. D. (1965). Microtine abundance and soil sodium levels. *J. Mammal.* **46,** 594–604.

Aumann, G. D., and Emlen, J. T. (1965). Relation of population density to sodium availability and sodium selection by microtine rodents. *Nature* **208,** 198–199.

Baker, B. E., Cook, H. W., and Teal, J. J. (1970). Muskox (*Ovibos moschatus*) milk. I. Gross composition, fatty acid, and mineral constitution. *Can. J. Zool.* **48,** 1345–1347.

Barney, G. H., Macapiniac, M. P., Pearson, W. N., and Darby, J. W. (1967). Parakeratosis of the tongue—a unique histopathologic lesion in the zinc-deficient squirrel monkey. *J. Nutr.* **93,** 511–517.

Beasom, S. L., and Pattee, O. H. (1978). Utilization of snails by Rio Grande turkey hens. *J. Wildl. Manage.* **42,** 916–919.

Belovsky, G. E. (1981). A possible population response of moose to sodium availability. *J. Mammal.* **62,** 631–633.

Belovsky, G. E., and Jordan, P. A. (1981). Sodium dynamics and adaptations of a moose population. *J. Mammal.* **62,** 613–621.

Bernhard, R. (1963). Specific gravity, ash, calcium and phosphorus content of antlers of Cervidae. *Can. Field-Nat.* **90,** 310–322.

Beyers, R. J., Smith, M. H., Gentry, J. B., and Ramsey, L. L. (1971). Standing crops of elements and atomic ratios in a small mammal community. *Acta Theriol.* **16,** 203–211.

Bintz, G. L. (1969). Sodium-22 retention as a function of water intake by *Citellus lateralis. In* "Physiological Systems in Semiarid Environments" (C. C. Hoff and M. L. Riedesel, eds.), pp. 45–52. Univ. New Mexico Press, Albuquerque.

Blair-West, J. R., Coghlan, J. P., Denton, D. A., Nelson, J. F., Orchard, E., Scoggins, B. A., Wright, R. D., Myers, K., and Jungueira, C. L. (1968). Physiological, morphological and behavioral adaptation to a sodium deficient environment by wild native Australian and introduced species of animals. *Nature* **217,** 922–928.

Botkin, D. B., Jordan, P. A., Dominski, A. S., Lowendorf, H. S., and Hutchinson, G. E. (1973). Sodium dynamics in a northern ecosystem. *Proc. Nat. Acad. Sci. (USA)* **70,** 2745–2748.

Brady, P. S., Brady, L. J., Whetter, P. A., Ullrey, D. E., and Fay, L. D. (1978). The effect of dietary selenium and vitamin E on biochemical parameters and survival of young among white-tailed deer (*Odocoileus virginianus*). *J. Nutr.* **108,** 1439–1448.

Brambell, M. R., and Mathews, S. J. (1976). Primates and carnivores at Regent's Park. *Symp. Zool. Soc. London* **40,** 147–165.

Bronner, F. (1964). Dynamics and function of calcium. *In* "Mineral Metabolism," Vol. II, Part A (C. L. Comar and F. Bronner, eds.), pp. 341–444. Academic Press, New York.

Calef, G. W., and Lortie, G. M. (1975). A mineral lick of the barren-ground caribou. *J. Mammal.* **56,** 240–242.

Case, R. M., and Robel, R. J. (1974). Bioenergetics of the bobwhite. *J. Wildl. Manage.* **38,** 638–652.

Chaddock, T. T. (1940). Chemical analysis of deer antlers. *Wisc. Cons. Bull.* (Madison) **5,** 42.

Chadwick, D. P., May, J. C., and Lorenz, D. (1979). Spontaneous zinc deficiency in marmosets, *Saguinus mystax. Lab. Anim. Sci.* **29,** 482–485.

Chambers, G. D., Sadler, K. C., and Breitenbach, R. P. (1966). Effects of dietary calcium levels on egg production and bone structure of pheasants. *J. Wildl. Manage.* **30,** 65–73.

Chapman, J. (1980). University of Maryland, Frostburg. Personal communication.

Christian, J. J. (1964). Potassium deficiency: A factor in mass mortality of sika (*Cervus nippon*)? *Wildl. Dis.* **37.**

Christian, J. J., Flyger, V., and Davis, D. E. (1960). Factors in a mass mortality of sika deer. *Chesapeake Sci.* **1,** 79–95.

Church, D. C. (1979). "Digestive Physiology and Nutrition of Ruminants," Vol. II. O. and B. Books, Corvallis, Oreg.

Colgrove, G. S. (1978). Suspected transportation associated myopathy in a dolphin. *J. Am. Vet. Med. Assoc.* **173,** 1121–1123.

Combs, C. F., Jr., and Scott, M. L. (1977). Nutritional interrelationships of vitamin E and selenium. *BioScience* **27,** 467–473.

Conlogue, G. J., Foreyt, W. J., Hanson, A. L., and Ogden, J. A. (1979). Juvenile rickets and hyperparathyroidism in the arctic fox. *J. Wildl. Dis.* **15,** 563–567.

Cook, H. W., Pearson, A. M., Simmons, N. M., and Baker, B. E. (1970a). Dall sheep (*Ovis dalli dalli*) milk. I. Effects of stage of lactation on the composition of the milk. *Can. J. Zool.* **48,** 629–633.

Cook, H. W., Rausch, R. A., and Baker, B. E. (1970b). Moose (*Alces alces*) milk. Gross composition, fatty acid, and mineral constitution. *Can. J. Zool.* **48,** 213–215.

Cowles, R. B. (1967). Fire suppression, faunal changes and condor diets. *Proc. Tall Timbers Fire Ecol. Conf.* **7,** 217–224.

Crocker, A. D., Cronshaw, J., and Holmes, W. N. (1974). The effect of a crude oil on intestinal absorption in ducklings (*Anas platyrhynchos*). *Environ. Pollut.* **7,** 165–177.

Dale, F. H. (1954). Influence of calcium on the distribution of the pheasant in North America. *Trans. North Am. Wildl. Conf.* **19,** 316–323.

Dieterich, R. A., and Van Pelt, R. W. (1972). Juvenile osteomalacia in a coyote. *J. Wildl. Dis.* **8,** 146–148.

Encke, W. (1962). Hand-rearing Cape hunting dogs (*Lycaon pictus*) at the Krefeld Zoo. *Int. Zoo Ybk.* **4,** 292–293.

Erdman, J. A., Ebens, R. J., and Case, A. A. (1978). Molybdenosis: A potential problem in ruminants grazing on coal mine spoils. *J. Range Manage.* **31,** 34–36.

Fleming, W. J., Haschek, W. M., Gutenmann, W. H., Caslick, J. W., and Lisk, D. J. (1977). Selenium and white muscle disease in woodchucks. *J. Wildl. Dis.* **13,** 265–268.

Flieg, G. M. (1973). Nutritional problems in young ratites. *Int. Zoo Ybk.* **13,** 158–163.

Flynn, A., Franzmann, A. W., and Arneson, P. D. (1976). Molybdenum–sulfur interactions in the utilization of marginal dietary copper in Alaskan moose. *In* "Molybdenum in the Environment" (W. R. Chappell and K. K. Petersen, eds.), pp. 115–124. Dekker, New York.

Flynn, A., Franzmann, A. W., Arneson, P. D., and Oldemeyer, J. L. (1977). Indications of copper deficiency in a subpopulation of Alaskan moose. *J. Nutr.* **107,** 1182–1189.

Fontenot, J. P. (1972). Magnesium in ruminant animals and grass tetany. *In* "Magnesium in the Environment" (J. B. Jones, Jr., M. C. Blount, and S. R. Wilkinson, eds.), pp. 131–151. Taylor County Printing Co., Reynolds, Ga.

Forbes, G. B. (1962). Sodium. *In* "Mineral Metabolism," Vol. II, Part B (C. L. Comar and F. Bronner, eds.), pp. 1–72. Academic Press, New York.

Fore, H., and Morton, R. A. (1952). Manganese in rabbit tissues. *Biochem. J.* **51**, 600–603.

Fowler, M. E. (1978a). Metabolic bone disease. In "Zoo and Wild Animal Medicine" (M. E. Fowler, ed.), pp. 55–76. Saunders, Philadelphia.

Fowler, M. E. (1978b). Nutritional value of hard water. *J. Zoo Anim. Med.* **9**, 96–98.

Franzmann, A. W., Oldemeyer, J. L., and Flynn, A. (1975). Minerals and moose. *Proc. Am. Moose Workshop and Conference* **11**, (Mimeo).

Fraser, D., and Reardon, E. (1980). Attraction of wild ungulates to mineral-rich springs in Central Canada. *Holarctic Ecol.* **3**, 36–40.

Fraser, D., Reardon, E., Dieken, F., and Loescher, B. (1980). Sampling problems and interpretation of chemical analysis of mineral springs used by wildlife. *J. Wildl. Manage.* **44**, 623–631.

French, C. E., McEwen, L. C., Magruder, N. D., Ingram, R. H., and Swift, R. W. (1956). Nutrient requirements for growth and antler development in the white-tailed deer. *J. Wildl. Manage.* **20**, 221–232.

Frieden, E. (1972). The chemical elements of life. *Sci. Am.* **227**, 52–60.

Gandal, C. P. (1966). White muscle disease in a breeding herd of Nyala antelope (*Traglophus angasi*) at New York Zoo. *Int. Zoo Ybk.* **6**, 277–278.

Geraci, J. R. (1972). Hyponatremia and the need for dietary salt supplementation in captive pinnipeds. *J. Am. Vet. Med. Assoc.* **161**, 618–623.

Gist, C. S., and Whicker, F. W. (1971). Radioiodine uptake and retention by the mule deer thyroid. *J. Wildl. Manage.* **35**, 461–468.

Gorham, J. R., Peckham, J. C., and Alexander, J. (1970). Rickets and osteodystrophia fibrosa in foxes fed a high horsemeat ration. *J. Am. Vet. Med. Assoc.* **156**, 1331–1333.

Green, B. (1978). Estimation of food consumption in the dingo, *Canis familiaris dingo,* by means of ^{22}Na turnover. *Ecology* **59**, 207–210.

Green, B., and Dunsmore, J. D. (1978). Turnover of tritiated water and ^{22}sodium in captive rabbits (*Oryctolagus cuniculus*). *J. Mammal.* **59**, 12–17.

Green, B., and Eberhard, I. (1979). Energy requirements and sodium and water turnovers in two captive marsupial carnivores: The Tasmanian devil, *Sarcophilus harrisii,* and the native cat, *Dasyurus viverrinus. Aust. J. Zool.* **27**, 1–8.

Green, B., Dunsmore, J., Bults, H., and Newgrain, K. (1978). Turnover of sodium and water by free-living rabbits, *Oryctolagus cuniculus. Aust. Wildl. Res.* **5**, 93–99.

Grimshaw, H. M., Ovington, J. D., Betts, M. M., and Gibb, J. A. (1958). The mineral content of birds and insects in plantations of *Pinus silvestris* L. *Oikos* **9**, 26–34.

Hall-Martin, A. J., Skinner, J. B., and Smith, A. (1977). Observations on lactation and milk composition of the giraffe *Giraffa camelopardalis. S. Afr. J. Wildl. Res.* **7**, 67–71.

Hanson, H. C., and Jones, R. L. (1974). An inferred sex differential in copper metabolism in Ross geese (*Anser rossii*): Biogeochemical and physiological considerations. *Arctic* **27**, 111–120.

Harbers, L. H., Callahan, S. L., and Ward, G. M. (1980). Release of calcium oxalate crystals from alfalfa in the digestive tracts of domestic and zoo animals. *J. Zoo Anim. Med.* **11**, 52–56.

Harper, J. A., and Labisky, R. F. (1964). The influence of calcium on the distribution of pheasants in Illinois. *J. Wildl. Manage.* **28**, 722–731.

Harthoorn, A. M., and Young, E. (1974). A relationship between acid-base balance and capture myopathy in zebra (*Equus burchelli*) and an apparent therapy. *Vet. Rec.* **95**, 337–342.

Havera, S. (1978). Nutrition, supplemental feeding, and body composition of the fox squirrel, *Sciurus niger,* in central Illinois. Doctoral dissertation, Univ. Illinois, Urbana-Champaign.

Henderson, B. M., and Winterfield, R. W. (1975). Acute copper toxicosis in the Canada goose. *Avian Dis.* **19**, 385–387.

Hebert, D. M., and Cowan, I. McT. (1971a). Natural salt licks as a part of the ecology of the mountain goat. *Can. J. Zool.* **49**, 605–610.

Hebert, D. M., and Cowan, I. McT. (1971b). White muscle disease in the mountain goat. *J. Wildl. Manage.* **35,** 752–756.

Henkin, R. I. (1979). "Zinc." Univ. Park Press, Baltimore.

Hidiroglou, M. (1980). Zinc, copper and manganese deficiencies and the ruminant skeleton: A review. *Can. J. Anim. Sci.* **70,** 579–590.

Hinkson, R. S., Jr., Smith, L. T., and Kese, A. G. (1970). Calcium requirements of the breeding pheasant hen. *J. Wildl. Manage.* **34,** 160–165.

Hinkson, R. S., Jr., Gardiner, E. E., Kese, A. G., Reddy, D. N., and Smith, L. T. (1971). Calcium requirement of the pheasant chick. *Poult. Sci.* **50,** 35–41.

Holm, E. R., and Scott, M. L. (1954). Studies on the nutrition of wild waterfowl. *New York Fish Game J.* **1,** 171–187.

Hume, I. D., and Dunning, A. (1979). Nitrogen and electrolyte balance in the wallabies *Thylogale thetis* and *Macropus eugenii* when given saline drinking water. *Comp. Biochem. Physiol.* **63A,** 135–139.

Hyvarinen, H., Helle, T., Nieminen, M., Vayrynen, P., and Vayrynen, R. (1977a). The influence of nutrition and seasonal conditions on mineral status in the reindeer. *Can. J. Zool.* **55,** 648–655.

Hyvarinen, H., Kay, R. N. B., and Hamilton, W. J. (1977b). Variation in the weight, specific gravity and composition of the antlers of red deer (*Cervus elaphus L.*). *Brit. J. Nutr.* **38,** 301–311.

Irving, J. T. (1964). Dynamics and function of phosphorus. *In* "Mineral Metabolism," Vol. II, Part A (C. L. Comar and F. Bronner, eds.), pp. 249–313. Academic Press, New York.

Jarrett, W. H. F., Jennings, J. W., Murray, M., and Harthoorn, A. M. (1964). Muscular dystrophy in wild Hunter's antelope. *E. Afr. Wild. J.* **2,** 158–159.

Joshua, I. G., and Mueller, W. J. (1979). The development of a specific appetite for calcium in growing broiler chicks. *Brit. Poult. Sci.* **20,** 481–490.

Karstad, L. (1967). Fluorosis in deer (*Odocoileus virginianus*). *Bull. Wildl. Dis. Assoc.* **3,** 42–46.

Kaufman, D. W., and Kaufman, G. A. (1975). Prediction of elemental content in the old-field mouse. *In* "Mineral Cycling in Southeastern Ecosystems" (F. G. Howell, J. B. Gentry, and M. H. Smith, eds.), pp. 528–535. ERDA Symp. Series Conf. 740513. National Technical Information Service, Springfield, Va.

Kaufman, G. A., and Kaufman, D. W. (1975). Effects of age, sex, and pelage phenotype on the elemental composition of the old-field mouse. *In* "Mineral Cycling in Southeastern Ecosystems" (F. G. Howell, J. B. Gentry, and M. H. Smith, eds.), pp. 518–527. ERDA Symp. Series Conf. 740513. National Technical Information Service, Springfield, Va.

Kopischke, E. D. (1966). Selection of calcium and magnesium-bearing grit by pheasants in Minnesota. *J. Wildl. Manage.* **30,** 276–279.

Kopischke, E. D., and Nelson, M. M. (1966). Grit availability and pheasant densities in Minnesota and South Dakota. *J. Wildl. Manage.* **30,** 269–275.

Korschgen, L. J., Chambers, G. D., and Sadler, K. C. (1965). Digestion rate of limestone force-fed to pheasants. *J. Wild. Manage.* **29,** 820–823.

Krapu, G. L., and Swanson, G. A. (1975). Some nutritional aspects of reproduction in prairie nesting pintails. *J. Wildl. Manage.* **39,** 156–162.

Krausman, P. R., and Bissonette, J. A. (1977). Bone-chewing behavior of desert mule deer. *Southwest. Nat.* **22,** 149–150.

Kreulen, D. (1975). Wildebeest habitat selection on the Serengeti plains, Tanzania, in relation to calcium and lactation: A preliminary report. *E. Afr. Wildl. J.* **13,** 297–304.

Kubota, J. (1975). Areas of molybdenum toxicity to grazing animals in the western states. *J. Range Manage.* **28,** 252–256.

Kubota, J., and Allaway, W. H. (1972). Geographic distribution of trace element problems. *In* "Micronutrients in Agriculture," (J. J. Mortvedt, P. M. Giordano, and W. L. Lindsay, eds.), pp. 525–554. Soil Sci. Soc. Am., Madison, Wisc.

Langman, V. A. (1978). Giraffe pica behavior and pathology as indicators of nutritional stress. *J. Wildl. Manage.* **42,** 141–147.

Leininger, W. C., Taylor, J. E., and Wambolt, C. L. (1977). Poisonous range plants of Montana. *Cooperative Extension Service, Montana State Univ., Bull. 348.*

Leon, B., and Belonje, P. C. (1979). Calcium, phosphorus, and magnesium excretion in the rock hyrax, *Procavia capensis. Comp. Biochem. Physiol.* **64A,** 67–72.

Loh, T. T., and Kaldor, I. (1973). Iron in milk and milk fractions of lactating rats, rabbits and quokkas. *Comp. Biochem. Physiol.* **44B,** 337–346.

Luick, J. R., White, R. G., Gau, A. M., and Jenness, R. (1974). Compositional changes in the milk secreted by grazing reindeer. I. Gross composition and ash. *J. Dairy Sci.* **57,** 1325–1333.

Macapinlae, M. P., Barney, G. H., Pearson, W. N., and Derby, W. J. (1967). Production of zinc deficiency in the squirrel monkey (*Saimisi sciureus*). *J. Nutr.* **93,** 499–510.

McCullough, D. R. (1969). The tule elk: It's history, behavior, and ecology. *Univ. Calif. Publ. Zool. Vol. 88.*

McDonald, P., Edwards, R. A., and Greenhalgh, J. F. D. (1973). "Animal Nutrition." Longman, New York.

McEwen, L. C., French, C. E., Magruder, N. D., Swift, R. W., and Ingram, R. H. (1957). Nutrient requirements of the white-tailed deer. *Trans. North Am. Wildl. Conf.* **22,** 119–132.

MacLean, S. F., Jr. (1974). Lemming bones as a source of calcium for Arctic sandpipers (*Calidris spp*). *Ibis* **116,** 552–557.

Magruder, N. D., French, C. E., McEwen, L. C., and Swift, R. W. (1957). Nutritional requirements of white-tailed deer for growth and antler development. *Pennsylvania St. Univ., University Park, Bull.* **628.**

Mahon, C. L. (1969). Mineral deficiencies in desert bighorns and domestic livestock in San Juan County. *Trans. Desert Bighorn Coun.* **13,** 27–32.

Mattfeld, G. F., Wiley, J. E., III, and Behrend, D. F. (1972). Salt versus browse-seasonal baits for deer trapping. *J. Wildl. Manage.* **36,** 996–998.

Miller, R. B. (1961). The chemical composition of rainwater at Taita, New Zealand, 1956–1958. *N. Z. J. Sci.* **4,** 844–853.

Miller, W. J., Britton, W. M., and Ansari, M. S. (1972). Magnesium in livestock nutrition. *In* "Magnesium in the Environment" (J. B. Jones, Jr., M. C. Blount, and S. R. Wilkinson, eds.), pp. 109–130. Taylor County Printing Co., Reynolds, Ga.

Morgulis, S. (1931). Studies on the chemical composition of bone ash. *J. Biol. Chem.* **93,** 455–466.

Morris, J. G. (1980). Assessment of sodium requirements of grazing beef cattle: A review. *J. Anim. Sci.* **50,** 145–152.

Mundy, P. J., and Ledger, J. A. (1976). Griffon vultures, carnivores and bones. *S. Afr. J. Sci.* **72,** 106–110.

Murray, M. (1967). The pathology of some diseases found in wild animals in East Africa. *E. Afr. Wildl. J.* **5,** 37–45.

Myers, K. (1967). Morphological changes in the adrenal glands of wild rabbits. *Nature* **213,** 147–150.

Nagy, K. A., and Milton, K. (1979). Aspects of dietary quality, nutrient assimilation and water balance in wild howler monkeys (*Alouatta palliata*). *Oecologia* **39,** 249–258.

Nagy, K. A., Shoemaker, V. H., and Costa, W. R. (1976). Water, electrolyte, and nitrogen budgets of jackrabbits (*Lepus californicus*) in the Mojave Desert. *Physiol. Zool.* **49,** 351–363.

National Research Council (NRC). (1966). Nutrient requirements of rabbits. Publ. No. 1194. Nat. Acad. Sci., Washington, D.C.

National Research Council (NRC). (1968). Nutrient requirements of mink and foxes. Publ. No. 1676. Nat. Acad. Sci., Washington, D.C.

National Research Council (NRC). (1970). Nutrient requirements of beef cattle. Nat. Acad. Sci., Washington, D.C.

National Research Council (NRC). (1971a). Atlas of nutritional data on United States and Canadian feeds. Nat. Acad. Sci., Washington, D.C.

National Research Council (NRC). (1971b). Nutrient requirements of poultry. Publ. No. 1861, Nat. Acad. Sci., Washington, D.C.

National Research Council (NRC). (1972). Nutrient requirements of laboratory animals. Publ. No. 2028, Nat. Acad. Sci., Washington, D.C.

National Research Council (NRC). (1975). Nutrient requirements of sheep. Publ. No. 2212, Nat. Acad. Sci., Washington, D.C.

National Research Council (NRC). (1977a). Nutrient requirements of poultry. Publ. No. 2725, Nat. Acad. Sci., Washington, D.C.

National Research Council (NRC). (1977b). Nutrient requirements of rabbits. Publ. No. 2607, Nat. Acad. Sci., Washington, D.C.

National Research Council (NRC). (1978a). Nutrient requirements of laboratory animals. Publ. No. 2767, Nat. Acad. Sci., Washington, D.C.

National Research Council (NRC). (1978b). Nutrient requirements of nonhuman primates. Publ. No. 2786, Nat. Acad. Sci., Washington, D.C.

Nelson, M. M., Moyle, J. B., and Farnham, A. L. (1966). Cobalt levels in foods and livers of pheasants. *J. Wildl. Manage.* **30,** 423–425.

Nestler, R. B. (1949). Nutrition of bobwhite quail. *J. Wildl. Manage.* **13,** 342–358.

Nestler, R. B., DeWitt, J. B., Derby, J. V., and Moschler, M. (1948). Calcium and phosphorus requirements of bobwhite quail chicks. *J. Wildl. Manage.* **12,** 32–36.

Newberne, P. M. (1970). Syndromes of nutritional deficiency disease in nonhuman primates. *In* "Feeding and Nutrition of Nonhuman Primates" (R. S. Harris, ed.), pp. 205–232. Academic Press, New York.

Newman, J. R., and Yu, M. (1976). Fluorosis in black-tailed deer. *J. Wildl. Dis.* **12,** 39–41.

Obeck, D. K. (1978). Galvanized caging as a potential factor in the development of the "fading infant" or "white monkey" syndrome. *Lab. Anim. Sci.* **28,** 698–704.

Ojanen, M., Haarakangas, H., and Hyvarinen, H. (1975). Seasonal changes in bone mineral content and alkaline phosphatase activity in the house sparrow (*Passer domesticus L.*). *Comp. Biochem. Physiol.* **50A,** 581–585.

Reid, T. C., McAllum, H. J. F., and Johnstone, P. D. (1980). Liver copper concentrations in red deer (*Cervus elaphus*) and wapiti (*C. canadensis*) in New Zealand. *Res. Vet. Sci.* **28,** 261–262.

Reynnells, R. D., and Flegal, C. J. (1979a). Dietary calcium and available phosphorus requirements of adult ring-necked pheasants. *Poult. Sci.* **58,** 1098.

Reynnells, R. D., and Flegal, C. J. (1979b). Dietary calcium and available phosphorus requirements for growing pheasants. *Poult. Sci.* **58,** 1097–1098.

Rings, R. W., Doyle, R. E., Hooper, B. E., and Kraner, K. L. (1969). Osteomalacia in the golden-mantled ground squirrel (*Citellus lateralis*). *J. Am. Vet. Med. Assoc.* **155,** 1224–1227.

Robbins, C. T., Moen, A. N., and Reid, J. T. (1974). Body composition of white-tailed deer. *J. Anim. Sci.* **38,** 871–876.

Romanoff, A. L., and Romanoff, A. J. (1949). "The Avian Egg." Wiley, New York.

Rush, W. M. (1932). "Northern Yellowstone Elk Study." Montana Fish Game Comm., Helena.

Sandstead, H. H., Strobel, D. A., Logan, G. M., Jr., Marks, E. O., and Jacob, R. A. (1978). Zinc deficiency in pregnant rhesus monkeys: Effects on behavior of infants. *Am. J. Clin. Nutr.* **31,** 844–849.

Schlesinger, W. H., and Hasey, M. M. (1980). The nutrient content of precipitation, dry fallout, and intercepted aerosols in the chaparral of southern California. *Am. Midl. Nat.* **103,** 114–122.

Scoggins, B. A., Blair-West, J. R., Coghlan, J. P., Denton, D. A., Myers, K., Nelson, J. F., Orchard, E., and Wright, R. D. (1970). The physiological and morphological response of mammals to changes in their sodium status. *In* "Hormones and the Environment" (G. K. Benson, and J. G. Phillips, eds.), pp. 577–602. Cambridge Univ. Press, Cambridge.

Scott, M. L. (1963a). The nutrition of pheasants and quail. *Game Bird Breeders Gazette* **12**, 16–19.

Scott, M. L. (1963b). Symposium—Specialty feed formulation—The nutrition of pheasants and quail. *Proc. Cornell Nutr. Conf. Feed Manufacturers,* 127–130.

Scott, M. L. (1977–1978). Twenty-five years of research in game bird nutrition. *World Pheasant Assoc. J.* **3**, 31–45.

Scott, M. L., and Holm, E. R. (1964). *Nutrition of Wild Waterfowl. Proc. Cornell Nutr. Conf. Feed Manufacturers,* 149–155.

Scott, M. L., van Teinhoven, A., Holm, E. R., and Reynolds, R. E. (1960). Studies on the sodium, chlorine and iodine requirements of young pheasants and quail. *J. Nutr.* **71**, 282–288.

Scott, P. P. (1968). The special features of nutrition of cats, with observations on wild felidae nutrition in the London Zoo. *Symp. Zool. Soc. London* **21**, 21–36.

Seal, U. S., and Jones, W. (1979). Goiter in a white-tailed deer population. Unpublished manuscript.

Seastedt, T. R., and MacLean, S. F., Jr. (1977). Calcium supplements in the diet of nestling Lapland longspurs *Calcarius lapponicus* near Barrow, Alaska. *Ibis 119,* 531–533.

Sekulic, R., and Estes, R. D. (1977). A note on bone chewing in the sable antelope in Kenya. *Mammalia* **41**, 537–539.

Shrift, A. (1972). Selenium toxicity. *In* "Phytochemical Ecology" (J. B. Harborne, ed.), pp. 145–161. Academic Press, New York.

Skar, H. J., Hagvar, S., Hagen, A., and Ostbye, E. (1975). Food habits and body composition of adult and juvenile meadow pipit *(Anthus pratensis (L.)). In* "Fennoscandian Tundra Ecosystems. Ecol. Studies" Vol. 17, Part 2 (F. E. Wielgolaski, ed), pp. 160–169. Springer-Verlag, New York.

Slusher, R., Bistner, S. I., and Kircher, C. (1965). Nutritional secondary hyperparathyroidism in a tiger. *J. Am. Vet. Med. Assoc.* **147**, 1109–1115.

Smith, M. C., Leatherland, J. F., and Myers, K. (1978). Effects of seasonal availability of sodium and potassium on the adrenal cortical function of a wild population of snowshoe hares, *Lepus americanus. Can. J. Zool.* **56**, 1869–1876.

Smith, S. E., Gardner, R. W., and Swanson, G. A. (1963). A study of the adequacy of cobalt nutrition in New York deer. *New York Fish Game J.* **10**, 225–227.

Staaland, H., White, R. W., Luick, J. R., and Holleman, D. F. (1980). Dietary influences on sodium and potassium metabolism of reindeer. *Can. J. Zool.* **58**, 1728–1734.

Stoszek, M. J., Willmes, H., and Kessler, W. B. (1980). Trace-mineral metabolism of Idaho pronghorn antelope. *NW Sec. Wildl. Soc. Ann. Meeting,* Banff, Alberta.

Stout, F. M., Oldfield, J. E., and Adair, J. (1960). Aberrant iron metabolism and the "cotton-fur" abnormality in mink. *J. Nutr.* **72**, 46–52.

Tomson, F. N., and Lotshaw, R. R. (1978). Hyperphosphatemia and hypocalcemia in lemurs. *J. Am. Vet. Med. Assoc.* **173**, 1103–1106.

Trost, R. E. (1981). Dynamics of grit selection and retention in captive mallards. *J. Wildl. Manage.* **45**, 64–73.

Uete, T., and Venning, E. H. (1962). Interplay between various adrenal cortical steroids with respect to electrolyte excretion. *Endocrinology* **71**, 768–778.

Ullrey, D. E., Youatt, W. G., Johnson, H. E., Fay, L. D., Schoepke, B. L., Magee, W. T., and Keahey, K. K. (1973). Calcium requirements of weaned white-tailed deer fawns. *J. Wildl. Manage.* **37**, 187–194.

Ullrey, D. E., Youatt, W. G., Johnson, H. E., Cowan, A. B., Fay, L. D., Covert, R. L., Magee, W. T., and Keahey, K. K. (1975). Phosphorus requirements of weaned white-tailed deer fawns. *J. Wildl. Manage.* **39**, 590–595.

Underwood, E. J. (1977). "Trace Elements in Human and Animal Nutrition." Academic Press, New York.

Van Pelt, R. W., and Caley, M. T. (1974). Nutritional secondary hyperparathyroidism in Alaskan red fox kits. *J. Wildl. Dis.* **10**, 47–52.

Wacker, W. E. C., and Vallee, B. L. (1964). Magnesium. *In* "Mineral Metabolism," Vol. II, Part A (C. L. Comar and F. Bronner, eds.), pp. 483–521. Academic Press, New York.

Wallach, J. D. (1970). Nutritional diseases of exotic animals. *J. Am. Vet. Med. Assoc.* **157,** 583–599.

Wallach, J. D. (1971). Nutritional problems in zoos. *Proc. Cornell Nutr. Conf. Feed Manufacturers,* 10–19.

Wallach, J. D., and Flieg, G. M. (1967). Nutritional secondary hyperparathyroidism in captive psittacine birds. *J. Am. Vet. Med. Assoc.* **151,** 880–883.

Wallach, J. D., and Flieg, G. M. (1969). Nutritional secondary hyperparathyroidism in captive birds. *J. Amer. Vet. Med. Assoc.* **155,** 1046–1051.

Wallach, J. D., and Flieg, G. M. (1970). Cramps and fits in carnivorous birds. *Int. Zoo Ybk.* **10,** 3–4.

Ward, G. M., and Nagy, J. G. (1976). Molybdenum and copper in Colorado forages, molybdenum toxicity in deer, and copper supplementation in cattle. *In* "Molybdenum in the Environment," Vol. I (W. R. Chappell and K. K. Petersen, eds.), pp. 97–113. Dekker, New York.

Watkins, B. E. (1980). Iodine status and thyroid activity of white-tailed deer (*Odocoileus virginianus borealis*). Doctoral dissertation, Michigan St. Univ.

Weeks, H. P., Jr. (1978). Characteristics of mineral licks and behavior of visiting white-tailed deer in southern Indiana. *Am. Midl. Nat.* **100,** 384–395.

Weeks, H. P., Jr., and Kirkpatrick, C. M. (1976). Adaptations of white-tailed deer to naturally occurring sodium deficiencies. *J. Wildl. Manage.* **40,** 610–625.

Weeks, H. P., Jr., and Kirkpatrick, C. M. (1978). Salt preferences and sodium drive phenology in fox squirrels and woodchucks. *J. Mammal.* **59,** 531–542.

Weir, J. S. (1972). Spatial distribution of elephants in an African National Park in relation to environmental sodium. *Oikos* **23,** 1–13.

Weir, J. S. (1973). Exploitation of water soluble sodium by elephants in Murchison Falls National Park, Uganda. *E. Afr. Wildl. J.* **11,** 1–7.

Widdowson, E. M., and Dickerson, J. W. T. (1964). Chemical composition of the body. *In* "Mineral Metabolism," Vol. II, Part A (C. L. Comar and F. Bronner, eds.), pp. 1–247. Academic Press, New York.

Wiener, J. G., Brisbin, I. L., Jr., and Smith, M. H. (1975). Chemical composition of white-tailed deer: Whole-body concentrations of macro- and micronutrients. *In* "Mineral Cycling in Southeastern Ecosystems" (F. G. Howell, J. B. Gentry, and M. H. Smith, eds.), pp. 536–541. ERDA Symp. Series Conf. 740513. National Technical Information Service, Springfield, Va.

Wilde, W. S. (1962). Potassium. *In* "Mineral Metabolism," Vol. II, Part B (C. L. Comar and F. Bronner, eds.), pp. 73–107. Academic Press, New York.

Wilkinson, S. R., Stuedemann, J. A., Jones, J. B., Jr., Jackson, W. A., and Dobson, J. W. (1972). Environmental factors affecting magnesium concentrations and tetanigenicity of pastures. *In* "Magnesium in the Environment" (J. B. Jones, Jr., M. C. Blount, and S. R. Wilkinson, eds.), pp. 153–175. Taylor County Printing Co., Reynolds, Ga.

Wilson, D. E., and Hirst, S. M. (1977). Ecology and factors limiting roan and sable antelope populations in South Africa. *Wildl. Monog.* **54.**

Wilson, H. R., Holland, M. W., Jr., and Harms, R. H. (1972). Dietary calcium and phosphorus requirements for bobwhite chicks. *J. Wildl. Manage.* **36,** 965–968.

Zumpt, I. F., and Heine, E. W. P. (1978). Some veterinary aspects of bontebok in the Cape of Good Hope Nature Reserve. *S. Afr. J. Wildl. Res.* **8,** 131–134.

6

Vitamins

Vitamins are organic compounds that usually occur in food in minute amounts and are distinct from carbohydrates, fat, and protein. They are essential for health, maintenance, and production. When vitamins are absent from the diet or are not properly absorbed or used, characteristic deficiency diseases appear. Although many vitamins cannot be synthesized by the animal and therefore must be obtained from the diet, bacterial synthesis of some vitamins in the gastrointestinal tract can reduce the requirements. Free-ranging wildlife must continually adapt their food habits to avoid diets deficient in one or more of the vitamins or precursors.

Vitamins may be classified as fat soluble (vitamins A, D, E, and K) or water soluble (thiamin, B_1; riboflavin, B_2; nicotinic acid or niacin; pyridoxine, B_6; pantothenic acid; biotin; folic acid; cyanocobalamin, B_{12}; choline; ascorbic acid, C; and *myo*-inositol). Fat-soluble vitamins are absorbed in the intestinal tract with fat and are often stored in large concentrations in the body to be used during periods of dietary inadequacy. Excesses of fat-soluble vitamins, primarily vitamins D and A, can be highly toxic. Conversely, water-soluble vitamins, with the exception of the liver storage of vitamin B_{12} and riboflavin, are generally not stored in large amounts and thus are needed in a rather constant dietary supply. Since excesses of water-soluble vitamins are readily excreted in the urine, toxic overdoses seldom, if ever, occur.

I. FAT-SOLUBLE VITAMINS

A. Vitamin A

Vitamin A has been of interest to wildlife biologists in North America since Nestler and his coworkers (Nestler, 1946; Nestler *et al.*, 1949a, 1949b) reported on its importance to bobwhite quail. Quail survival, egg production, egg hatchability, and growth generally increased as vitamin A was added to a deficient basal diet (Fig. 6.1). However, requirement estimates have varied dramatically between and within species (Table 6.1).

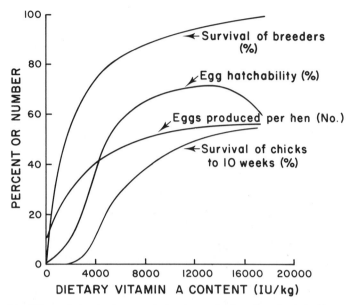

Fig. 6.1. Effects of vitamin A on survival and reproduction of captive bobwhite quail. (From Nestler, 1946, courtesy of the Wildlife Management Institute.)

Acute deficiencies of vitamin A produce debilitating signs because of its importance in vision and the maintenance and growth of epithelial derivatives (Table 6.2). Vitamin A deficiencies are far more common in captive than in free-ranging wildlife because of the feeding of inadequate diets. For example, captive carnivores fed exclusively on carcass meat without liver, cod liver oil, or synthetic sources of vitamin A and granivorous birds are prone to deficiencies (Warner, 1961; Heywood, 1967; P. P. Scott, 1968; Wallach, 1970). Yellow corn is one of the few commercially available grains having significant amounts of provitamin A. Coupled with the inadequacy of a pure meat diet, a higher requirement may also exist for cats, polar bears, and perhaps other specialized carnivores than for other species (Heywood, 1967; Foster, 1981). Hand-reared armadillos (Block, 1974) and polar bear cubs (Wortman and Larue, 1974) developed several abnormalities suggestive of vitamin A deficiencies. Similarly, a captive white-tailed deer being fed winter browse very low in carotene became blind and partially paralyzed (Silver and Colovos, 1957).

Few, if any, cases of acute vitamin A deficiencies have been observed in free-ranging wildlife (Notini, 1941; Cowan and Fowle, 1944; Nestler *et al.*, 1949b). Liver vitamin A levels vary seasonally in free-ranging herbivores and omnivores as intake and requirements fluctuate. Incipient vitamin A deficiencies based on liver concentrations relative to domestic animal standards were reported in 2–3%

TABLE 6.1

Fat-Soluble Vitamin Requirements or Allowances of Several Wildlife Species per Kilogram of Diet

Species	A (IU)	D_3 (IU)	E (IU)	K (mg)	Reference
Birds					
Bobwhite quail					
Growth	3,000	900			NRC (1977a)
Growth	10,000–13,000	1,500	10–15	0.5–1.0	M. L. Scott (1977–1978)
Egg production	3,000	900			NRC (1977a)
Egg production	13,000	1,500	25	0.5	M. L. Scott (1977–1978)
Ducks and geese					
(domestic)					
Growth	1,500–4,000	200			NRC (1977a)
Growth	2,000–11,000	1,500	10–20	0.5–1.0	M. L. Scott (1977–1978)
Egg production	4,000	200–500			NRC (1977a)
Egg production	11,000	1,500	25	0.5	M. L. Scott (1977–1978)
Pheasant					
Growth (0–6 weeks)	3,000	1,200			NRC (1977a)
Growth (6–20 weeks)	3,000	900			NRC (1977a)
Growth	2,000–11,000	1,500	10–15	0.5–1.0	M. L. Scott (1977–1978)
Egg production	11,000	1,500	25	0.5	M. L. Scott (1977–1978)
Turkey					
(domestic)					
Growth (0-3 weeks)	4,000	900	12	1.0	NRC (1977a)
Growth (8–16 weeks)	4,000	900	10	0.8	NRC (1977a)
Growth	4,000–11,000	1,500	10–20	0.5–1.0	M. L. Scott (1977–1978)
Egg production	4,000	900	25	1.0	NRC (1977a)
Egg production	11,000	1,500	25	0.5	M. L. Scott (1977–1978)
Mammals					
Fox	2,410				NRC (1968)
Guinea pig					
Growth	23,000	1,000		5.0	NRC (1978a)
Hamster					
Growth	6,700	2,484	3000	4.0	NRC (1978a)
Laboratory mouse					
Growth and reproduction	500	150	20	3.0	NRC (1978a)
Laboratory rat					
Growth and reproduction	4,000	1,000	30	0.05	NRC (1978a)
Mink	3,500		25		NRC (1968)
Nonhuman primate	10,000–15,000	2,000	50		NRC (1978b)
Rabbit					
(domestic)					
Growth	580		40		NRC (1977b)
Gestation	>1,160		40	0.2	NRC (1977b)

TABLE 6.2
General Functions and Reported Deficiency Symptoms of the Fat-Soluble Vitamins

Vitamin	Major functions	Deficiency signs
A (Retinol, retinal, and retinoic acid)	Major constituent of visual pigment (rhodopsin), maintenance, differentation and proliferation of epithelial tissue, glycoprotein synthesis	Nervous disorders, reduced fertility or sterility, birth defects, reduced egg hatchability and chick survival, reduced growth or loss of weight, oral and nasal pustules, weakness, night blindness, impaired eyesight because of none or copious lacrimination, corneal degeneration, eye infections and eyelid adhesion, bone and teeth abnormalities, lack of alertness, visceral gout, unsteady gait and incoordination, ruffled-droopy appearance
D (D_2-ergocalciferol) (D_3-cholecalciferol)	Necessary for active calcium absorption from the gut, calcium metabolism and resorption from bone, HPO_4^{2-} reabsorption from kidney tubules	Rickets, osteomalacia, nervous disorders
E (tocopherol)	Antioxidant and useful in several different metabolic systems	Yellow fat disease (steatitis), sudden death when subjected to stress, dystrophic

(continued)

of deer killed by cars during winter in Michigan (Youatt *et al.*, 1976). Although Nestler *et al.* (1949b) suggested that free-ranging game birds might be limited directly by vitamin A, later researchers disagreed (Schultz, 1948, 1959; Jones, 1950; Thompson and Baumann, 1950). Vitamin A sources, primarily green vegetation, were always in sufficient quantities to prevent complete depletion in Gambel's quail in Arizona, even in the driest years (Hungerford, 1964). However, when average body reserves fell below 550 μg/liver or 175 μg/g of liver, sex organs regressed even during the normal breeding period.

Thus, it has often been postulated that vitamin A synchronizes reproduction in desert birds and mammals with rains and the emergence of green vegetation (Hungerford, 1964; Reichman and Van De Graaf, 1975). However, Leopold *et al.* (1976) and Labov (1977) have suggested that vitamin A may be less impor-

TABLE 6.2 *Continued*

Vitamin	Major functions	Deficiency signs
		lesions of intercostal and myocardial muscles, orange or brownish yellow discoloration of body fat or body organs, hemorrhages and increased blood cell volume and mean cell hemoglobin, hypersensitivity, decreased activity and depression, anorexia, fever, lumpiness of subcutaneous fat, severe edema, exudative diathesis or fluid accumulation in the pleural and abdominal cavity and body tissues, nutritional muscular dystrophy, severe hemolytic anemia, weight loss, abnormal pelage molt, unsteady gait, blood in urine, fur discoloration, reproductive failure, ataxia, electrolyte imbalances
K (Phylloquinone and menaquinone)	Necessary for blood clotting	Stillbirth or death of neonates soon after birth, hemorrhaging

tant than other plant constituents to desert mammals and birds as an actual regulator of reproduction. Phytoestrogens, substances produced by plants having estrogenic effects, were more abundant in the stunted vegetation growing during dry years but were nearly absent in the luxuriant growth of wet years. The reproductive effort of free-ranging quail populations was inversely proportional to plant estrogen content, and captive feeding trials further demonstrated the inhibitory effects of plant estrogens on quail reproduction. Thus, while acute vitamin A deficiencies may be extremely rare in free-ranging populations, evidence does exist indicating that very marginal deficiencies may occasionally occur and, in conjunction with other plant constituents, may be an important mechanism controlling reproduction, particularly in desert-dwelling species.

Vitamin A is not found in plants and therefore must be of animal origin.

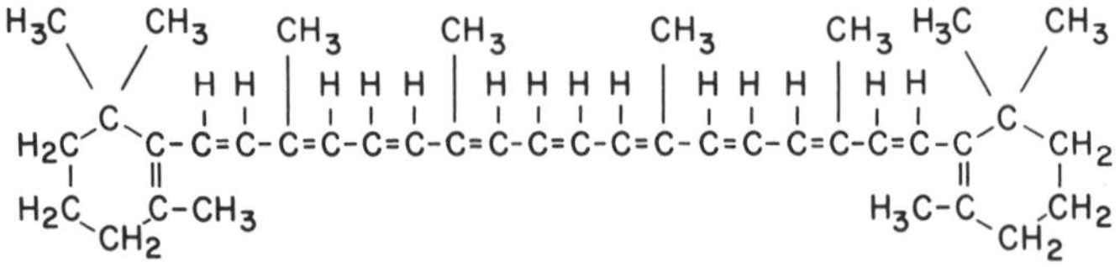

$$H_3C \quad CH_3 \quad CH_3 \qquad CH_3 \qquad\quad CH_3 \qquad\quad CH_3 \ H_3C \quad CH_3$$

$$
\begin{array}{c}
H_2C \\
| \\
H_2C \\
\searrow CH_2
\end{array}
\overset{C}{\underset{C-CH_3}{\|}}
\!\!\! C{-}C{=}C{-}C{=}C{-}C{=}C{-}C{=}C{-}C{=}C{-}C{=}C{-}C{=}C{-}C{=}C{-}C
\begin{array}{c}
H_3C{-}C \\
\nearrow\searrow \\
H_3C{-}C \\
\searrow CH_2
\end{array}
$$

β-Carotene

$$
\begin{array}{c}
H_3C \quad H_3C \quad CH_3 \quad CH_3 \\
H_2C \\
| \\
H_2C \searrow \\
\quad CH_2
\end{array}
C{-}C{=}C{-}C{=}C{-}C{=}C{-}C{=}C{-}CH_2OH
$$

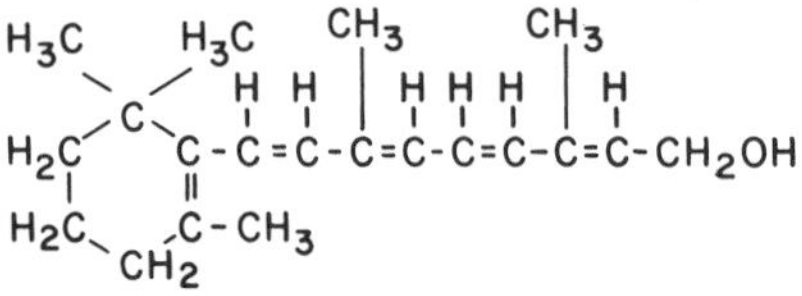

Vitamin A

Fig. 6.2. Chemical structures of β-carotene and vitamin A.

Fortunately, β-carotene and other provitamin A sources are widely distributed in plants and insects and can be hydrolyzed to retinal and subsequently retinol (Fig. 6.2) in the intestinal wall (Haugen and Hove, 1960) and liver (McDonald *et al.*, 1973). Retinol and retinal are interchangeable forms of the vitamin that satisfy all requirements. Retinoic acid can meet many of the requirements but can not function in vision or reproduction. Thus, animals consuming vitamin A-deficient diets supplemented with retinoic acid will be blind and sterile, but otherwise healthy (DeLuca, 1979; Goodman, 1979; Smith and Goodman, 1979).

Although it appears theoretically possible to produce two units of vitamin A from each unit of β-carotene consumed (Fig. 6.2), the efficiency of absorption and conversion in herbivores and granivores is usually well below the maximum possible (Nestler, 1946). Since carnivores normally ingest either retinol or reti-nyl esters from the prey, metabolic systems for the production of vitamin A in carnivores may not even exist. Although mink have retained a minimal capability to convert β-carotene to vitamin A (each unit of true vitamin A must be replaced by six times as much β-carotene), cats have not and therefore require true vitamin A (Warner *et al.,* 1963; P. P. Scott, 1968). Many marine birds and mammals may have similar conversion inefficiencies, since their requirements can be met by the vitamin A contained in marine invertebrates and other verte-brates (Kon and Thompson, 1949).

However, red muscle meat and many species of food fish are lacking in vitamin A (P. P. Scott, 1968; Wallach, 1970). The primary site of vitamin A and β-carotene storage in animals, and therefore a major dietary source in carnivores, is the liver (Table 6.3). Smaller amounts can be stored in the kidneys, body fat, lungs, and adrenal glands (Thompson and Baumann, 1950; P. P. Scott, 1968).

TABLE 6.3
Concentrations of Vitamin A in the Liver of Several Wildlife Species[a]

Species	Concentration		Reference
	μg/g	IU/g	
Polar bear	9807	32,650	Russell (1967)
		13,000–18,000	Rodahl and Moore (1943)
		24,300	Rodahl (1949)
Seal (*Phoca barbata*)		13,000	Rodahl and Moore (1943)
Soup fin shark	400	1,330	Russell (1967)
Mule deer			
Males	118–289	393–962	Anderson *et al.* (1972)
Females	116–371	588–1,235	
Pronghorn antelope		1,024–2,200	Weswig (1956)
Dog fish	180	600	Russell (1967)
White-tailed deer	50–143	167–477	Youatt *et al.* (1976)
Gambel's quail	52–927	173–3,087	Hungerford (1964)
Bobwhite quail	6–227	20–756	Nestler (1946)
		42–3,668	Schultz (1959)
Felids (captive)	Nil–384	Nil–1,280	Heywood (1967)

[a]One IU of vitamin A activity = 0.3 μg of crystalline retinol.

One must be very careful in supplementing diets of captive animals with vitamin A because it can be toxic if fed in excess. Martin (1975) suggested that excesses of vitamin A are a relatively common cause of ill health and death in captive wildlife. Typical signs of hypervitaminosis A are internal hemorrhaging, yellow discoloration of the liver and fat deposits, weight loss, deformed embryos and neonates, bone fractures, and reduced reproduction (D. W. Friend and Crampton, 1961; Martin 1975). Because of very high liver storage capabilities of marine mammals for vitamin A (Table 6.3), some of the classical cases of hypervitaminosis A have occurred when humans have consumed polar bear liver (Rodahl and Moore, 1943).

B. Vitamin D

Vitamin D is well recognized as the antirachitic vitamin necessary for proper calcium absorption and metabolism. Vitamin D is required for the formation of a protein necessary for the active transport of calcium across the intestinal mucosa and into the body (Wasserman *et al.*, 1966). Although very small amounts of dietary calcium can be absorbed by simple diffusion, active transport requiring vitamin D is necessary if calcium homeostasis is to occur. However, abundant

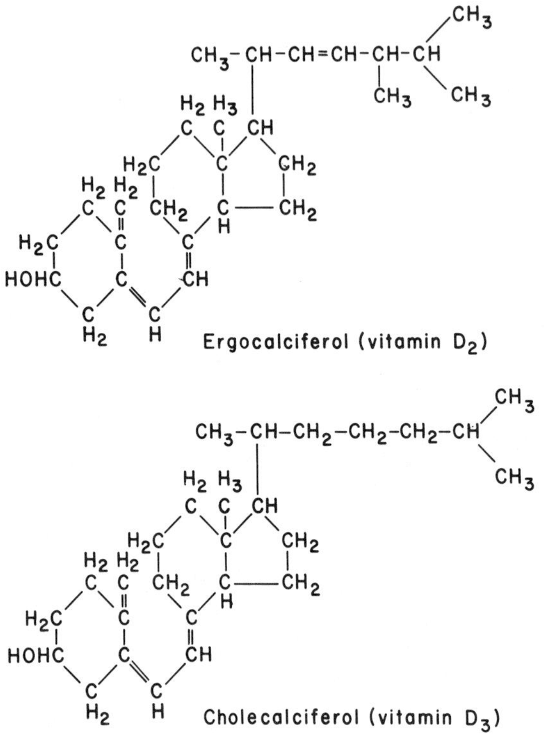

Fig. 6.3. The two major forms of vitamin D.

vitamin D cannot overcome the effects of excess phosphorus or other elements that reduce the total availability of ionized calcium for absorption. Apparently, vitamin D does not affect calcium deposition in the bones once it is absorbed since injected calcium is deposited equally well in normal and vitamin D-deficient chicks (Keane *et al.*, 1956). However, calcium resorption induced by parathyroid hormone is a vitamin D-dependent process. Thus, signs of vitamin D deficiency reflect improper or inadequate calcium metabolism (Table 6.2).

The antirachitic vitamin occurs in two major forms (vitamin D_2 = ergocalciferol and vitamin D_3 = cholecalciferol) (Fig. 6.3). Humans, cattle, Old World primates, and rats can utilize both forms of the vitamin effectively, whereas New World primates utilize vitamin D_2 less efficiently than vitamin D_3 (Portman, 1970). For example, vitamin D_3 is at least eight times more effective than vitamin D_2 in preventing rickets in squirrel monkeys (Lehner *et al.*, 1966). Birds, reptiles, amphibians, and fish require vitamin D_3 and cannot utilize vitamin D_2 efficiently (Hay and Watson, 1977). The specificity of the two forms of the vitamin in the various animal groups is partially due to the binding capacity and transport by blood proteins. The blood proteins of 11 orders of birds were found to be specific for vitamin D_3 (Hay and Watson, 1977). Mammals are much more variable in their capabilities to transport and use D_2 and D_3 (Table 6.4). The echidna, an egg-laying, very primitive mammal, preferentially utilizes vitamin D_3, as do reptiles and birds. In approximately 25% of all the proteins examined that transport vitamin D in mammals, D_3 is bound preferentially to D_2, but to a much lesser degree than observed in the nonmammalian vertebrates. One should be aware of these differences when supplementing the diets of captive animals.

TABLE 6.4
Mammalian Species Examined for Plasma Protein Vitamin D Transport Specificity[a,b]

Monotremes
Tasmanian echidna (*Tachyglossus setosus*)*

Marsupials
Bennett's wallaby (*Protemnodon rufogrisea*)* (10%)

Insectivores
European hedgehog (*Erinaceus europaeus*)* (15%)
Common Eurasian mole (*Talpa europaea*)

Bats
Bat-noctules (*Nyctalus noctula*)

Primates
(a) Prosimians
Large tree-shrew (*Tupaia tana*)* (15%)
Thick-tailed bushbaby (*Galago crassicaudatus*)
(b) New-World monkeys
Owl monkey (*Aotus trivirgatus*)
White-throated capuchin (*Cebus capucinu*)
White-fronted capuchin (*Cebus albifrons*)
Brown capuchin (*Cebus apella*)
Common marmoset (*Callithrix jaccus*)
(c) Old-World monkeys
Rhesus macaque (*Macaca mulatta*)
Olive baboon (*Papio anubis*)
Patas monkey (*Erythrocebus patas*)
(d) Apes
White-handed gibbon (*Hylobates lar*)

Rodents
Brush-tailed porcupine (*Atherurus*)
Agoutis (*Dasyprocta*)* (30%)

Cetaceans
Pacific dolphin (*Delphinus bairdi*)
Killer whale (*Orcinus orca*)

(*continued*)

TABLE 6.4 *Continued*

Carnivores
  Lion (*Panthera leo*)* (10%)
  Tiger (*Panthera tigris*)* (20%)
  Small Indian mongoose (*Herpestes auropunctatus*)
  Giant panda (*Ailuropoda melanoleuca*)* (30%)
Elephants
  Indian elephant (*Elephas maximus*)
Ungulates
  Odd-toed
    White rhinoceros (*Diceros simus*)
  Even-toed
    Llama (*Lama glama*)
    Guanaco (*Lama guanicoe*)
    Bactrian camel (*Camelus bactrianus*)* (20%)
    Red deer (*Cervus elaphus*)
    Eland (*Taurotragus oryx*)
    Yak (*Bos grunniens*)
    European bison (*Bison bonasus*)
    Goat (*Capra hircus*)* (10%)

[a]Reprinted with permission from *Comparative Biochemistry and Physiology*, Vol. 56B, A. W. M. Hay and G. Watson, Vitamin D_2 in vertebrate evolution, Copyright 1977, Pergamon Press, Ltd.

[b]Species in which the efficiency of binding of 25-OH-D_3 was greater than 25-OH-D_2 are marked with an asterisk. Figures in brackets represent additional quantity of 25-OH-D_2 required to obtain the same displacement of [^{3}H]-25-OH-D_3 as obtained with 25-OH-D_3. The properties of the echidna proteins are similar to those of the reptiles. Those species not marked exhibited equivalence of binding of 25-OH-D_3 and 25-OH-D_2.

Transport mechanisms are critical to vitamin D metabolism, since several chemical changes must occur before either cholecalciferol or ergocalciferol is effective in calcium metabolism. Vitamin D is the only vitamin that can be produced simply by the irradiation of a naturally occurring body sterol (7-dehydrocholesterol) on or within the skin surface (Fig. 6.4). The conversion of 7-dehydrocholesterol to cholecalciferol is a two-step process driven initially by ultraviolet radiation to form previtamin D_3, which is then very slowly converted thermally to vitamin D_3. Since the plasma-binding protein has an affinity for vitamin D that is 1000 times greater than for the previtamin form, the delay in the thermal conversion process provides the means for a very gradual, more continuous release of the vitamin following exposure to sunlight and thus may be important in preventing vitamin D toxicity due to prolonged sun exposure (Holick and Clark, 1978). The absorbed cholecalciferol or ergocalciferol must then be transported in succession to the liver and kidney with accompanying molecular alterations before becoming metabolically active. Liver enzymes, and possibly to a lesser extent kidney and small intestine enzymes in some species,

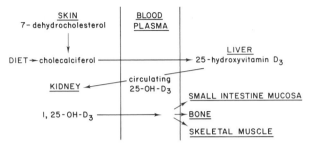

Fig. 6.4. Intermediary metabolism of vitamin D_3. Dietary ergocalciferol (vitamin D_2) must also undergo the same hydroxylations to become metabolically active. (From Fraser and Kodicek (1970), reprinted by permission from *Nature*, Vol. 228, pp. 764–766. Copyright © 1970, Macmillan Journals Limited.)

convert vitamin D_3 to 25-hydroxyvitamin D_3, which at physiological concentrations is not active in calcium metabolism. The subsequent metabolism of 25-OH-D_3 to 1,25-hydroxyvitamin D_3 by the kidney is stimulated by parathyroid hormone and reduced cellular phosphate levels during hypocalcemia and is the active form of the vitamin stimulating bone resorption and intestinal calcium absorption. Conversely, high dietary calcium reduces the production of 1,25-OH-D_3. Thus, the amount of metabolically active vitamin D is controlled through a feedback regulation system to maintain calcium homeostasis. Metabolic aberrations anywhere within the process will produce a functional vitamin D and calcium deficiency.

The divergence of mammals from the other vertebrate taxa in the utilization of vitamin D_2 has been suggested as an evolutionary response in primitive insectivorous mammals to being nocturnally active (Hay and Watson, 1977). Diurnal concealment would reduce the opportunities for the production of adequate vitamin D_3 by irradiation, consequently requiring the utilization of a dietary source. Many invertebrates when irradiated contain an antirachitic factor having ergosterol-like activity that may be important to nocturnally active insectivores, such as bats.

Since vitamin D is normally not stored in the body of terrestrial vertebrates, vitamin D deficiency symptoms can appear within several days to weeks if dietary or irradiation sources are inadequate (Mann, 1970). Although vitamin D deficiencies should be extremely rare in free-ranging animals because of irradiation and dietary intake, captive animals not exposed to sunlight or artificial ultraviolet light sources and consuming inadequate diets can experience deficiencies. Ring-necked pheasants without dietary vitamin D_3 or exposure to sunlight developed rickets starting the fifth day after hatching (Millar *et al.*, 1977). Nonsupplemented domestic cow's milk and, perhaps, the milk of many other species whose young normally experience sunlight shortly after birth are notoriously deficient in vitamin D and hence require either dietary supplementation if the animal is bottle-raised or exposure of the young animal to sunlight. For example, captive maternal-nursed snow leopards and Cape hunting dogs developed rickets when not allowed to exercise outside (Encke, 1962; Wallach, 1970). The exposure of Cape hunting dogs to an artificial ultraviolet light source for only 3 min per day prevented vitamin D-deficient rickets. The milks of species normally raising the young in burrows or other darkened places must contain moderate levels of the vitamin.

Although irradiation is probably the main source of vitamin D for most animals, many plants contain ergosterol, which under ultraviolet irradiation becomes ergocalciferol (D_2) (one IU of vitamin D activity $= 0.025$ μg of crystalline vitamin D_3). Seeds and their by-products are practically devoid of the vitamin (Maynard and Loosli, 1969). Among the largest naturally occurring sources of vitamin D_3 available for dietary supplementation are fish oils, particularly liver oils. Supplementation from such sources is relatively unusual, since few animal products contain any significant quantity of vitamin D_3. Of course, purified vitamin D_2 and D_3 are also commercially available. Thus, supplementation of the diets of captive animals can be achieved easily, although overdoses should be avoided since they can be quite detrimental. Excess vitamin D_2 and D_3 retards bone growth, increases bone resorption, produces bone deformities, increases fracture susceptibility, and promotes cellular degeneration and calcification of soft tissues, including the circulatory system, urinary tract, and respiratory tract (M. L. Scott et al., 1976; Millar et al., 1977).

Several plants also contain toxic levels of $1,25-OH-D_3$-glycoside. When these plants are consumed, the carbohydrate unit is cleaved in the gastrointestinal tract and the metabolically active form of the vitamin is absorbed. Consequently, when these plants are consumed in conjunction with adequate calcium and phosphorus, the normal vitamin D feedback control at the kidney is bypassed. Calcium and phosphorus absorption and deposition proceed out of control. Of the three plants identified thus far that produce vitamin D toxicity, two are in the Solanaceae (potato family, Solanum malacoxylon and Cestrum diurnum) and one is in the Graminae (grass family, Trisetum flavescens). S. malacoxylon occurs in South America; C. diurnum occurs in the southeastern United States, Jamaica, and Hawaii; and T. flavescens occurs in Germany (Wasserman, 1975; Wasserman et al., 1976).

C. Vitamin E

The primary function of vitamin E is to maintain the functional integrity of cellular and subcellular membranes by preventing lipid peroxidation of unsaturated fatty acids, although many other roles have been suggested (Nair and Kayden, 1972; Glavird, 1973; Combs and Scott, 1977; Machlin, 1980). Vitamin E and selenium both function in protecting biological membranes, since several of the deficiency diseases can be ameliorated by either selenium or vitamin E. Thus, the diagnosis of a vitamin E deficiency is complicated by many of the similar deficiency signs of selenium.

The most active and widely distributed form is α-tocopherol (Fig. 6.5), although many other tocopherol isomers exist and have various degrees of activity (Hjarde *et al.*, 1973; Combs and Scott, 1977; Engelhardt, 1977). The activity of the various isomers may in part be related to the form normally ingested and utilized (Engelhardt, 1977). As an example, the plasma level of ingested (+)-α-tocopherol in harp seals was 30 times higher than when a similar amount of the (±)-γ form was ingested, suggesting differences in either absorption, excretion, or metabolism (Losowsky *et al.*, 1972; Engelhardt, 1977). The apparent absorption of dietary α-tocopherol in captive mink ranged from 27 to 53% and decreased as dietary fat content increased (Eskeland and Rimeslatten, 1979). Absorption efficiencies may also decrease as the dietary vitamin E and polyunsaturated fatty acid concentration, particularly linoleic acid, increase, but the absorption efficiency is apparently unaltered by the relative vitamin E status of the animal (Losowsky *et al.*, 1972; Engelhardt, 1977; Eskeland and Rimeslatten, 1979).

Although vitamin E occurs in many common foods and feeds (Eskeland and Rimeslatten, 1979), numerous deficiencies in captive wildlife fed diets based on frozen marine fish have been reported (Warner, 1961; Fitch and Dinning, 1963; Bone, 1968; P. P. Scott, 1968; Wagner and Dietlein, 1970; Engelhardt, 1974; Campbell and Montali, 1980). Paradoxically, live or fresh fish are in general very good sources of the vitamin (Helgebostad and Ender, 1973). However, the combination of a relatively high concentration of polyunsaturated fatty acids in marine fish, the frozen storage of the fish for several months before being fed, and the evisceration of the fish before freezing contribute to producing the deficiency. These factors are important because (*a*) the antioxidant role of vitamin E increases the requirement as unsaturated fats or dietary fat rancidity increases; (*b*) vitamin E in fish is oxidized and inactivated even when stored at very cold temperatures (50% loss after 6 months of storage at −20°C); and (*c*)

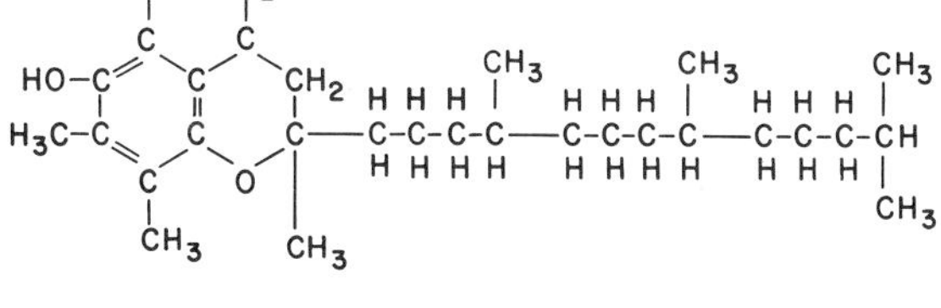

α-Tocopherol (vitamin E)

Fig. 6.5. Molecular structure of vitamin E.

the viscera of fish are important storage sites of the vitamin (Ackman and Cormier, 1967; Helgebostad and Ender, 1973; Engelhardt and Geraci, 1978). Thus, the feeding of frozen marine fish without a vitamin E supplement produces an ideal situation for the development of a deficiency. Similarly, the excessive utilization of cod liver oil or other oils high in unsaturated fatty acids and low in vitamin E as a vitamin A or D supplement can contribute to a vitamin E deficiency. Engelhardt and Geraci (1978) suggested that diets be supplemented with 100 IU of vitamin E per kilogram of fish to offset the oxidative loss of vitamin E during storage and the higher requirement relative to the polyunsaturated fat content of the diet. The sodium requirement of vitamin E deficient harp seals maintained in fresh water was double the requirement of the normal control seals.

Vitamin E deficiencies have also been reported in captive and free-ranging herbivores (Haugen and Hove, 1960; Kakulas, 1961; Anonymous, 1968; Higginson et al, 1973; Decker and McDermid, 1977; Brady et al., 1978). Although low dietary selenium may have been a contributing factor in many of the cases, tocopherol degradation by gastrointestinal microflora may be important when animals consume the concentrated diets often fed in captivity (Oksanen, 1973; Sorensen, 1973). For example, the destruction of vitamin E increased from 8 to 42% as the dietary content of corn increased from 20 to 80%.

Deficiency signs can appear within 1–4 weeks after an animal begins consuming a deficient diet. Similarly, most symptoms can be readily reversed, with repair of even severe muscle damage occurring within a similar span (West and Mason, 1958; Olson and Carpenter, 1967; Chan and Hegarty, 1977; Dahlin et al., 1978). As with selenium, the treatment of the muscle damage resulting from an acute deficiency is usually possible only in captive animals. Therapy is quite effective if muscle damage is not too extensive. Stress associated with confinement may also markedly increase the requirement (MacKenzie and Fletcher, 1979).

Vitamin E is widely distributed in green plants and seeds and can be stored in relatively large concentrations in the animal body (Dicks, 1965; Kivimae and Carpena, 1973) (one IU of vitamin E activity = 1.0 mg of (±)-α-tocopherol acetate). Grass and forb tocopherol levels generally increase during early growth and are higher in leaves than in stems, but they normally decrease dramatically as development approaches seed maturity (Kivimae and Carpena, 1973). Reductions of from 80 to 90% in grasses are common. Although many animal tissues accumulate the vitamin, fat reserves are normally the site of greatest storage (Engelhardt, 1974; Engelhardt et al., 1975). Protein-rich animal feeds, such as meat or fish meals, skim milk, and bone meal, and fat-extracted plant products are very poor sources (Sorensen, 1973; Kivimae and Carpena, 1973). Since vitamin E production is limited to chlorophyll-containing plants and microorganisms (J. Green et al., 1959), phytoplankton is likely a major source of the vitamin for the marine food web. Thus, while dietary sources of the vitamin are relatively common, deficiencies can occur and should always be a concern when animals are consuming diets high in polyunsaturated fatty acids, diets stored for a considerable time prior to feeding, or such diets as mature, leached forages or red meat, which may have minimal amounts of the vitamin.

D. Vitamin K

Vitamin K is well known as the antihemorrhagic vitamin necessary for the liver synthesis of requisite proteins in the blood-clotting process, although other roles in calcium metabolism and bone development may exist (Lian et al., 1978; Almquist, 1979; Gallop et al., 1980). The most important part of the vitamin K molecule is the menadione ring structure (Fig. 6.6), since it as well as many other molecules of various side chain lengths are biologically active. The two naturally occurring forms of the vitamin are phylloquinone (K_1), available from green plants, and menaquinone (K_2), available from synthesis by microorganisms or from animal tissue.

Because of the ubiquitous distribution of vitamin K in plants and animals and its synthesis by gastrointestinal microorganisms, it is exceedingly doubtful if free-ranging or captive wildlife consuming relatively natural diets ever suffer a deficiency. However, relatively purified diets not containing green plant matter, oil seed meals, or animal or fish meals should be supplemented with vitamin K.

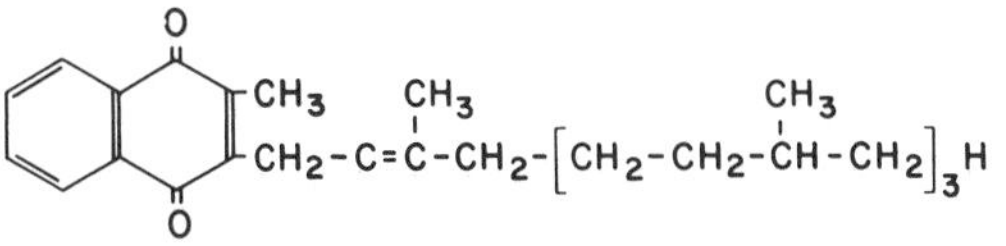

Phylloquinone (vitamin K_1)

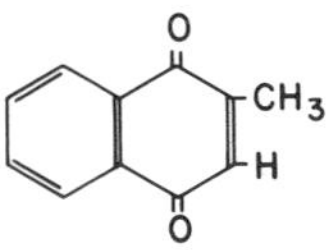

Menadione (vitamin K_3)

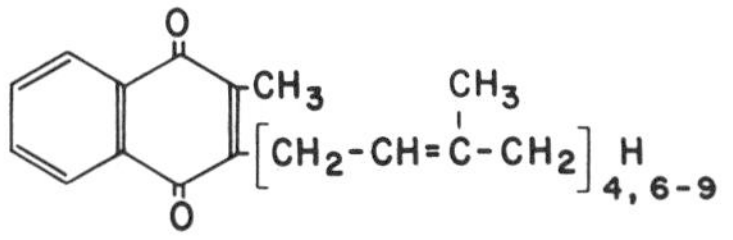

Menaquinone (vitamin K_2)

Fig. 6.6. Molecular structures of the three major forms of vitamin K. The side chain of vitamin K_2 can range from four to nine isoprene units.

Similarly, several vitamin K antagonists, such as the rodenticide warfarin or dicoumarol found in moldy sweet clover, can markedly increase the vitamin K requirement. The excessive consumption of vitamin K antagonists will produce death due to hemorrhaging. The prolonged use of ingestible drugs that reduce gastrointestinal microbial activity may also require supplemental vitamin K for those species dependent on bacterially produced vitamin K.

Warfarin resistance in wild brown rats (*Rattus norvegicus*) is a dominant characteristic controlled by a single gene. Heterozygous and homozygous warfarin-resistant rats have a 2 and 10 times higher vitamin K requirement than the more common homozygous susceptible individuals. Microbial synthesis of vitamin K and reingestion of cecal droppings would enable the homozygous susceptible and heterozygous resistant rats to meet their requirements. However, intestinal synthesis is inadequate for the homozygous resistant rats. Thus, depending on the vitamin K content of the food source, homozygous resistant animals may die out when warfarin is withdrawn and leave the more fit homozygous and heterozygous rats (Partridge, 1980).

II. WATER-SOLUBLE VITAMINS

Our understanding of the requirements and metabolism of water-soluble vitamins in wildlife is totally inadequate relative to their importance in animal functioning. The paucity of studies is, in part, due to the rarity of B-vitamin deficiencies observed in free-ranging wildlife. Because of synthesis by intestinal microbes and the general availability of dietary sources, B-vitamin deficiencies in free-ranging animals would be uncommon, with the potential exception of vitamin B_{12} because of its structural necessity for cobalt (P. P. Scott, 1968).

A. Thiamin

Thiamin has been studied far more frequently than all other water-soluble vitamins needed by wildlife (Table 6.5, 6.6, and 6.7). Thiamin deficiencies have been reported in captive Atlantic bottle-nose dolphins (White, 1970), herring gulls (M. Friend and Trainer, 1969; Gilman, 1978), seals (Hubbard, 1969; Geraci, 1972), polar bear cubs (Hess, 1971), mink (Warner, 1961; Travis, 1963), peregrine falcon (Ward, 1971), western grebes (Ratti, 1977), and sea lions (Rigdon and Drager, 1955), which in most cases were consuming diets composed of raw, dead fish, such as herring, mackerel, blue runner, alewives, smelt, carp, and whiting. Many species of uncooked fish, particularly freshwater species of the Cyprinidae, contain a thiaminase that destroys thiamin in the fish or in other foods fed during the same meal (R. C. Green *et al.*, 1942; Harris, 1951). Thiaminase is not found in the muscle but is confined in relatively large concentrations in the viscera and trimmings, including the head, skin, fins, and

TABLE 6.5
General Functions and Reported Deficiency Signs of the Water-Soluble Vitamins[a]

Vitamin	Major functions	Deficiency signs
Thiamin (vitamin B$_1$)	Necessary coenzyme in carbohydrate metabolism, possibly important in nerve or neuromuscular impulse transmission	Anorexia, weakness, diarrhea, lethargy, seizures and other neurological disorders, weight loss, leg weakness, ruffled feathers, unsteady gait, death, ''Chastek paralysis,'' liver degeneration, impaired digestion
Riboflavin (vitamin B$_2$)	Functions in two coenzymes— flavin adenine dinucleotide (FAD) and flavin mononucleotide	Curled toe paralysis, anorexia, loss of weight, weakness, perosis, reduced growth, death, reduced efficiency of feed utilization, rough hair coat, corneal vascularization, atrophy of the skin, heart muscle damage, diarrhea, leg paralysis, reduced fertility
Niacin (nicotinic acid)	Functions in two coenzymes— nicotinamide adenine dinucleotide (NAD) and nicotinamide adenine dinucleotide phosphate (NADP)	Reduced growth, enlarged hocks, poor feathering, death, erythema, anorexia, diarrhea, drooling, death, dermatitis, tongue discoloration
Vitamin B$_6$ (pyridoxine, pyridoxamine, and pyridoxal)	Functions in enzyme systems of protein metabolism	Testicular atrophy, sterility, anorexia, retarded growth, muscular incoordination, roughness and thinning of the hair coat, convulsions and neurological disorders, death, enlarged kidneys and adrenals, anemia, abnormal excitability
Pantothenic acid	A component of coenzyme A, necessary for fat, carbohydrate, and amino acid metabolism	Death, perosis, anorexia, skin lesions, crusty scabs about the beak and eyes, emaciation, weakness, reduced growth, leg disorders, poor feathering, intestinal hemorrhages, enlarged adrenals, enlarged fatty degeneration of the liver, enlarged and congested kidneys, reproductive failure
Biotin	Functions as a coenzyme in carbon dioxide fixation and decarboxylation	Fur discoloration, hair loss, degenerative changes in the hair follicles, thickened and scaly

(continued)

TABLE 6.5 *Continued*

Vitamin	Major functions	Deficiency signs
		skin, conjunctivitis, fatty infiltration of the liver, eye infections, dermatitis
Folacin (folic acid)	Transfer of single carbon units in molecular transformations	Retarded growth, anorexia, decreased activity, weakness, diarrhea, profuse salivation, convulsions, death, adrenal hemorrhages, fatty infiltration of the liver, reduced hemoglobin, hematocrit, and leukocytes
Vitamin B_{12} (cyanocobalamin)	Functions as a coenzyme in single carbon metabolism and in carbohydrate metabolism	Anorexia, loss of weight, fatty degeneration of the liver, neurological and locomotion disorders
Choline	Nerve functioning (acetylcholine) and the phospholipid lecithin as a necessary constituent of cells and tissues throughout the body	Diffuse fatty infiltration of the liver, rupture and hemorrhaging of the liver, enlarged hocks, slipped tendons, reduced growth of the leg bones, awkward gait, growth retardation, muscular weakness, lowered hematocrit, pale kidneys
Vitamin C (ascorbic acid)	May reduce infections, necessary for phagocytic activity, bone and collagen formation, functions in hydroxylation reactions	Scurvy, severe necrotic stomatitis, anorexia, weight loss, gingivitis, glossitis, pharyngitis, hemorrhages throughout the body, weakness, stiffened hind legs, lowered body temperature, diarrhea, bone fractures, abnormal behavior, enteritis, retarded growth, death

[a]Deficiency signs vary between species and depend on the severity of the deficiency.

skeleton (R. C. Green *et al.,* 1942). The captive peregrine falcon was fed hatchling cockerels, pigeon and quail muscle, beef, chicken gizzards, and hearts (Ward, 1971). As suspected from the earlier example, thiaminase is also found in the heart and spleen of warm-blooded animals (M. L. Scott *et al.,* 1976) and newly hatched chickens (Ward, 1971). Bracken fern *(Pteris aquilinum)* and horsetail *(Equisetum arvense)* also contain thiaminase (Harris, 1951).

TABLE 6.6
Chemical Structure of the Water-Soluble Vitamins

Vitamin	Structure
Thiamin	
Riboflavin	
Niacin	
Pyridoxine	
Pantothenic acid	
Biotin	
Folacin	

(*continued*)

TABLE 6.6 *Continued*

Vitamin	Structure

B$_{12}$

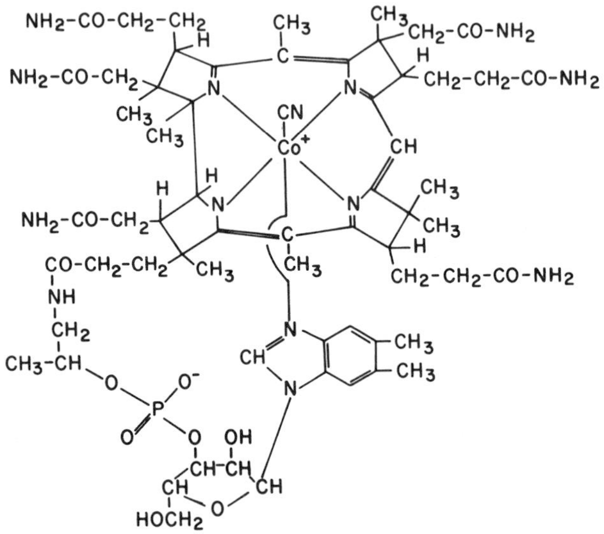

NH$_2$-CO-CH$_2$-CH$_2$　　　CH$_3$　　CH$_3$-CH$_2$-CO-NH$_2$

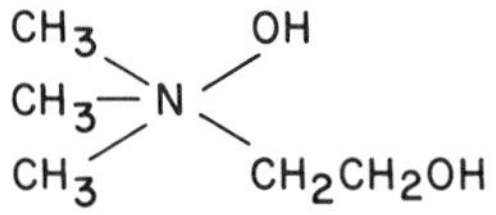

NH$_2$-CO-CH$_2$　　　　　　C　　　　CH$_2$-CH$_2$-CO-NH$_2$

CH$_3$　CH$_3$　　N　CN　N　　CH

　　　　　Co$^+$

H　H　　　　　　　CH$_3$

NH$_2$-CO-CH$_2$　N　　　N　CH$_3$

CO-CH$_2$-CH$_2$　CH$_3$　CH$_3$　CH$_2$-CH$_2$-CO-NH$_2$

NH

CH$_2$

CH$_3$-CH　O　O$^-$

　　　O　P　O$^-$

CH$_3$-CH$_2$　　N　CH$_3$

　　　CH　N　CH$_3$

O　OH

CH-CH

CH　　CH

HOCH$_2$　O

Choline

CH$_3$　　OH

CH$_3$-N

CH$_3$　　CH$_2$CH$_2$OH

C

O=C

HO-C

HO-C 　 O

H-C

HO-C-H

CH$_2$OH

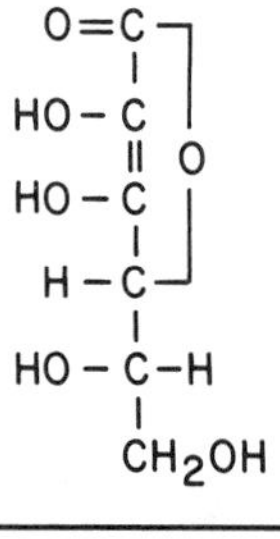

It is, perhaps, surprising that fish diets have produced major thiamin deficiencies in captive wildlife since many free-living marine and terrestrial predators and scavengers feed entirely on fish. Thiaminase is apparently sequestered in an inactive form in the living animal and stimulated or activated by the breakdown products associated with death and storage (Deolalkar and Sohonie, 1957a, 1957b; White, 1970). Thiamin destruction is virtually complete by 90 min when dead smelt are incubated at 37°C (Geraci, 1972). Similarly, since ingested

TABLE 6.7

Water-Soluble Vitamin Requirements or Allowances for Several Wildlife Species per Kilogram of Diet

Species	Riboflavin (mg)	Niacin (mg)	Choline (mg)	Thiamin (mg)	Biotin (mg)	Folic acid (mg)	Pantothenic acid (mg)	Pyridoxine (mg)	B$_{12}$ (μg)	Ascorbic acid (mg)	Reference
Birds											
Bobwhite quail											
Growth	3.8	31	1500	—	—	—	13	—	—	—	NRC (1977a)
Growth	3.5–4.0	60–70	1100–1500	—	—	—	8–10	—	3–5	—	M. L. Scott (1977–1978)
Egg production	4.0	20	1000	—	—	—	15	—	—	—	NRC (1977a)
Egg production	3.5	60	1100	—	—	—	15	—	3	—	M. L. Scott (1977–1978)
Ducks and geese (domestic)											
Growth	2.5–4.0	35–55	—	—	—	—	11	2.6	—	—	NRC (1977a)
Growth	3.5–4.0	60–70	1100–1500	—	—	—	8–11	—	3–5	—	M. L. Scott (1977–1978)
Egg production	4.0	20–40	—	—	—	—	10	3.0	—	—	NRC (1977a)
Egg production	3.5	60	1100	—	—	—	15	—	5	—	M. L. Scott (1977–1978)
Pheasant											
Growth (0–6 weeks)	3.5	60	1500	—	—	—	10	—	—	—	NRC (1977a)
Growth (6–20 weeks)	2.6	40	1000	—	—	—	10	—	—	—	NRC (1977a)
Growth	3.5–4.0	60–70	1000–1500	—	—	—	8–10	—	3–5	—	M. L. Scott (1977–1978)
Egg production	3.5	60	1100	—	—	—	15	—	5	—	M. L. Scott (1977–1978)
Turkey (domestic)											
Growth (0–3 weeks)	3.6	70	1900	2.0	0.2	1.0	11	4.5	3	—	NRC (1977a)
Growth (8–16 weeks)	3.0	50	1100	2.0	0.1	0.8	9	3.5	3	—	NRC (1977a)
Growth	3.5–4.0	60–70	1500–1980	—	—	—	8–11	—	3–5	—	M. L. Scott (1977–1978)
Egg production	4.0	30	1000	2.0	0.15	1.0	16	4.0	3	—	NRC (1977a)
Egg production	3.5	60	1500	—	—	—	16	—	5	—	M. L. Scott (1977–1978)

(continued)

TABLE 6.7 *Continued*

Species	Riboflavin (mg)	Niacin (mg)	Choline (mg)	Thiamin (mg)	Biotin (mg)	Folic acid (mg)	Pantothenic acid (mg)	Pyridoxine (mg)	B_{12} (µg)	Ascorbic acid (mg)	Reference
Mammals											
Fox	2.6–4.0	10	—	0.5–1.0	—	0.2	8	2.0	—	—	NRC (1968)
Guinea pig											
Growth	3.0	10	1000	2.0	0.3	4.0	20	3.0	10	200	NRC (1978a)
Hamster											
Growth	15.0	90	2000	20.0	0.6	2.0	40	6.0	10	—	NRC (1978a)
Laboratory mouse											
Growth and repro-duction	7.0	10	600	5.0	0.2	0.5	10	1.0	10	—	NRC (1978a)
Laboratory rat											
Growth and repro-duction	3.0	20	1000	4.0	—	1.0	3	6.0	50	—	NRC (1978a)
Mink	1.5	20	—	1.2	—	0.5	6	1.1	—	—	NRC (1968)
Nonhuman primate	5	50	—	—	0.1	0.2	15	2.5	—	100	NRC (1978b)
Rabbit (domestic)											
Growth	—	180	1200	—	—	—	—	39.0	—	—	NRC (1977b)

thiaminase is readily inactivated by peptic digestion and stomach acidity and therefore does not destroy internal body thiamin, the consumption of thiaminase-containing fish does not always produce a deficiency. The thiamin requirement of either captive or free-ranging animals can be met by ingesting the needed thiamin in a meal or supplement temporally isolated from foods containing thiaminase. Gilman (1978) suggested that a thiamin deficiency in free-ranging adult herring gulls was unlikely, since they have a relatively low requirement and use alternative thiamin-rich food sources. Paradoxically, whereas captive herring gulls fed fresh smelt develop a deficiency, those fed rotten smelt for 10 weeks to simulate feeding during a fish die-off did not develop the deficiency. Young, parental-fed chicks consuming thiaminase-rich food boluses may be quite susceptible to a deficiency resulting in high, early mortality. Although many wild ruminants may ingest sizable quantities of bracken fern, the destruction of the thiaminase and simultaneous production of thiamin by the microorganisms of the rumen prevent a deficiency in these animals. Similarly, predatory species consuming the entire animal may ingest additional thiamin associated with the gastrointestinal contents of prey species (Ward, 1971).

Thiamin deficiencies can usually be rapidly and easily cured even in near terminally sick animals by the addition of adequate thiamin to the diet or by injecting thiamin hydrochloride. The symptoms of thiamin deficiency began disappearing in Atlantic bottle-nosed dolphin 6 hr after an injection of thiamin and in harp seals 12 hr after an injection (White, 1970; Geraci, 1972). Thiamin-deficient western grebes exhibiting massive neurological disorders completely recovered within 48 hr of oral thiamin supplementation (Ratti, 1977). The daily thiamin requirement is, of course, very dependent on dietary thiaminase activity, temporal sequence of food ingestion, and, particularly, dietary fat content. Since thiamin is necessary for the synthesis of fatty acids, diets high in fats spare thiamin. Animals consuming diets high in fat but deficient in thiamin remain healthier for a much longer period of time than those receiving the same amount of thiamin but less dietary fat. Although the average thiamin requirement of animals consuming diets devoid of thiaminase may be as little as 1–2 mg/kg of diet (Table 6.7), the effective thiamin requirement of seals and foxes consuming thiaminase containing herring or carp was 25–35 and 147 mg/kg of diet, respectively, with the difference due to thiaminase content (R. C. Green et al., 1942; Geraci, 1972).

Infection and nutrition are often closely related processes. Thiamin-deficient herring gulls consuming alewives were highly susceptible to the respiratory fungal disease aspergillosis (M. Friend and Trainer, 1969). Antifungal agents were generally ineffective in combating the disease. When the diet was supplemented with thiamin rather than the antifungal therapeutic agents, aspergillosis mortality ceased and most of the sick birds immediately recovered.

B. Other B Vitamins

Riboflavin, niacin, pantothenic acid, choline, folacin, and pyridoxine metabolism and requirements have been studied in pheasants and quail (M. L. Scott *et al.*, 1959, 1964; Serafin, 1974), mink, foxes, guinea pigs, primates, and virtually all common laboratory animals (data summarized in NRC publications). Riboflavin is not produced to any significant extent by animals and therefore must initially come from plants or microorganisms. Even though animal tissue is a very good source of riboflavin, a riboflavin deficiency was reported in a free-ranging golden eagle (Stauber, 1973). The deficiency may have been precipitated by the consumption of environmental toxicants that interfered with riboflavin metabolism (Stauber, personal communication). Although riboflavin can be readily excreted, the biological half-life of the vitamin in well-nourished rhesus monkeys was 3.2 days but lengthened during the deficient state to 9.5 days as the animals responded by conserving body stores (Greenberg, 1970). Riboflavin deficiencies in many captive birds (such as the ostrich) have also been noted (Wallach, 1970). While curled-toe paralysis associated with riboflavin deficiency in birds is a useful diagnostic tool, it is not to be confused with the laterally curled toes produced when birds are maintained simply on smooth surfaces (Wallach, 1970). Biotin deficiencies can occur if the diet is high in raw egg white, which contains a compound called avidin that binds dietary biotin into a nonavailable form.

C. Vitamin C

Vitamin C, or ascorbic acid, is one of the exceptions to the general definition of vitamins in that it can be produced by many animals and thus is not always a dietary requirement. The development of the synthetic capability and its transition within different organs of the body follows an interesting phylogenetic scale. Invertebrates and fish lack the synthetic capability and are thus dependent on dietary sources. Amphibians, reptiles, most birds, and monotremes synthesize ascorbic acid in the kidney. Approximately one-half of the Passeriformes (particularly numerous Southeast Asian species, such as bulbuls, Asian paradise-flycatcher, black-hooded oriole, pale-billed flowerpecker, and crimson sunbird, as well as some shrikes and the barn swallow) are unable to synthesize the vitamin and are therefore totally dependent on a dietary source. The remaining Passeriformes synthesize ascorbic acid in the liver (such as, house sparrow, jungle myna, Asian pied starling, and magpie robin) or in both the liver and kidney (Indian house crow and common myna). Like the advanced taxa of birds, most mammals synthesize ascorbic acid in the liver, but a few, including primates, bats, guinea pigs, and possibly cetaceans, lack a synthetic ability. Thus, the ability to synthesize ascorbic acid appeared initially in the kidney with the evolution of terrestrial, vertebrate life forms and was later either transferred to the liver or lost (Roy and Guha, 1958; Miller and Ridgeway, 1963; Ridgeway, 1965; Elliot *et al.*, 1966; Yess and Hegsted, 1967; Hubbard, 1969; Chaudhuri and Chatterjee, 1969; Chatterjee, 1973; Birney *et al.*, 1976, 1980).

However, even those species that synthesize ascorbic acid may require a dietary supply at critical times in the life cycle. For example, in willow ptarmigan, a species with the ability to synthesize vitamin C in the kidney, chicks that were fed a commercial poultry diet containing 265 mg vitamin C per kilogram of diet developed scurvy and died by 4 weeks of age (Hanssen *et al.*, 1979). The deficiency was cured by vitamin C supplementation or the provision and consumption of blueberry or other plants normally found in the ptarmigan's diet. Although a diet containing 750 mg vitamin C per kilogram of diet was adequate, many plants normally consumed by free-ranging ptarmigan contain 2000–5000 mg vitamin C per kilogram. The capabilities to synthesize vitamin C are sufficient for adult ptarmigan since they thrive in captivity on diets devoid of the vitamin. However, the dietary requirement of the chick is approximately 125–150 mg ascorbic acid per kilogram body weight, which is higher than the natural synthetic capabilities and much higher that the 5 mg/kg body weight required by the newborn guinea pig. Snowshoe hares may be similar in having a synthetic capability but are nevertheless dependent on dietary sources (Jenness *et al.*, 1978). Thus, vitamin C should not be ignored even in those species able to synthesize it, since rapid early growth, stress, or organochloride pesticides may increase the ascorbic acid requirement beyond the capabilities of endogenous synthesis (M. L. Scott, 1975; Street and Chadwick, 1975; Chatterjee *et al.*, 1975; Hanssen *et al.*, 1979). Deficiency signs can appear within 2–4 days in ptarmigan chicks, 3 weeks in guinea pigs, to 3–4 months in humans (Chatterjee *et al.*, 1975; Hanssen *et al.*, 1979).

Vitamin C is not as widespread as many of the other water-soluble vitamins. The richest sources are green plants and fleshy fruits, such as rose hips and berries, but it is virtually absent in eggs, seeds, grains, and most bacteria and protozoa. Animal tissues are, in general, poor sources of the vitamin, although the adrenal cortex and liver (300 mg/kg) are relatively good sources.

III. CONCLUSIONS

Even though vitamins are usually found in very minute amounts, they are crucially important to animal health. Deficiencies produce signs that the wildlife nutritionist should be encouraged to recognize. While most of the deficiencies have been detected in captive animals, free-ranging wildlife may also suffer.

REFERENCES

Ackman, R. G., and Cormier, M. G. (1967). α-Tocopherol in some Atlantic fish and shellfish with particular reference to live-holding without food. *J. Fish. Res. Bd. Can.* **24,** 357–373.

Almquist, H. J. (1979). Vitamin K: Discovery, identification, synthesis, functions. *Fed. Proc.* **38,** 2687–2689.

Anderson, A. E., Medin, D. E., and Bowden, D. C. (1972). Carotene and vitamin A in the liver and blood serum of a Rocky Mountain mule deer, *Odocoileus hemionus hemionus,* population. *Comp. Biochem. Physiol.* **41B,** 745–758.

Anonymous. (1968). Avitaminosis E in a white-tailed deer *(Odocoileus virginianus). J. Small Anim. Pract.* **9,** 130–131.

Birney, E. C., Jenness, R., and Ayaz, K. M. (1976). Inability of bats to synthesize L-ascorbic acid. *Nature* **260,** 626–628.

Birney, E. C., Jenness, R., and Hume, I. D. (1980). Evolution of an enzyme system: Ascorbic acid biosynthesis in monotremes and marsupials. *Evolution* **34,** 230–239.

Block, J. A. (1974). Hand-rearing seven-banded armadillos *(Dasypus septemcinctus)* at the National Zoological Park, Washington. *Int. Zoo Ybk.* **14,** 210–214.

Bone, W. J. (1968). Pansteatitis in a lion *(Felis leo). J. Am. Vet. Med. Assoc.* **153,** 791–792.

Brady, P. S., Brady, L. J., Whetter, P. A., Ullrey, D. E., and Fay, L. D. (1978). The effect of dietary selenium and vitamin E on biochemical parameters and survival of young among white-tailed deer *(Odocoileus virginianus). J. Nutr.* **108,** 1439–1448.

Campbell, G., and Montali, R. J. (1980). Myodegeneration in captive brown pelicans attributed to vitamin E deficiency. *J. Zoo Anim. Med.* **11,** 35–40.

Chan, A. C., and Hegarty, P. V. J. (1977). Morphological changes in skeletal muscles in vitamin E-deficient and refed rabbits. *Brit. J. Nutr.* **38,** 361–370.

Chatterjee, I. B. (1973). Evolution and biosynthesis of ascorbic acid. *Science* **182,** 1271–1272.

Chatterjee, I. B., Majumder, A. K., Nandi, B. K., and Subramanian, N. (1975). Synthesis and some major functions of vitamin C in animals. *Ann. New York Acad. Sci.* **258,** 24–47.

Chaudhuri, C. R., and Chatterjee, I. B. (1969). L-Ascorbic acid synthesis in birds: Phylogenetic trend. *Science* **164,** 435–436.

Combs, G. F., Jr., and Scott, M. L. (1977). Nutritional interrelationships of vitamin E and selenium. *BioScience* **27,** 467–473.

Cowan, I. McT., and Fowle, C. D. (1944). Visceral gout in a wild ruffed grouse. *J. Wildl. Manage.* **8,** 260–261.

Dahlin, K. J., Chan, A. C., Benson, E. S., and Hegarty, P. V. J. (1978). Rehabilitation effect vitamin E therapy on the ultrastructural changes in skeletal muscles of vitamin E deficient rabbits. *Am. J. Clin. Nutr.* **31,** 94–99.

Decker, R. A., and McDermid, A. M. (1977). Nutritional myopathy in a young camel. *J. Zoo Anim. Med.* **8,** 20–21.

DeLuca, H. F. (1979). Retinoic acid metabolism. *Fed. Proc.* **38,** 2519–2523.

Deolalkar, S. T., and Sohonie, K. (1957a). Studies on thiaminase from fish. I. Properties of thiaminase. *Indian J. Med. Res.* **45,** 571–586.

Deolalkar, S. T., and Sohonie, K. (1957b). Studies on thiaminase from fish. II. Effect of certain compounds on thiaminase activity. *Indian J. Med. Res.* **45,** 587–592.

Dicks, M. W. (1965). Vitamin E content of foods and feeds for human and animal consumption. *Wyoming Agric. Exp. Sta. Bull.* **435.**

Elliot, O., Yess, N. J., and Hegsted, D. M. (1966). Biosynthesis of ascorbic acid in the tree shrew and the slow loris. *Nature* **212,** 739–740.

Encke, W. (1962). Hand-rearing Cape hunting dogs *(Lycaon pictus)* at the Krefeld Zoo. *Int. Zoo Ybk.* **4,** 292–293.

Engelhardt, F. R. (1974). Aspects of vitamin E deprivation in the harp seal, *Pagophilus groenlandicus*. Doctoral dissertation, Univ. Guelph.

Engelhardt, F. R. (1977). Plasma and tissue levels of dietary radiotocopherols in the harp seal, *Phoca groenlandica*. *Can. J. Physiol. Pharmacol.* **55**, 601–608.

Engelhardt, F. R., and Geraci, J. R. (1978). Effects of experimental vitamin E deprivation in the harp seal, *Phoca groenlandica*. *Can. J. Zool.* **56**, 2186–2193.

Engelhardt, F. R., Geraci, J. R., and Walker, B. L. (1975). Tocopherol distribution in the harp seal, *Pagophilus groenlandicus*. *Comp. Biochem. Physiol.* **52B**, 561–562.

Eskeland, B., and Rimeslatten, H. (1979). Studies on the absorption of labelled and dietary α-tocopherol in mink as influenced by some dietary factors. *Acta Agric. Scandinavica* **29**, 75–80.

Fitch, C. D., and Dinning, J. S. (1963). Vitamin E deficiency in the monkey. *J. Nutr.* **79**, 69–78.

Foster, J. W. (1981). Dermatitis in polar bears—A nutritional approach to therapy. Woodland Park Zoo, Seattle, Washington.

Fraser, D. R., and Kodicek, E. (1970). Unique biosynthesis by kidney of a biologically active vitamin D metabolite. *Nature* **228**, 764–766.

Friend, D. W., and Crampton, E. W. (1961). The adverse effect of raw whale liver on the breeding performance of female mink. *J. Nutr.* **73**, 317–320.

Friend, M., and Trainer, D. O. (1969). Aspergillosis in captive herring gulls. *Bull. Wildl. Dis. Assoc.* **5**, 271–275.

Gallop, P. M., Lian, J. B., and Hauschka, P. V. (1980). Carboxylated calcium-binding proteins and vitamin K. *New England J. Med.* **302**, 1460–1466.

Geraci, J. R. (1972). Experimental thiamin deficiency in captive harp seals, *Phoca groenlandica* induced by eating herring, *Clupea harengus,* and smelt, *Osmerus mordax. Can. J. Zool.* **50**, 170–195.

Gilman, A. P. (1978). Natural and induced thiamin deficiency in the herring gull, *Larus argentatus.* Doctoral dissertation, Univ. Guelph.

Glavind, J. (1973). The biological antioxidant theory and the function of vitamin E. *Acta Agric. Scandinavica Suppl.* **19**, 105–112.

Goodman, D. S. (1979). Vitamin A and retinoids: Recent advances. *Fed. Proc.* **38**, 2501–2503.

Green, J., Price, S. A., and Gare, L. (1959). Tocopherols in microorganisms. *Nature* **184**, 1339.

Green, R. C., Carlson, W. E., and Evans, C. A. (1942). The inactivation of vitamin B_1 in diets containing whole fish. *J. Nutr.* **23**, 165–174.

Greenberg, L. D. (1970). Nutritional requirements of Macaque monkeys. *In* "Feeding and Nutrition of Nonhuman Primartes" (R. S. Harris, ed.), pp. 117–142. Academic Press, New York.

Hanssen, K., Grav, H. J., Steen, J. B., and Lysnes, H. (1979). Vitamin C deficiency in growing willow ptarmigan *(Lagopus lagopus lagopus).* *J. Nutr.* **109**, 2260–2278.

Harris, R. S. (1951). Thiaminase. *In* "The Enzymes," Vol. I, Part 2 (J. B. Sumner and K. Myrback, eds.), pp. 1186–1206. Academic Press, New York.

Haugen, A. O., and Hove, E. L. (1960). Vitamin A and E in deer blood. *J. Mammal.* **41**, 410–411.

Hay, A. W. M., and Watson, G. (1975). Binding of 25-hydroxyvitamin D_2 to plasma protein in New World monkeys. *Nature* **256**, 150.

Hay, A. W. M., and Watson, G. (1977). Vitamin D_2 in vertebrate evolution. *Comp. Biochem. Physiol.* **56B**, 375–380.

Helgebostad, A., and Ender, F. (1973). Vitamin E and its function in the health and disease of fur-bearing animals. *Acta Agric. Scandinavica Suppl.* **19**, 79–83.

Hess, J. K. (1971). Hand-rearing polar bear cubs (*Thalarctos maritimus*) at St. Paul Zoo. *Int. Zoo Ybk.* **11**, 102–107.

Heywood, R. (1967). Vitamin A in the liver and kidney of some felidae. *Brit. Vet. J.* **123**, 390–396.

Higginson, J. A., Julian, R. J., and van Drummel, A. A. (1973). Muscular dystrophy in zebra foals. *J. Zoo Anim. Med.* **4**, 24–27.

Hjarde, W., Leerbeck, E., and Leth, L. (1973). The chemistry of vitamin E (including its chemical determination). *Acta Agric. Scandinavica Suppl.* **19,** 87–96.

Holick, M. F., and Clark, M. B. (1978). The photobiogenesis and metabolism of vitamin D. *Fed. Proc.* **37,** 2567–2574.

Hubbard, R. C. (1969). Chemotherapy in captive marine mammals. *Bull. Wildl. Dis. Assoc.* **5,** 218–230.

Hungerford, C. R. (1964). Vitamin A and productivity in Gambel's quail. *J. Wildl. Manage.* **28,** 141–147.

Jenness, R., Birney, E. C., and Ayaz, K. L. (1978). Ascorbic acid and L-gulonolactone oxidase in lagomorphs. *Comp. Biochem. Physiol.* **61B,** 395–399.

Jones, G. E. (1950). A study of vitamin A storage in bobwhite quail in Cleveland County, Oklahoma with comparative data from other counties. Master's thesis, Univ. Oklahoma.

Kakulas, B. A. (1961). Myopathy affecting the Rottnest quokka *(Setonix brachyurus)* reversed by α-tocopherol. *Nature* **191,** 402–403.

Keane, K. W., Collins, R. A., and Gillis, M. B. (1956). Isotopic tracer studies on the effect of vitamin D on calcium metabolism in the chick. *Poult. Sci.* **35,** 1216–1222.

Kivimae, A., and Carpena, C. (1973). The level of vitamin E content in some conventional feeding stuffs and the effects of genetic variety; harvesting; processing and storage. *Acta Agric. Scandinavica Suppl.* **19,** 161–168.

Kon, S. K., and Thompson, S. Y. (1949). Preformed vitamin A in northern krill. *Biochem. J.* **45,** XXXI–XXXII.

Labov, J. B. (1977). Phytoestrogens and mammaliam reproduction. *Comp. Biochem. Physiol.* **57A,** 3–9.

Lehner, N. D. M., Bullock, B. C., Clarkson, T. B., and Lofland, H. B. (1966). Biological activity of vitamins D_2 and D_3 fed to squirrel monkeys. *Fed. Proc.* **25,** 533.

Leopold, A. S., Erwin, M., Oh, J., and Browning, B. (1976). Phytoestrogens: Adverse effects on reproduction in California quail. *Science* **191,** 98–100.

Lian, J. B., Hauschka, P. V., and Gallop, P. M. (1978). Properties and biosynthesis of a vitamin K-dependent calcium binding protein in bone. *Fed. Proc.* **37,** 2615–2620.

Losowsky, M. S., Kelleher, J., Walker, B. E., Davies, T., and Smith, C. L. (1972). Intake and absorption of tocopherol. *Ann. New York Acad. Sci.* **203,** 212–222.

McDonald, P., Edwards, R. A., and Greenhalgh, J. F. D. (1973). "Animal Nutrition." Longman, New York.

Machlin, L. J. (1980). "Vitamin E: A Comprehensive Treatise." Dekker, New York.

MacKenzie, W. F., and Fletcher, K. (1979). Megavitamin E responsive myopathy in Goodfellow tree kangaroos associated with confinement. *In* "The Comparative Pathology of Zoo Animals" (R. J. Montali and G. Migaki, eds.), pp. 35–39. Smithsonian Inst. Press, Washington, D.C.

Mann, G. V. (1970). Nutritional requirements of *Cebus* monkeys. *In* "Feeding and Nutrition of Nonhuman Primates" (R. S. Harris, ed.), pp. 143–157. Academic Press, New York.

Martin, R. D. (1975). General principles for breeding small mammals in captivity. *In* "Breeding Endangered Species in Captivity" (R. D. Martin, ed.), pp. 143–166. Academic Press, New York.

Maynard, L. A., and Loosli, J. K. (1969). "Animal Nutrition." McGraw-Hill, New York.

Millar, R. I., Smith, L. T., and Wood, J. H. (1977). The study of the dietary vitamin D_3 requirement of ringnecked pheasant chicks. *Poult. Sci.* **56,** 1739 (abstract).

Miller, R. M., and Ridgeway, S. H. (1963). Clinical experience with dolphins and whales. *Small Anim. Clin.* **3,** 189–193.

Nair, P. P., and Kayden, H. J. (eds.). (1972). International conference on vitamin E and its role in cellular metabolism. *Ann. New York Acad. Sci.* **203.**

National Research Council (NRC). (1968). Nutrient requirements of mink and foxes. Publ. No. 1676, Nat. Acad. Sci., Washington, D.C.

National Research Council (NRC). (1977a). Nutrient requirements of poultry. Publ. No. 2725, Nat. Acad. Sci., Washington, D.C.

National Research Council (NRC). (1977b). Nutrient requirements of rabbits. Publ. No. 2607, Nat. Acad. Sci., Washington, D.C.

National Research Council (NRC). (1978a). Nutrient requirements of laboratory animals. Publ. No. 2767, Nat. Acad. Sci., Washington, D.C.

National Research Council (NRC). (1978b). Nutrient requirements of nonhuman primates. Publ. No. 2786, Nat. Acad. Sci., Washington, D.C.

Nestler, R. B. (1946). Vitamin A, vital factor in the survival of bobwhites. *Trans North Am. Wildl. Conf.* **11,** 176–195.

Nestler, R. B. (1949). Nutrition of bobwhite quail. *J. Wildl. Manage.* **13,** 342–358.

Nestler, R. B., Derby, J. V., and DeWitt, J. B. (1949a). Vitamin A and carotene content of some wildlife foods. *J. Wildl. Manage.* **13,** 271–274.

Nestler, R. B., DeWitt, J. B., and Derby, J. V., Jr. (1949b). Vitamin A storage in wild quail and its possible significance. *J. Wildl. Manage.* **13,** 265–271.

Notini, G. (1941). Verksamheten vid andkarantanen. *Svensk Jakt.* **79,** 155–159.

Oksanen, H. E. (1973). Aspects of vitamin E deficiency in ruminants. *Acta Agric. Scandinavica Suppl.* **19,** 22–28.

Olson, R. E., and Carpenter, P. C. (1967). Regulatory function of vitamin E. *Advances in Enzyme Regulation* **5,** 325–334.

Partridge, G. G. (1980). The vitamin K requirements of wild brown rats *(Rattus norvegicus)* resistant to warfarin. *Comp. Biochem. Physiol.* **66A,** 83–87.

Portman, O. W. (1970). Nutritional requirements of squirrel and woolly monkeys. *In* "Feeding and Nutrition of Nonhuman Primates" (R. S. Harris, ed.), pp. 159–173. Academic Press, New York.

Ratti, J. T. (1977). Reproductive separation and isolation mechanisms between dark and light-phase western grebes. Doctoral dissertation, Utah State Univ., Logan.

Reichman, O. J., and Van De Graaf, K. M. (1975). Association between ingestion of green vegetation and desert rodent reproduction. *J. Mammal.* **56,** 503–506.

Ridgeway, S. H. (1965). Medical care of marine mammals. *J. Am. Vet. Med. Assoc.* **147,** 1077–1085.

Rigdon, R. H., and Drager, G. A. (1955). Thiamine deficiency in sea lions *(Otaria californiana)* fed only frozen fish. *J. Am. Vet. Med. Assoc.* **127,** 453–455.

Rodahl, K. (1949). Toxicity of polar bear liver. *Nature* **164,** 530.

Rodahl, K., and Moore, T. (1943). The vitamin A content and toxicity of bear and seal liver. *Biochem. J.* **37,** 166–168.

Roy, R. N., and Guha, B. C. (1958). Species difference in regard to the biosynthesis of ascorbic acid. *Nature* **182,** 319–320.

Russell, F. E. (1967). Vitamin A content of polar bear liver. *Toxicon* **5,** 61–62.

Schultz, V. B. (1948). Vitamin A as a survival factor of the bobwhite quail *(Colinus v. virginianus)* in Ohio during the winter of 1946–47. *J. Wildl. Manage.* **12,** 251–263.

Schultz, V. B. (1959). Vitamin A and Ohio bobwhite quail during the winter of 1947–48. *J. Wildl. Manage.* **23,** 322–327.

Scott, M. L. (1975). Environmental influences on ascorbic acid requirements in animals. *Ann. New York Acad. Sci.* **258,** 151–155.

Scott, M. L. (1977–1978). Twenty-five years of research in game bird nutrition. *World Pheasant Assoc. J.* **3,** 31–45.

Scott, M. L., Holm, E. R., and Reynolds, R. E. (1959). Studies on the niacin, riboflavin, choline, manganese, and zinc requirements of young ring-necked pheasants for growth, feathering, and prevention of leg disorders. *Poult. Sci.* **38,** 1344–1350.

Scott, M. L., Holm, E. R., and Reynolds, R. E. (1964). Studies on the pantothenic acid and unidentified factor requirements of young ring-necked pheasants and bobwhite quail. *Poult. Sci.* **43**, 1534–1539.

Scott, M. L., Nesheim, M. C., and Young, R. J. (1976). "Nutrition of the Chicken." M. L. Scott and Assoc., Ithaca, New York.

Scott, P. P. (1968). The special features of nutrition of cats, with observations on wild felidae nutrition in the London Zoo. *Symp. Zool. Soc. London* **21**, 21–36.

Serafin, J. A. (1974). Studies on the riboflavin, niacin, pantothenic acid and choline requirements of young bobwhite quail. *Poult. Sci.* **53**, 1522–1532.

Silver, H., and Colovos, N. F. (1957). Nutritive evaluation of some forage rations of deer. *New Hampshire Fish Game Dept., Tech. Circular No. 15.*

Smith, J. E., and Goodman, D. S. (1979). Retinol-binding protein and the regulation of vitamin A transport. *Fed. Proc.* **38**, 2504–2509.

Sorensen, P. H. (1973). Basic principles involved in the supplementation of compound feeds with vitamin E. *Acta Agric. Scandinavica* **19**, 177–180.

Stauber, E. (1973). Suspected riboflavin deficiency in a golden eagle. *J. Am. Vet. Med. Assoc.* **163**, 645–646.

Street, J. C., and Chadwick, R. W. (1975). Ascorbic acid requirements and metabolism in relation to organochloride pesticides. *Ann. New York Acad. Sci.* **258**, 132–143.

Thompson, D. R., and Baumann, C. A. (1950). Vitamin A in pheasants, quail, and muskrats. *J. Wildl. Manage.* **14**, 42–49.

Travis, H. F. (1963). Symposium—Specialty feed formulation—Some considerations in the formulation. *Proc. Cornell Nutr. Conf. Feed Manufacturers,* 122–126.

Wagner, J. E., and Dietlein, D. R. (1970). Steatitis in an American white pelican. *Int. Zoo Ybk.* **10**, 174.

Wallach, J. D. (1970). Nutritional diseases of exotic animals. *J. Am. Vet. Med. Assoc.* **157**, 583–599.

Ward, F. P. (1971). Thiamine deficiency in a peregrine falcon. *J. Am. Vet. Med. Assoc.* **159**, 599–601.

Warner, R. G. (1961). Recent developments in mink nutrition. *Proc. Cornell Nutr. Conf. Feed Manufacturers,* 96–100.

Warner, R. G., Travis, H. F., Bassett, C. F., Krook, P., and McCarthy, B. (1963). Utilization of carotene by growing mink kits. Mink Farmers Research Foundation Progress Report.

Wasserman, R. H. (1975). Active vitamin D-like substances in *Solanum malacoxylon* and other calcinogenic plants. *Nutr. Rev.* **33**, 1–5.

Wasserman, R. H., Taylor, A. N., and Kallfelz, F. A. (1966). Vitamin D and transfer of plasma calcium to intestinal lumen in chicks and rats. *Am. J. Physiol.* **211**, 419–423.

Wasserman, R. H., Henion, J. D., Haussler, M. R., and McCain, T. A. (1976). Calcinogenic factor in *Solanum malacoxylon:* Evidence that 1,25-dihydroxyvitamin D_3-glycoside. *Science* **194**, 853–855.

West, W. T., and Mason, K. E. (1958). Histopathology of muscular dystrophy in the vitamin E-deficient hamster. *Am. J. Anat.* **102**, 323–363.

Weswig, P. H. (1956). Vitamin A storage. *Oregon St. Game Bull.* **11**, 7.

White, J. R. (1970). Thiamine deficiency in an Atlantic bottle-nosed dolphin *(Tursiops truncatus)* on a diet of raw fish. *J. Am. Vet. Med. Assoc.* **157**, 559–562.

Wortman, J. D., and Larue, M. D. (1974). Hand-rearing polar bear cubs *(Thalarctos maritimus)* at Topeka Zoo. *Int. Zoo Ybk.* **14**, 215–218.

Yess, N. J., and Hegsted, D. M. (1967). Biosynthesis of ascorbic acid in the acouchi and agouti. *J. Nutr.* **92**, 331–333.

Youatt, W. G., Ullrey, D. E., and Magee, W. T. (1976). Vitamin A concentration in livers of white-tailed deer. *J. Wildl. Manage.* **40**, 172–173.

7

Estimation of Energy and
Protein Requirements

The essential continuity between an animal and its environment is revealed most
obviously in energy exchanges and transformations.

—King, 1974, p. 4

I. INTRODUCTION

Studies of wildlife energy transactions have been popular since the early 1950s
because of the recognition that the animal is often intimately coupled to its
environment via energy flow and must ultimately balance its energy ledger. The
nutritionist is often required to estimate energy and protein requirements of
wildlife as a prelude to estimating necessary food intake. Before examining some
of the requirement estimates, it is essential to understand the general types and
suitability of the methods employed.

For the captive animal in which the ingested food can be easily measured,
changes in tissue balance relative to feed intake can be used to estimate requirements for many whole-body processes. For example, if the energy or material
state of the animal is constant, the observed intake is the maintenance requirement. Although the maintenance requirement is often defined as the intake at
which a weight change does not occur, one must be careful in using weights as
the sole criterion of energy and matter balances, because compensatory changes
in body constituents also take place. When a range of intakes (X) are graphically
compared to the body changes produced (Y), the X intercept at zero gain or loss is
again the maintenance requirement, while the slope of the line is the gain or loss
per unit of additional intake (Fig. 7.1). Intake per animal as well as the dependent variable are often expressed as a function of body weight, since rarely are
the experimental animals perfectly matched and the use of such a common
denominator provides a means for extrapolating or comparing the results to other
animals of the same or differing species. Additional productive requirements can
be determined by feeding at differing levels and measuring the rate (such as milk

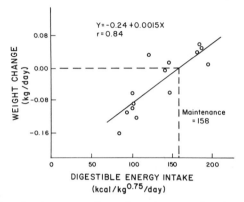

Fig. 7.1. Estimation of maintenance and growth requirements of captive white-tailed deer as a function of metabolic body weight as determined by feeding many different levels of digestible energy. (From Ullrey *et al.*, 1970, courtesy of the *Journal of Wildlife Management*.)

production) or success (such as reproduction) of the specific process being examined. Thus, all the requirements can be estimated by a direct feeding trial approach, although such estimates may be applicable only to the conditions in which the measurement occurred.

The use of captive animal feeding trials to estimate the requirements of the free-ranging animal poses many philosophical questions. As an example, the maintenance energy requirements of captive animals are often below what might be expected under free-ranging conditions since activity and thermoregulatory requirements may be minimized. The captive animal rarely needs to search for food, move between cover and food patches, or flee from predators. For example, the energy expenditure of a free-ranging male black duck during the summer was approximately 15% higher than the captive nonflying bird swimming and feeding in a large pond (Wooley and Owen, 1978). Similar results have been observed for nonflying versus flying blue-winged teal (Owen, 1969, 1970). Free-ranging wild and domestic ungulates may expend 25–100% more energy than confined animals (Osuji, 1974; Holleman *et al.*, 1979). The expenditures by wild herbivores may be even higher in heavily grazed areas requiring extensive food searching activities, in areas where disturbance by human beings or predators is extensive, or in severe continental climates requiring an increased thermoregulatory expenditure. Thus, the captive feeding trial approach to estimating energy requirements, while very applicable to understanding other captive animals, may not provide the type of information necessary to understand the free-ranging animal because it simplifies the animal– environment interaction. If one is unable to discern underlying interspecific similarities, a feeding trial approach requires that animals of n productive states in n environments be used to estimate all possible requirement levels.

The questioning by ecologists of feeding trials using captive animals has provided the impetus to examine wildlife requirements from a very basic perspective suitable for predicting requirements under many circumstances. The general approach has been to divide whole-body maintenance and productive processes into their basic components. Since the relative magnitudes of many of the components are not unique to the species or living state, interspecific comparisons can provide the basis for estimating requirements for species that have not been studied or for those living in markedly different environments. In either approach, the prediction of daily requirements must be based on an ecological awareness of the animal's daily life. Because of the number of biotic and abiotic variables affecting the animal's daily existence, the complexity of predicting requirements can be formidable (Fig. 7.2). Overzealous efforts to initially sim-

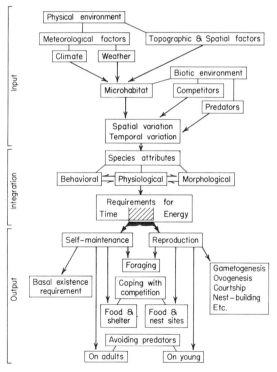

Fig. 7.2. Some of the interactions affecting the energy requirements in free-living birds. (From King, 1974, courtesy of the Nuttall Ornithological Club.)

plify without the necessary basic underpinning knowledge of the animal–environment interaction should be avoided.

II. Energy Expenditure for Maintenance

Maintenance energy expenditure is defined as the necessary chemical energy ingested to maintain basic body functioning or basal metabolism, to support activity costs, and to thermoregulate by balancing heat loss with heat production. The chemical energy used in these three processes as fats, proteins, or carbohydrates are oxidized is converted to heat. For example, if the animal oxidizes 1 mole of glucose, 6 moles of carbon dioxide and water are produced as 673 kcal of heat are liberated. The stoichiometric relationships between heat production and gas exchanges are constant for the complete oxidation of any organic compound. Therefore, the energy used in the individual maintenance processes can be determined by measuring either heat production (direct calorimetry) or gas production (indirect calorimetry).

Since atmospheric air is a relatively stable 20.939% oxygen, 0.031% carbon dioxide, and 79.03% nitrogen, indirect calorimetry requires only the determination of the expired air volume and composition. Respired air sampling in indirect calorimetry is usually accomplished either by confining the animal to a chamber through which atmospheric air is pumped, by placing on the experimental animal a face mask having a valve system for routing expired air directly into sensors or indirectly via sampling bags, or by surgically providing other gas collecting ports, such as tracheotomies (Mautz and Fair, 1980) (Fig. 7.3). The method of air sampling used depends on specific considerations relative to the test animals and hypothesis being examined.

The chamber continues to be popular for birds and small mammals because of the difficulty of placing masks on such animals and the need for visual isolation if many of the animals are to remain calm. The use of a mask depends on considerations of mask weight, air resistance, and dead space, if a passive system, and ease of animal habituation. Although light weight masks have been used successfully on flying parakeets and gulls (Tucker, 1969), masks have largely been confined to the larger mammals in which weight and dead space are relatively minor considerations. Since the energy equivalent of consumed oxygen is less variable as different substrates are metabolized (4.69 to 5.05 kcal/liter or $\pm$ 3.6% of the mean) than the respired carbon dioxide (5.05–6.69 kcal/liter or $\pm$ 16.3% of the mean), the measurement of oxygen consumption and the use of appropriate energy equivalents in indirect calorimetry to estimate maintenance requirements is preferred to relying solely on carbon dioxide production measurements (Tables 7.1 and 7.2).

Fig. 7.3. Different types of gas-collecting procedures used in calorimetric measurements: (top) a white-crowned sparrow running on a treadmill in which air is pumped through the plexiglass box (courtesy of David P. Mack, Washington State University, Pullman) and (bottom) a passive mask system with one-way valves on a mule deer in which air is inspired on the open side and expired through the long tube.

TABLE 7.1

Thermal Equivalents of Oxygen and Carbon Dioxide and the Corresponding Percentage of Fat and Carbohydrates Oxidized for Different Respiratory Quotients in Mammals[a]

Nonprotein RQ	O_2 kcal/liter	CO_2 kcal/liter	kcal/g	Percentage O_2 consumed by Carbohydrates	Fat	Percentage heat produced by oxidation of Carbohydrates	Fat
0.70	4.686	6.694	3.408	0	100	0	100
0.72	4.702	6.531	3.325	4.4	95.6	4.8	95.2
0.74	4.727	6.388	3.252	11.3	88.7	12.0	88.0
0.76	4.752	6.253	3.183	18.1	81.9	19.2	80.8
0.78	4.776	6.123	3.117	24.9	75.1	26.3	73.7
0.80	4.801	6.001	3.055	31.7	68.3	33.4	66.6
0.82	4.825	5.884	2.996	38.6	61.4	40.3	59.7
0.84	4.850	5.774	2.939	45.4	54.6	47.2	52.8
0.86	4.875	5.669	2.886	52.2	47.8	54.1	45.9
0.88	4.900	5.568	2.835	59.0	41.0	60.8	39.2
0.90	4.924	5.471	2.785	65.9	34.1	67.5	32.5
0.92	4.948	5.378	2.738	72.7	27.3	74.1	25.9
0.94	4.973	5.290	2.693	79.5	20.5	80.7	19.3
0.96	4.997	5.205	2.650	86.3	13.7	87.2	12.8
0.98	5.022	5.124	2.609	93.2	6.8	93.6	6.4
1.00	5.047	5.047	2.569	100	0	100	0

[a]From Brody, 1945, p. 310.

TABLE 7.2

Thermal Equivalents of Oxygen and Carbon Dioxide for Various Substrates Metabolized and the Corresponding Respiratory Quotient in Birds[a]

RQ	Substrate	Thermal equivalents (kcal/liter) O_2	CO_2
0.71	Fat	4.686	6.694
0.72	Protein	4.750	6.597
1.00	Carbohydrate	5.047	5.047

[a]From King and Farner, 1961; Romijn and Lokhorst, 1961.

A. Basal Metabolism

1. Measurement Criteria

The concept of basal metabolism as some lower level of metabolism necessary for basic body functioning has pervaded thought and studies of energy metabolism for many years. Basal metabolic rate has been defined as the energy expenditure of an animal (a) in muscular and psychic repose although not sleeping; (b) in a thermoneutral environment; and (c) in a postabsorptive state (Brody, 1945; Kleiber, 1961; Blaxter, 1962; King, 1974). The values determined are not necessarily minimal, since sleeping, prolonged starvation, and torpidity, estivation, or hibernation can further reduce energy expenditure below basal metabolic rate levels (Passmore and Durnin, 1955; Kayser, 1961; Bartholomew and Hudson, 1962; Lasiewski, 1963; Brown and Bartholomew, 1969; Toutain et al., 1977; Westerterp, 1977; Walker et al., 1979). However, since the animal is not expending energy to thermoregulate, process or metabolize food, or increase the activity state, basal metabolic rate does represent a baseline on which the additional energy expenditures of the active animal can be added. Although a portion of basal metabolism is due to breathing and blood circulation, approximately 80% of the energy is expended in maintaining electrochemical gradients across cell membranes and in the synthetic processes involved in replacing proteins and other macromolecules (Passmore, 1971).

The postabsorptive state occurs when an animal fasts until it is no longer directly metabolizing ingested feed and is therefore dependent on mobilized body tissue for meeting its energy requirements. Since it is difficult to directly measure physiological emptying of metabolizable material from the gastrointestinal tract, indirect indices—primarily the absence of methane production associated with gastrointestinal fermentation; a stabilized, reduced metabolic rate relative to the fed animal; or a stoichiometric gas relationship indicating fat metabolism—are often used to determine when the animal has reached basal conditions.

The stoichiometric ratios between carbon dioxide produced and atmospheric oxygen consumed by the animal are termed the *respiratory quotient* (RQ) and normally range from 0.70 to 1.00. Since the fatty acids of metabolized fat are relatively poor in molecular oxygen and carbohydrates are relatively rich, the RQ of fats is lower than that of carbohydrates because additional atmospheric oxygen must be respired relative to the carbon dioxide produced (Table 7.1). As the animal fasts, the observed RQ decreases from levels characteristic of carbohydrate metabolism (1.0) to a nonprotein corrected RQ of 0.7 associated with fat metabolism (Fig. 7.4). Occasionally, an RQ outside the range of 0.7–1.0 occurs. A very high RQ can occur during the initial phases of exercise as carbon dioxide is flushed from the animal, during intense fattening, or during fermentation in the nonfasted ruminant, whereas an RQ lower than 0.7 may be due to the retention of

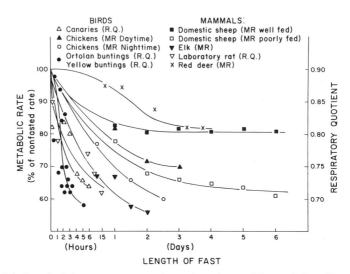

Fig. 7.4. Length of time necessary to reach postabsorptive conditions as indicated by a minimal RQ or stable metabolic rates in several animal species. (Data from Benedict and Fox, 1933; Barott *et al.*, 1938; Marston, 1948; Wallgren, 1954; Westerterp, 1977; Robbins, unpublished; Simpson *et al.*, 1978b.)

carbon dioxide, excretion of bicarbonates, or the preferential or partial metabolism of long-chain fatty acids. The RQ for protein metabolism is different between birds (0.72) and mammals (0.82) and reflects the differing metabolic approaches between the two groups of organisms in excreting excess nitrogen (Tables 7.1 and 7.2). The uric acid of bird excreta contains 29% oxygen and 35% nitrogen, whereas urea contains 27% oxygen and 47% nitrogen.

The necessary length of fasting to achieve basal conditions is largely dependent on the size of the animal, the rate of food passage through the gastrointestinal tract, and the level of previous feeding (Marston, 1948). Although numerous small birds, such as ortolan and yellow buntings and canaries, achieve postabsorptive conditions in 3–6 hr (Stevenson, 1933; Brenner and Malin, 1965), shrews began dying after 2 hr of fasting because of their small size and very intense metabolism (Gebczynski, 1971), whereas desert rodents required more than 6 hr to become postabsorptive (Hudson and Deavers, 1976). The required length of fasting increases to more than 48 hr for larger ruminants because of their relatively slow rate of food passage (Silver, 1968; Wesley *et al.*, 1973; Simpson *et al.*, 1978b) (Fig. 7.4).

Unfortunately, the fasting of some animals can make it very difficult to meet the condition of a calm, resting attitude. Hummingbirds often enter torpor when fasted as a physiological and ecological adaptation to reduced energy availability, whereas many rodents and birds become hyperactive (Ketterson and King, 1977). The experimental difficulties in achieving the postabsorptive state in a female Indian elephant were described by Benedict (1936, pp. 233–234):

> . . . when she was not fed as frequently as she wished. She became almost frantically restless, moved from side to side, turned not a little, trumpeted, and gave every evidence of extreme discomfort and irritability. . . . At night when the keeper was sleeping and [she] became hungry, she would gather up refuse from the floor and throw it at his cot to wake him to go get hay. If this did not waken him, she would search around until she found a stone. On several occasions stones as large as hens eggs were removed from his bed . . . . One night she actually . . . . crashed through the doors of the barn, went immediately to another building 150 meters away where she knew hay was stored, ate what hay she wanted, and returned to the barn . . . therefore, it was apparent that the measurement of the basal metabolism of the elephant in the true post-absorptive condition was impracticable . . . .

Thus, wildlife present special problems in meeting all conditions of basal metabolic rate measurements. Perhaps one of the most difficult conditions to judge is whether the animal is in a calm, resting state without excess apprehension and muscle tonus. Extensive training and habituation may be necessary. Similarly, social animals may need the psychophysiological stimulation of adjacent conspecifics to reach basal conditions (Martin *et al.,* 1980). Basal metabolic rate in diurnally active birds can often be measured in the dark to reduce all activity. Avian ecologists have accepted sleeping as a component of basal metabolic rate determinations, while measurements made during the day on quiescent but awake fasted birds in a thermoneutral environment are termed *fasting metabolic rates.* Thus, additional terminology has arisen, such as *standard metabolic rate* (SMR) and *resting metabolic rate* (RMR). In general, all the metabolic rate terminology has been so ''mutilated and confused'' (King, 1974, p. 30) that the student should read the methodology of each article carefully to determine the conditions of the measurements.

2. Interspecific Comparisons

The extensive use of basal metabolic rate measurements has become popular because of the general interspecific predictability and unifying nature of the measurements to understanding energy metabolism of diverse biological systems (Brody, 1945; Kleiber, 1961; Schmidt-Nielsen, 1970). Historically, basal metabolic rates of different species have been compared to an exponential function of body weight. Kleiber's (1947) determination that basal metabolic rates are a function of the 0.75 power of body weight has been widely accepted (Fig. 7.5), with the basal metabolic rate of eutherian or placental mammals described by the equation:

$$BMR = 70 \ W^{0.75}$$

where *BMR* is the basal metabolic rate in kilocalories per day and *W* is the body weight in kilograms.

The Kleiber equation is a very broad generalization, and many groups of mammals have significantly lower or higher metabolic rates than predicted. One of the major differences occurs in neonatal mammals whose metabolic rate changes relative to age. Neonates often have a metabolic rate similar to adults, but it subsequently rises within the first few days of life to levels as high as twice the adult level before gradually falling with maturity (Wesley *et al.*, 1973; Poczopko, 1979; Ashwell-Erickson *et al.*, 1979).

Many adult desert rodents, fossorial rodents, arboreal folivores, and desert carnivores have lower than expected basal metabolic rates relative to the Kleiber equation (Hart, 1971; Hudson *et al.*, 1972; Scott *et al.*, 1972; McNab, 1974; Hudson and Deavers, 1976; McNab, 1978b; McNab, 1979; Noll-Banholzer, 1979; Vleck, 1979; Knight and Skinner, 1981), while marine mammals, many mustelids, lagomorphs, shrews, and microtine rodents often have elevated basal metabolic rates (Irving, 1972; Wang *et al.*, 1973; McNab, 1974; Newman and Rudd, 1978). The lower metabolic rate of desert-dwelling species, which can be further reduced by dehydration (Reese and Haines, 1978; Ward and Armitage, 1978), could have significant ecological advantages in conserving body water and stretching limited food resources. Wild ungulates often have higher fasting metabolic rates than predicted by the Kleiber equation, although the psychological and activity state of the animals can be questioned in some of the observations (Silver *et al.*, 1959; Brockway and Maloiy, 1967; Rogerson, 1968; Silver, 1968; McEwan, 1970; Wesley *et al.*, 1973; Weiner, 1977).

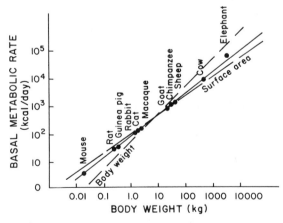

Fig. 7.5. Relationships between basal metabolic rate (0.75 power of body weight), body weight (power of 1), and surface area (0.67 power) in several adult eutherian mammals. (From Kleiber, 1947, courtesy of the American Physiological Society.)

While the basal metabolic rates of most animals are a function of body weight to the 0.75 power (Schmidt-Nielsen, 1970; Poczopko, 1979), several entire taxonomical categories have constants significantly different from 70 (Kleiber, 1947). For example, the constant for marsupials is 48.6 (T. J. Dawson and Hulbert, 1970; McNab, 1978a)—that is, 70% of the eutherian level—whereas the basal metabolic rate of the even more primitive echidna (monotreme) is 27.5% of placental mammals (T. J. Dawson et al., 1979).

The lowest resting metabolic rates in mammals occur during hibernation and estivation. This enables the free-ranging animal in a predictable environment containing an abundant, seasonal food supply the necessary life strategy to stagger the accumulation and use of energy and matter reserves and thereby occupy a niche that could not support an active animal year round. The metabolic rates of hibernators are, in part, dependent on environmental temperature and therefore body temperature (Bartholomew and Hudson, 1962; South and House, 1967; Brown and Bartholomew, 1969; Cranford, 1978; Vogt and Robert, 1978). Many small mammals are able to lower their body temperature to 2°–3°C during hibernation in cold environments. However, the body temperature in larger hibernators can be much higher than in the smaller species, because increasing fur thickness and decreasing surface area to volume ratios limit the rate of cooling. For example, the body temperature of bears does not drop below 31°C (Morrison, 1960; Kayser, 1961; Watts et al., 1981). The minimal metabolic rates (Kayser, 1964) of hibernators, excluding arousal periods and not corrected for differing body temperatures, are expressed by the following equation:

$$\text{Metabolic rate} = 3.2 \ W^{1.03}$$

where metabolic rate is in kilocalories per day and W is the body weight in kilograms.

Thus, unlike basal metabolism, energy expenditure during hibernation is a linear function of body weight (Kayser, 1961). The hibernation metabolic rate converges toward the interspecific basal metabolic rate in larger hibernators as body temperature remains high. However, if energy storage, particularly fat, is a linear function of body weight, the unity function of the hibernation metabolic rate indicates that the various hibernators are equally adapted for prolonged fasting (Morrison, 1960). Although small animals consume very little energy during the inactive phase of hibernation, the physiological processes of arousal are very costly. For example, ground squirrels expended 90% of all the energy consumed during the total hibernation period in the arousal periods, even though 93% of the total time was spent in actual hibernation. However, even with the very costly arousals, the energy savings of hibernation in ground squirrels when compared to remaining active averaged 88% (Wang, 1979). Thus, the greatly reduced metabolic rate of small hibernators represents an extreme example of reduced energy expenditure and savings.

The best fit equations for the basal metabolic rate of adult birds have been widely debated (King and Farner, 1961; Lasiewski and Dawson, 1967; Zar, 1968, 1969; Aschoff and Pohl, 1970a; Poczopko, 1971; King, 1974). King and Farner (1961) first suggested that all birds might not be adequately represented by one equation, since larger birds (over 125 g) were apparently different from the smaller species. Lasiewski and Dawson (1967) later pointed out that it was not purely a question of bird size, but rather that passerines and nonpasserines were represented by differing equations.

Although the BMR of larger nonpasserines is indistinguishable from the generalized adult eutherian equation, passerines have metabolic rates 30–70% higher than the generalized mammalian equation (Calder and King, 1974). Prinzinger and Hänssler (1980) have proposed that the BMR of small nonpasserines is also equally high. The basal metabolic rates of birds measured during the nonactive phases of the daily cycle are approximately 24% lower than those measured during the active phase on resting birds (Aschoff and Pohl, 1970a) (Fig. 7.6). The differences may be due in part to sleeping during the nonactive phase or to a true difference in metabolic cycles.

The metabolic rate of nestling birds follows a trend during growth similar to that observed in neonatal mammals of an initial rise followed by a decline to adult levels (Poczopko, 1979). However, the initial rate is below the adult level in waterfowl (ranging from 65–90% of the predicted adult level) and gallinaceous birds (40–72%) but virtually identical in larids (W. R. Dawson et al., 1976).

Thus, basal metabolic rate estimations should not be regarded as immutable constants (Thonney et al., 1976), particularly when attempting by a summative approach to estimate the energy requirements of free-ranging animals. Not only are there major differences between taxonomical groups, but basal metabolic rates vary with season in many wildlife species that remain active year round

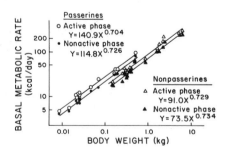

Fig. 7.6. Basal metabolic rates in birds measured during the active and nonactive phases of the daily cycle. (From Aschoff and Pohl, 1970b, courtesy of the *Journal für Ornithologie*.)

(Silver, 1968; West, 1972; Hinds, 1973; Zervanos, 1975; Chappel and Hudson, 1978b; Moen, 1978; Newman and Rudd, 1978; Ringberg, 1979). Seasonally depressed metabolic rates, even in active homeotherms, are adaptive and often observed during periods of food scarcity.

B. Activity

Since animals cannot exist indefinitely under basal conditions but must engage in additional energy-demanding activities, the determination of the energy expenditure for numerous activities has become an increasingly popular field of investigation as allometric equations have been used to evaluate energy costs of specific activities for various groups of organisms (Taylor et al., 1970, 1972; Tucker, 1970; Schmidt-Nielsen, 1972; Berger and Hart, 1974; Fedak et al., 1974; Wunder, 1975; Baudinette et al., 1976; Taylor, 1977; Cohen et al., 1978). Energy expenditures for activity have been used to answer basic physiological questions (Taylor et al., 1972), interpret life strategies (Chassin et al., 1976; Powell, 1979), or estimate animal requirements in carrying capacity determinations relative to management programs (Moen, 1973). The summative estimation of the energy requirements for activity is dependent on the quantification of the energy expenditure for each activity per unit time multiplied by the length of time the animal is involved in that activity.

1. Standing

For virtually all animals, except possibly members of the genus *Equus,* which can passively lock the legs and do not need to maintain muscle contractions to remain upright, standing is energy demanding. The cost of standing in wild animals is often higher than the 9% increase over lying reported for human beings, cattle, and sheep (Brody, 1945) and used by Moen (1973) for wild ungulates. Measurements of the cost of standing or the extrapolation of the cost of locomotion to zero velocity to estimate the cost of standing in eight mammals ranging in size from kangaroo rats to moose have averaged $24.8 \pm 11.0\%$ higher than the cost of lying (Taylor et al., 1970; Wesley et al., 1973; Weiner, 1977; Renecker et al., 1978; Gates and Hudson, 1978, 1979; Chappel and Hudson, 1979). The cost of standing in birds ($N = 10$) is extremely variable but averages 13.6% higher than the cost of resting (Taylor et al., 1970; Fedak et al., 1974; Brackenbury and Avery, 1980). The very large range in the cost of standing for the different species may simply represent experimental difficulties or indeed a true difference that can be related to skeletal or muscular morphology (Chappel and Hudson, 1979).

2. Terrestrial Locomotion

a. Horizontal Movement. The energy costs of terrestrial transport (such as walking, trotting, and running) have been determined for a number of animals by exercising them either on treadmills or on various ground surfaces and measuring oxygen consumption. For most of the animals studied thus far, the cost has been expressed as a linear function of the speed of travel, with the y intercept corresponding to the standing expenditure (Fig. 7.7). Notable exceptions to the preceding generalization about linearity do exist and to some extent reflect the differing efficiencies of transport by differing gaits and speeds within those gaits (T. J. Dawson and Taylor, 1973; T. J. Dawson, 1976; Chassin *et al.*, 1976; Alexander *et al.*, 1980). For example, at very slow speeds, the kangaroo uses a pentapedal type of locomotion in which the body is partially supported by the tail. Since pentapedal locomotion is very costly compared to the faster bipedal hopping, the transition between the two gaits is not reflected in increasing costs. Similarly, the costs of transport by differing gaits and speeds in more traditional bipeds and quadrupeds are a series of slightly curvilinear relationships (Hoyt and Taylor, 1981) (Fig. 7.8). The animal switches between gaits when the cost of continuing in the same gait is higher than that of moving into the next appropriate gait. However, the error in using the linear expression is so minimal that an assumption of linearity is acceptable for most studies.

The slope of the regression between the cost of horizontal movement and speed is the net cost expressed in kilocalories per kilogram per unit of distance moved and becomes increasingly steep as the weight of the animal becomes smaller (Figs. 7.7 and 7.9). Thus, the same increase in speed produces a far greater rise in the metabolic rate of a mouse than in that of the larger dog (Fig. 7.7). Earlier studies by Taylor *et al.* (1970) and Fedak *et al.* (1974) of the net cost of locomotion

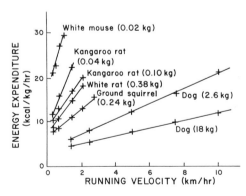

Fig. 7.7. Energy expenditure as a function of running velocity in a range of animals. (From Taylor *et al.*, 1970, courtesy of the American Physiological Society.)

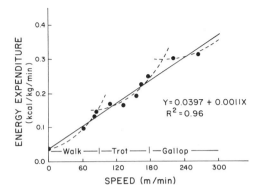

Fig. 7.8. Energetic cost of differing gaits in an elk calf. (From Parker, 1981.)

suggested that bipeds and quadrupeds were represented by differing equations relative to body weight (Table 7.3). However, more recent studies have demonstrated that these conclusions were based on inadequate samples and that most bipeds and quadrupeds can now be represented by one line (Fedak and Seeherman, 1979; Paladino, 1979). The increasing cost at decreasing weights is a predictable function of approximately the one-third power of body weight (Table 7.3). The growing animal also experiences the same change in costs as a function of increasing body weight. For example, the net cost of horizontal movement (Y, kilocalories per kilogram per kilometer) in growing elk and mule deer decreases as body weight (W in grams) increases ($Y = 31.10W^{-0.34}$) (Parker, 1981). Thus, there is not a single slope or cost for a given species, but an entire range of values dependent on the individual's body weight.

One of the most costly forms of horizontal locomotion is the waddling of

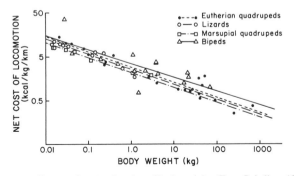

Fig. 7.9. Net cost of locomotion as a function of body weight. (From Paladino, 1979, courtesy of F. Paladino, Department of Zoology, Miami University, Ohio.)

TABLE 7.3
Net Cost of Horizontal Locomotion (kcal/kg/km) as a Function of Body Weight (g) in Several Groups of Animals

Group	Equation[a]	R^2	Reference
Lizards	$Y = 26.50 \, W^{-0.34}$	0.88	Bakker (1972)
Birds	$Y = 11.81 \, W^{-0.20}$		Fedak et al. (1974)
Bipeds (birds and apes)	$Y = 20.24 \, W^{-0.32}$	0.59	Paladino (1979)
Marsupials	$Y = 22.80 \, W^{-0,34}$		Baudinette et al. (1976)
Eutherian mammals	$Y = 40.61 \, W^{-0.40}$		Taylor et al. (1970)
	$Y = 30.48 \, W^{-0.34}$		Cohen et al. (1978)
Eutherian quadrupeds	$Y = 31.34 \, W^{-0.32}$	0.90	Cohen et al. (1978)
All together (birds, mammals, lizards)	$Y = 25.92 \, W^{-0.32}$	0.85	Paladino (1979)
	$Y = 18.77 \, W^{-0.28}$	0.84	Fedak and Seeherman (1979)

[a]Kcal = 4.8 × liters of oxygen consumed when only oxygen consumption data were presented.

penguins and waterfowl, for which the net cost is approximately twice that for the other groups of animals ($54.29 \, W^{-0.33}$) (Pinshow et al., 1977). Specializations for aquatic locomotion have occurred at the energetic expense of efficient terrestrial locomotion. For many aquatic species that spend little time on land, the additional cost may be insignificant. However, penguins make long treks, as far as 120 km, to and from their rookeries, while fasting, and the additional costs may have serious effects on breeding success during seasons when rookeries are separated from the sea by unusually wide ice (Le Maho, 1977; Pinshow et al., 1977).

The average cost for an animal to walk or run 1 km can be estimated by summing the net cost for locomotion in kilocalories per kilogram per kilometer (Table 7.3) and the overhead costs of basal metabolism and standing:

$$\text{Net Cost + Overhead Cost}$$
$$Y_{\text{kcal/hr}} = 25.92 W^{0.68} + \frac{I(K)W^{0.75}}{V},$$

where W is body weight in grams, V is speed of movement in kilometers per hour, I is increment due to standing, and K is the basal metabolic rate constant on a per hour basis. As one would expect from the logarithmic functions, it is much more costly per unit of body weight for a mouse to move a given distance than for an elephant, even though the net cost per unit of distance for each animal is independent of speed. Note that this is not the case for the entire animal, as the elephant will expend far more total energy than will the mouse. However, the total cost per kilometer—that is, the net cost plus the overhead cost of the posture of locomotion—is not independent of speed and is a curvilinear function that decreases at increasing speeds until a minimal asymptotic cost per unit of distance is achieved (Taylor *et al.*, 1970). The decreasing cost per unit of distance with increasing speed can be visualized in that the net cost remains constant, a function of body weight only, whereas the overhead cost is both weight and time dependent and becomes geometrically less important as speed of movement increases.

Since the cost of movement either per unit distance or per unit time is an entire range of values that is largely dependent on speed, one would not want initially to use a single value or increment as the cost of walking or running. However, observations of free-ranging animals suggest that the speed of movement is subjected to an optimization process since groups of animals often move within a fairly restricted range of speeds. For example, migrating African animals chose a restricted range of speeds (Pennycuick, 1975) that could conceivably be chosen to minimize energy expenditure per unit of distance covered (Hoyt and Taylor, 1981). However, the foraging animal may attempt to maximize the net rate of energy gain by choosing a speed that provides the opportunity for the greatest food intake at the least metabolic cost (Pyke, 1981). Thus, the estimation of locomotion costs should ultimately depend on an understanding of the constraints acting on the choice of movement speed.

Since the preceding discussion is based on measurements of animals exercising on hard surfaces, one must realize that locomotion on uneven surfaces or on softer surfaces in which sinking occurs will be far more costly. For example, reindeer expended 12% more energy when walking on dry tundra than on a packed-dirt road, and 25% more when walking on wet tundra than on the road (White and Yousef, 1977). Unfortunately, the cost of locomotion in snow has almost been neglected, even though it is critically important to temperate and arctic wildlife (Telfer and Kelsall, 1979). Energy expenditure of white-tailed deer in snow significantly increased as snow depth increased. The increase was curvilinear with the greatest effect occurring at a sinking depth in excess of 25–30 cm or when snow crusts formed that were unable to support locomotion (Mattfeld, 1974). The snow depths associated with markedly increased energy expenditures in deer were similar to snow depths reported to exclude deer use in many field situations (Hepburn, 1959; Telfer, 1967; Kelsall and Prescott, 1971).

b. Vertical Movement.   Animals rarely move solely on flat surfaces but are often required to climb up and down hills, trees, or cliffs. Upward movements can be quite costly because of the energy expended in working against gravity, whereas downslope locomotion can be very inexpensive because the potential energy stored in moving upslope is recovered as kinetic energy during descent (Taylor *et al.*, 1972). Several different ways are available to express the cost of vertical locomotion. Since we have already estimated horizontal costs, one of the simpler methods will be to divide the cost of moving on a slope into its horizontal

TABLE 7.4

Mean Energetic Cost of Lifting One Kilogram of Body Weight One Vertical Meter for Several Animal Species[a]

Species	Body weight (kg)	Angle of slopes	Cost (kcal/kg/m)	Reference
Red squirrel	0.25	0–37°	7.69	Wunder and Morrison (1974)
Domestic dog	12.8	0–11.5°	7.71	Raab et al. (1976)
Chimpanzee	17.5	0–15°	3.54	Taylor et al. (1972)
Mule deer	14–24	0–14.3°	5.99	Parker (1981)
Domestic sheep	40.8	0–5.1°	6.36	Clapperton (1964)
Lion cubs	53.5	0–17.1°	3.27	Chassin et al. (1976)
Elk	28–145	0–14.3°	5.73	Parker (1981)
Red deer	68.3	0–14°	5.13	Brockway and Gessaman (1977)
Reindeer	96.5	0–5.1°	6.04	White and Yousef (1977)
Domestic cattle	120–191	0–6°	6.32	Ribeiro et al. (1977)
Burro	253.5	0–9.6°	7.20	Yousef et al. (1972)
		Mean = 5.91 ± 1.47		

[a]For those species in which individual body weights varied extensively, a weight range is reported. For those species comprising test animals of similar weights, a mean body weight is reported.

and vertical vectors. Animals expend an average of 6 kcal/kg per vertical meter climbed (Table 7.4). The net cost is independent of the species weight and angle of ascent within the more moderate slope angles examined thus far. Since all animals expend the same amount of energy per unit of weight for uphill locomotion, while the smaller animal expends more energy per unit of weight for basal metabolism, upslope movement requires a much smaller energy expenditure increment for the smaller animal than for the larger one. Such a relationship may be an important reason why squirrels can literally run up trees, whereas larger animals, such as human beings, find it far more difficult (Taylor et al., 1972). Similarly, for different species traversing very rugged terrain in nonfeeding activities, the average angle of ascent decreases with increasing body size. However, as terrain becomes steeper, animals pick steeper ascents, since the more vertical path is still less costly than traveling a much greater total distance when using a shallower trail angle (Reichman and Aitchison, 1981). For the feeding animal consuming an evenly distributed resource in steep terrain in which there is no need to be at a different elevation at the end of a feeding bout, horizontal movement should be maximized and vertical displacement minimized in order to maximize the net rate of energy intake (Fig. 7.10).

The efficiency of recovering the potential energy stored in moving upslope is dependent on both the animal's weight and degree of incline (Table 7.5). Small animals are far more efficient at moving down moderate slopes than are the

TABLE 7.5
Energy Recovered during Downhill Movement in Several Species[a]

Species	Body weight (kg)	Incline angle	Energy recovered (kcal/kg/m)	Efficiency (%)	Reference
Mouse	0.03	15	2.05	87	Taylor *et al.* (1972)
Dog	12.8	4.2	2.63	100	Raab *et al.* (1976)
		6.7	2.32	99	
		11.5	1.43	61	
Chimpanzee	17.5	15	2.23	95	Taylor *et al.* (1972)
Mule deer	19	14.3	1.76	75	Parker (1981)
Reindeer	96	2.9	1.66	71	White and Yousef (1977)
		5.1	1.38	59	
Elk	50	14.3	1.26	54	Parker (1981)
	100	14.3	0.85	35	
	150	14.3	0.66	28	
Burro	253.5	5.7	1.12	48	Yousef *et al.* (1972)
		9.6	0.23	10	

[a]Per meter of vertical movement, 2.34 kcal of potential energy are stored.

Fig. 7.10. Feeding trails worn into a mountainside from years of domestic and wild ungulate feeding activities. Note that the very steep vertical profile is turned into a series of horizontal profiles. The more vertical trail has been worn by nonfeeding animals simply moving to the top of the hill.

larger animals. As the angle of the slope increases, the larger animal becomes less efficient at a lower slope angle than does the smaller animal. For example, the burro begins expending approximately the same amount of energy in controlling its movement as it gains in moving down a 10° slope. Hence, the cost to a burro in moving down a 10° slope is approximately the same per unit of distance as when moving on the horizontal. When moving down slopes steeper than 10°, the burro would expend more energy than it would recover, and thus such movement would be more costly than moving the same distance on the horizontal. However, the dog and mule deer would still recover far more energy than they would have to expend in controlling their acceleration when moving on the same 10° slope. Thus, one would expect that when animals of different weights descend a hill, the larger animal will either pick a less steep route or will not attempt to control its acceleration and will simply run down the hill if it is a relatively short hill. In conducting field studies with tame elk, we have observed that larger elk prefer to run down a relatively steep, short hill that as calves they had routinely walked down.

3. Burrowing

Burrowing, such as by the pocket gopher, is 360–3400 times more costly, depending on soil type, than simply walking the same distance. Burrowing in a cohesive clay averaged 11 times more costly than burrowing in loose soils (fine sand, gravelly sand, or sandy loam) (Vleck, 1979). However, the very high cost of burrowing per unit of time or distance may be a relatively small expenditure if the animal is able to occupy the burrow for an extensive period of time, and it is obviously essential for the exploitation of the subterranean niche. The actual cost of locomotion within an existing burrow should be the same as normal terrestrial costs unless friction against the walls or a necessitated altered type of locomotion within the confines of the burrow increase the costs.

4. Brachiation

Brachiation, while appearing as the analogue of walking, was approximately 1.48 times as costly as walking in the spider monkey whereas simply hanging was 1.57±0.27 more costly than resting in spider monkeys and slow loris. Although brachiation appears to be more costly than walking, it is essential for the exploitation of the arboreal environment, with the energetic comparison perhaps being one of the more minor considerations in understanding its evolution (Parsons and Taylor, 1977).

5. Flying

Flight, whether in birds, bats, or insects, is very economical per unit of distance covered in comparison to terrestrial locomotion. For example, a 10 g bird expends less than 1% of the energy that a 10 g mouse expends in moving

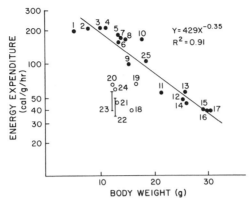

Fig. 7.11. Energy cost of bird flight as a function of body weight (1—*Calypte costae;* 2—
Amazilia fimbriata; 3—*Eulampis jugularis;* 4—*Spinus spinus;* 5 and 6—*Fringilla coelebs;* 7—
Fringilla montifringilla; 8—*Pyrrhula pyrrhula;* 9—*Melopsittacus undulatus;* 10—*Carpodacus
erytherinus;* 11—*Pluvialis dominica;* 12—*Larus artricilla;* 13—*Columba livia;* 14—*Larus dela-
warensis;* 15—*Larus marinus;* 16—*Anas platyrhynchos;* 17—*Anas rubripes;* 18—*Apus apus;* 19—
Progne subis; 20 and 24—*Hirundo rustica;* 21, 22, and 23—*Delichon urbica;* 25—*Sturnus vul-
garis*). The line is fitted for points 1–17 and 25. (Torre-Bueno and Larochelle, 1978; Hails 1979.)
(Adapted and reprinted with permission from *Comparative Biochemistry and Physiology,* Vol. 63A,
C. J. Hails, A comparison of flight energetics in Hirundines and other birds, Copyright 1979,
Pergamon Press, Ltd.)

1 km (Tucker, 1970; Schmidt-Nielsen, 1972; Thomas and Suthers, 1972; King,
1974). Although the energetic cost of bird flight per unit of body weight generally decreases as body size increases (Fig. 7.11), major differences in costs do
occur and can be related to differences in flight pattern, wing loading (i.e., body
weight to wing area), and wing conformation (i.e., wing area to wingspan).
While most actively flying birds expend energy at approximately 12 times BMR,
flight in martins and swallows is at an energetic cost of 2.9–5.7 times BMR. The
markedly lower cost of these aerial feeders is in part due to a relatively low wing
loading, a gliding flight pattern, and a very high wingspan to wing area ratio
(Hails, 1979). Even lower flight costs undoubtedly occur in soaring birds, such
as raptors and vultures. Similarly, since the energy expended in ascent can be
recovered very efficiently during descent, an undulating flight pattern is quite
similar, if not equal, in cost to a purely horizontal effort (Tucker, 1969; Schmidt-
Nielsen, 1972).

Unlike terrestrial locomotion, energy expenditure per unit of time during flight
is not a direct linear function of speed, since both gulls and starlings maintained a
relatively constant energy expenditure, whereas the cost to parakeets initially
decreased then increased as flight speeds doubled (Tucker, 1969; Torre-Bueno
and Larochelle, 1978). The constant energy expenditure in starlings was accom-

plished by maintaining a constant wing-beat rate but varying body attitude and wing-beat amplitude. Flight speeds in free-ranging birds are apparently compromises between maintaining a minimum metabolic rate per unit of time and per unit of ground distance covered, with efficient ground coverage being of overriding importance (Schnell and Hellack, 1979). However, as improved measurement techniques are developed, many of the theoretical considerations of bird flight may need to be reevaluated.

6. Swimming

Energy expenditures during surface and submerged swimming have been measured in the mallard duck (Prange and Schmidt-Nielsen, 1970) and the California sea lion (Costello and Whittow, 1975). Surface swimming by the mallard cost approximately 1.7 times the resting metabolic rate in water at the most efficient speed (0.50 m/sec) relative to distance covered. Higher swimming speeds increased the cost up to 3 times resting metabolic rate as a wake was created and pushed in front of the duck. Free-swimming ducks in a pond had a mean speed of 0.48 m/sec, suggesting that they indeed swim at the most efficient speed relative to distance covered.

Swimming at moderate speeds is far less costly than other forms of locomotion because of the beneficial effects of bouyancy and the fact that for many types of swimming, such as that of submerged penguins, thrust is generated over both halves of the wing stroke cycle (Schmidt-Nielsen, 1972; Clark and Bemis, 1979). However, the cost of submerged swimming is very difficult to estimate, as it is often independent of the speed of travel. While the oxygen consumption during the first 10 sec of a 60-sec swim by the sea lion rose to approximately 2 times the resting rate, the metabolic rate decreased during the next 50 sec to ultimately the resting level (Costello and Whittow, 1975). The uncoupling of metabolic rate and activity state is due to the diving reflex that is triggered during submerged swimming in virtually all aquatic animals and reduces blood flow and oxygen consumption in skeletal muscles as anaerobic metabolism predominates while maintaining cerebral and coronary circulation.

At very high speeds, swimming becomes increasingly costly since drag between the body surface and the water is proportional to the square of velocity. The leaping by many dolphins and fish when swimming at high speeds is a demonstration of the fact that less energy is expended per unit of distance covered by leaping than by remaining submerged. The swimming velocity at which this occurs represents a trade-off between the energy expended in becoming airborne versus that expended to overcome the increasing drag (Au and Weihs, 1980).

7. Feeding

The energy cost of feeding is the additional expense of ingesting and manipulating the food above the energy cost of the general activity state, such as

TABLE 7.6
Additional Cost of Eating above Standing in Several Ruminants[a]

Species	Energy expenditure (kcal/kg BW/hr)	Reference
Bighorn sheep	0.44	Chappel and Hudson (1978a)
Domestic cattle	0.44	Adam et al. (1979)
Domestic sheep	0.54	Graham (1964)
	0.52	Young (1966)
	0.83	Webster and Hayes (1968)
	0.44	Osuji (1974)
Moose	0.52	Renecker et al. (1978)

[a]From Wickstrom, 1981.

standing, walking, or flying. Feeding costs have generally not been measured, in part because of the common use of face masks, which preclude feeding in indirect calorimetry. Measurements of feeding costs should markedly increase in large herbivores as tracheotomies are further developed and used for expired air sampling (Mautz and Fair, 1980). The cost ranges from probably insignificant levels in animals ingesting food requiring minimal mastication (such as hummingbirds consuming nectar, or fish, grain, or insect feeders that easily and rapidly consume their food whole) to probably the highest levels in those animals extensively manipulating or masticating their food (such as herbivores and those animals that must separate edible and nonedible portions).

Feeding costs are dependent, therefore, on feed type, composition, and availability. The ingestion of very fibrous feeds by white-tailed deer often required more than 100 mastications per gram dry weight (Crawford and Whelan, 1973). The cost of eating in stall-fed ruminants per unit of time averages 0.53 ± 0.14 per kilogram of body weight per hour and is independent of both the animal species and type of food ingested (Table 7.6). However, the cost per unit of food ingested can vary widely because different foods can be consumed at different rates. Foraging costs of free-ranging herbivores can also increase as snow impedes access to normally available food. For example, caribou would theoretically expend 47–149 additional kilocalories per day (2–7% of the interspecific basal metabolic rate for a 90 kg caribou) in digging 90–140 feeding craters (Thing, 1977).

8. Other Activities

Wild animals can participate in many activities other than those discussed here. For example, self-grooming, conspecific grooming, and fighting in the house mouse cost 1.7, 1.4, and 3.0 times the sleeping metabolic rate (Malloy and Herreid, 1979). However, since the costs for very few of these activities have been examined in wildlife, students seeking energy expenditure determinations for other activities should consult the available information on domestic species (such as Graham, 1964; Young, 1966; Clark *et al.*, 1972; Osuji, 1974; Osuji *et al.*, 1975; Toutain *et al.*, 1977) while developing wildlife experiments of greater precision and ecological relevance.

9. Conclusions

If reasonable approximations of the costs for various activities are available, time–activity budgets for an animal's daily life can be measured and multiplied by the respective costs for each activity to estimate the daily energy expenditure for activity. Such estimates are for basal metabolism and activity and require an estimate of the thermoregulatory costs to approximate the daily maintenance requirement.

C. Thermoregulation

Birds and mammals maintain a relatively high, stable body temperature by balancing heat inputs with heat losses. The body temperature of birds is generally regulated between 40° and 44°C and mammals between 36° and 40°C (Morowitz, 1968). Temperature gradients between the animal and its environment of up to 100°C are possible in arctic environments. Since heat flux is driven by temperature gradients, numerous physiological, anatomical, and behavioral mechanisms must exist if endotherms even in less harsh environments are to regulate body temperature without exhausting metabolic heat production capabilities. The stability of body temperature must be viewed with caution, since heat balance may not be instantaneous and the temperature of the entire body or a part of it may either rise or fall from the normal set-point. However, if the endotherm cannot ultimately balance its heat budget by use of the heat produced in other metabolic processes or absorbed from the environment, a thermoregulatory cost must be added to basal metabolism and activity to estimate the cost of maintenance.

Thermoregulation is governed by the laws of thermodynamics, in that an accounting of the flow of heat must not violate the conservation of energy. Thermoregulation is a very recent field of investigation for the wildlife ecologist, since the impetus for much of the current effort traces to Scholander *et al.* (1950a, 1950b, 1950c). The conceptual definition of thermoregulation as a balancing between heat input and heat loss necessitates amplification and understanding of both sides of the implied equation in which

$$H_m \pm Q_r \pm Q_c \pm Q_k + Q_e = H_s,$$

where metabolic heat energy (H_m) plus or minus heat gained or lost by radiation (Q_r), convection (Q_c), conduction (Q_k), and evaporation (Q_e) equals heat stored (H_s).

1. Modes of Heat Loss

a. Radiation. Radiation is the flow of energy through space, including vacuums, from all objects above absolute zero ($-273°C$) as electromagnetic waves. The wave phenomenon of radiation gives rise to a spectrum of energy levels in biological systems ranging from visible and UV light to invisible infrared heat. These differences are due to the length of the wave in microns (λ_{max}), which is inversely proportional to the absolute temperature (T) of an emitting black body, or Wein's law:

$$\lambda_{max} = 2897/T.$$

Thus, all living objects are radiating energy in the relatively long wavelength infrared spectrum, while the sun's radiant wavelength is much shorter, with most falling within the infrared and visible spectrums.

Radiant energy striking an object can be absorbed, reflected, or transmitted. Whereas transmission of solar energy through the earth's atmosphere is essential for photosynthesis and the energy budget of the earth, absorption and reflectance are most immediate to an animal's energy budget, since transmission is confined to only the very outer surfaces of the pelage or plumage. The absorption of radiant energy by an animal's surface is dependent on both the physical characteristics of the surface and the energy wave. For example, the absorptivity of an animal's surface to infrared radiation is virtually 100% irrespective of color or whether it's covered by hair or feathers (Birkebak, 1966), but the absorptivity of shorter wavelengths is quite variable. The plumages of many dark-colored birds reflect 15–25% of the incident solar radiation as compared to over 50% in lighter colored birds (Birkebak, 1966; Walsberg et al., 1978).

The amount of energy radiated by an object is proportional to the fourth power of its absolute temperature:

$$Q_r = \xi\sigma T_s^4,$$

where Q_r is the radiant energy in kilocalories per square meter per hour, ξ is the emissivity ranging from zero to one, σ is the Stefan-Boltzmann proportionality constant (4.93×10^{-8} kcal/m²/hr), and T_s is the surface temperature in °K. Emissivity is a description of how well the surface of an object emits radiant energy at its characteristic wavelength in comparison to the surface of a theoretically perfect radiator, or "black body." Because emission equals absorption at a specified wavelength when transmission is zero, the emissivity of animal surfaces is virtually one since they are emitting in the infrared.

Just as the animal is absorbing and radiating heat, so are all objects in its environment. Consequently, the net radiation exchange describing the quantity and direction of the heat flux is the important quantity. The temperature of the air or space through which the energy is being radiated is unimportant except as it affects the temperature of the radiating surfaces.

b. Convection.   Convection is the flow of heat in a moving fluid, which is most often air or water in biological systems. If the movement is due to density or buoyancy differences between particle aggregations in the fluid, the convection is called free convection. However, if the convection is due to the animal moving through the fluid or to pressure differences often quite removed from the animal, such as wind, the heat transfer is called forced convection. Convective heat transfer occurs at the interface of a relatively motionless transitional boundary layer of either air or water surrounding the animal, or within the animal as in the respiratory tract, and can be described in the simplest cases by the equation:

$$Q_c = h_c A(T_s - T_a),$$

where Q_c is the quantity of heat transferred by convection from an object of area A. The direction of the heat flow is dependent on the temperature gradient between the surface (T_s) and the fluid (T_a). The convection coefficient, h_c, is a mathematical description of the rate at which heat moves across the temperature gradient and is dependent on the size, shape, and surface qualities of the object; the fluid velocity, viscosity, and turbulence in forced convection; and the magnitude of the temperature gradient in free convection.

Wind velocities above a surface range from zero at the surface as friction impedes molecular movement to a maximum velocity some distance from the surface (Fig. 7.12). Such an orderly progression of wind velocities with height is termed a wind profile. The characteristics of the wind profile are largely dependent on the roughness of the surface and the speed and turbulence of the fluid. Wind profiles offer an example of the complexity of describing convective heat loss from an animal. While the leading edge of the body of a small passerine

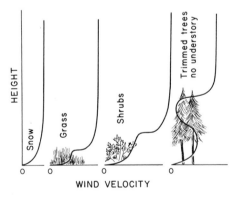

Fig. 7.12. Wind profile over snow or bare ground and different vegetation.

might be entirely within a given wind velocity, the larger deer can be within an entire spectrum of wind velocities requiring a computer analysis to describe even our simplest perception of convective heat loss. Similarly, since objects of small diameter are more efficient convectors of heat than are large ones, the legs of the standing animal have a different convection coefficient than do other body segments.

c. Conduction. Conduction is the flow of heat when oscillating molecules exchange kinetic energy without appreciably changing their position. Since conduction is the only means of heat transfer through opaque solids, it is an important means of transferring heat between an animal and other contacting surfaces, such as the ground, snow, vegetation, or motionless air, and through non-vascularized tissue segments. Heat flow through still air within the pelage or plumage is usually termed *conduction,* although free convection is far more important than is actual conduction (Cena and Monteith, 1975; Davis and Birkebak, 1975). Conduction is expressed by the equation

$$Q_k = h_k A(T_1 - T_2)/L,$$

where Q_k is heat transferred by conduction per unit of time and area (A). As with convection, conduction is driven by a temperature gradient ($T_1 - T_2$), but it is inversely proportional to the length of the conducting path (L), or the depth of insulation. The conductivity coefficient (h_k) describes the rate or efficiency with which the heat differential described by the temperature gradient can be conducted within or between the contacting surfaces. For the standing animal, conduction may be relatively unimportant since the bottoms of the feet are the only surfaces in which heat is being transferred solely by conduction. However, for the bedded animal, the resting or incubating bird, or the animal in which water of a temperature markedly different from body temperature contacts the skin surface, conduction can be the major route of energy transfer. Conduction losses can also be internal, such as the warming of ingested snow or water to body temperature and its subsequent loss via evaporation or urination (Young and Degen, 1980).

d. Evaporation. Evaporative heat loss is that heat removed from the animal as water changes from a liquid to a gas. Although the heat of vaporization is dependent on the temperature of the water, an average commonly used without significant error for animals is 580 cal per gram of water vaporized at 30°C. The rate of vaporization is dependent on the absolute humidity gradient, the interacting surface area, and the absolute amount of water available. Evaporation becomes the dominant means of heat loss as ambient temperature approaches or exceeds surface temperature. At very cold ambient temperatures, evaporative loss is small, since air, even at saturation, can hold very little water. Wind velocity is also important to evaporative heat loss, but its role is in maintaining a vapor pressure gradient across the boundary layer by removing more saturated air closer to the evaporative surface.

2. Surface Characteristics and Insulation Quality

a. Surface Area.   All four modes of heat loss are dependent on area. Unfortunately, the surface area of an animal is a very difficult quantity to define or measure. Most animals are multitudes of geometric shapes. Methods to measure area have included skinning the animal and measuring the area of the flattened skin, coating the hair or feather surface with various materials that can be subsequently removed and their area measured, or delineating the geometric components of the animal and measuring the necessary linear dimensions to calculate surface area. Depending on the method used and the subsequent application of the result, the measured surface area may not reflect the actual animal–environment interface in which heat exchange is occurring. For example, the area of the detached bird skin is expressed by the equation

$$A_{cm^2} = 10W^{0.667},$$

where body weight (W) is in grams. However, the external surface area exposed to heat interchange is approximately 20% less (Walsberg and King, 1978):

$$A_{cm^2} = 8.11W^{0.667}.$$

The difference between the two areas is due to the inclusion when the bird is skinned of skin surfaces that are not external surfaces relative to heat exchange— that is, the undersides of the folded wing and adjoining body surfaces, the skin of the retracted neck, and the upper parts of the legs that are covered by contour feathers. Similarly, the surface area of mammals is strongly dependent on body posture—that is, whether the animal is standing or lying. Even the surface area of the lying animal is affected by the extent to which the animal is curled and whether the legs are stretched out or are under the body. Many observations of subtle postural differences in wildlife can be interpreted from the alteration of surface area and corresponding heat loss. Thus, the thermally active surface of any animal is variable.

b. Insulation Quality.   The amount of heat moving through the pelage or plumage per unit area is dependent on the quality of insulation. The pelage of mammals is approximately one-third less insulative than the plumage on a similar sized bird (Aschoff, 1981). The term *conductance* is routinely used to describe insulation qualities, largely because of the method of measurement. A patch of skin with its hair or feathers is removed from the animal and placed on a hot plate to simulate body temperature. At the outer surface of the pelage or plumage is another plate at a different temperature, usually colder. The rate of heat flow within the established temperature gradient is measured and expressed

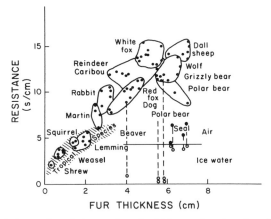

Fig. 7.13. Insulation quality relative to winter fur thickness of arctic and tropical mammals. For aquatic mammals, the measurements in 0°C air are connected by vertical dashed lines with the same measurements taken in ice water. (From Scholander *et al.*, 1950a, courtesy of the Marine Biological Laboratory.)

per unit of area and insulation depth—that is, the mathematical description of conductivity. However, since the majority of the heat moving through the dry coat is by free convection, with much smaller amounts by conduction and radiation along and between the hairs or feathers, the insulation quality is largely dependent on the extent to which air movement is prevented.

Insulation generally increases as fur depth increases (Fig. 7.13). The pelages of tropical mammals per unit of depth are similar in insulation quality to those of arctic animals (Scholander *et al.*, 1950a). The hair length on small arctic mammals is probably limited by the constraint that it not impede locomotion. Although the larger arctic mammals could conceivably carry longer pelages, their insulation and size already provide a very favorable thermoregulatory regime (Scholander *et al.*, 1950b, 1950c; Scholander, 1978). The variation in heat flux between pelts of similar depth suggests that other factors are also involved. One is simply the density of the hair. For example, red fox skins contained 3600 hairs per square centimeter as compared to 260 for the badger (Cena and Monteith, 1975). Similarly, hair depth and insulation quality vary dramatically with seasonal molt cycles and on different parts of the same animal (Jacobsen, 1980). Water within the pelage or plumage, such as sweat or rain, can markedly reduce the insulation, since the conductivity of water is 24 times greater than air at normal temperatures (Calder and King, 1974). Note the difference in insulation quality of the furs of the beaver, polar bear, and seal between measurements in air and in ice water (Fig. 7.13).

3. Estimation of Thermoregulatory Costs

While only a few of the problems inherent in quantifying the heat loss of a free-ranging animal have been suggested, it is obvious that some simplifications are currently necessary. Virtually all animal and environmental variables change during the daily cycle. The modes of heat loss are not simply independent and additive but are functioning concurrently, with the heat removed by one mode affecting the intensity of another. For example, heat removed by convection can lower the surface temperature and reduce radiant losses. An example of the interaction between radiation and convection in determining animal heat budgets was found by Walsberg *et al.* (1978), who addressed the seemingly paradoxical questions posed by the black plumages in desert birds (such as corvids and vultures) and white plumages in arctic birds (ptarmigan). The intuitive suggestion from human experiences and from animal studies in which forced convection was ignored is that these coat colors would be thermally maladapted. Nevertheless, the heat load generated by direct solar radiation was dependent on the interaction of plumage coloration, degree of plumage erection, and wind speed. Indeed, the black plumages did absorb more radiation, creating a thermal disadvantage at low wind speeds. However, as wind speed increased, the heat load on the erected black plumage fell below that of the white plumage as the radiant energy absorbed at the outer surface of the black plumage was convected away from the animal rather than inward toward the body surface. Radiation penetrated the white plumage much deeper before being absorbed and thereby reduced the effectiveness of forced convection. Thus, the black vulture and the white ptarmigan are at a thermal advantage in their respective environments if wind is common.

Because of the complexity of the physical and biological processes determining heat loss in the free-ranging animal, many investigators have relied on empirical measurements of heat loss from a variety of wildlife species in laboratory metabolism chambers. Scholander *et al.* (1950a, 1950b, 1950c) popularized such an approach, since many of the environmental and animal variables can be reduced or removed in an effort to broadly compare the thermoregulatory capabilities of a wide range of species. In most cases, ambient temperature has been the major environmental variable.

As ambient temperature decreases, the animal is initially able to thermoregulate by using only physical means to reduce heat loss, such as altering insulation or surface area. The temperature range in which thermoregulation can occur without increasing metabolic heat production is termed the thermoneutral zone and is bounded by the upper (T_{cu}) and lower (T_{cl}) critical temperatures (Fig. 7.14). Continued reductions in ambient temperature are met by the requisite need to increase metabolic heat production if body temperature is to remain constant. Eventually, an upper point in metabolic heat production (summit metabolism) is

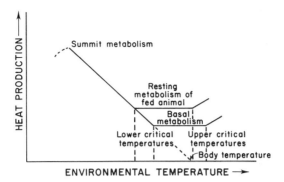

Fig. 7.14. The general Scholander model depicting a simplified thermoregulatory response.

reached where system exhaustion occurs and body temperature begins falling. Similarly, as ambient temperature rises above the upper critical temperature, heat production also increases because of the use of energy-demanding processes to dissipate heat, such as panting or sweating, and the Q_{10} effect on metabolism if body temperature rises (Whittow and Findlay, 1968; Brackenbury and Avery, 1980).

The slope of the relationship between metabolic heat production and ambient temperature below the lower critical temperature is a description of insulation quality. Theoretically, since physical insulation should be maximized at the lower critical temperature, the additional heat lost as ambient temperature decreases is a measure of whole-animal conductance. Conductances and lower critical temperatures are an entire range of values that generally decrease as animal size increases (Bradley and Deavers, 1980) (Fig. 7.15). Whole-animal conductances (C in kilocalories per kilogram per day per degree centigrade) in mammals weighing less than 5 kg are related to body weight (W in grams) by the following equation (Bradley and Deavers, 1980):

$$C = 87.84W^{-0.426}.$$

However, the Bradley–Deavers equation is based on a compilation of the results of Scholander type experiments, without regard for normal environment of the animals being compared. If one examines a series of small arctic mammals (lemmings, least weasel, varying hare, arctic fox, red fox, and porcupine), the equation relating conductance to body weight during winter becomes

$$C = 107.93W^{-0.54}.$$

Lower critical temperatures can be predicted using these equations by assuming that the slope of the line relating heat loss to ambient temperature extrapolates to body temperature at zero conductance (i.e., zero conductance when

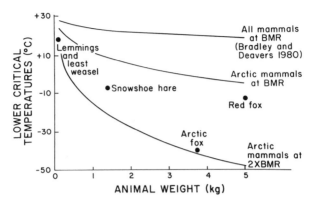

Fig. 7.15. Predicted (lines) and observed (data points) lower critical temperatures for mammals using average conductances (Bradley and Deavers, 1980), conductances specific for arctic mammals Casey *et al.*, 1979), and the interspecific BMR. The lower critical temperatures are predicted as a function of differing rates of heat production.

$T_a = T_b$). Lower critical temperatures predicted using the equation specific for arctic mammals are much lower than those predicted using the Bradley–Deavers equation (Fig. 7.15). The disagreement between the observed lower critical temperatures and those predicted from conductance may be due to (*a*) differences between actual basal metabolic rate and that predicted from the Kleiber interspecific equation or (*b*) a reduction in body temperature at colder ambient temperatures to minimize heat loss. Heterothermy is quite common in cold-adapted animals, particularly in appendages whose temperature may drop to only a few degrees above freezing.

In understanding the effects of human beings on wildlife energetics, several authors have used the Scholander model to evaluate energy expenditures by oil-contaminated muskrats, mallards, black ducks, and scaup (Hartung, 1967; McEwan and Koelink, 1973; McEwan *et al.*, 1974). Small amounts of oil matted the pelage or plumage and reduced its insulation and bouyancy. The magnitude of the increased heat production varied with the type of oil, the amount, and the length of time since exposure (Fig. 7.16). Although the animals were oiled by allowing them to swim in tanks with measured amounts of oil added, heat production measurements were on animals removed from the water. Heat loss by an oil-contaminated animal remaining in cold water would quickly exceed metabolic capacity. Heavily oiled muskrats remain out of water for weeks and oiled ducks often rest on shore and become isolated from food resources.

Critical temperatures and conductances determined using Scholander methods have a limited value in predicting thermoregulatory requirements of free-ranging animals because of the simplification of the thermal environment and possible restriction of the animal's responses. An additional step taken by many researchers has been to incorporate wind, artificial solar radiation, or other variables into metabolism chamber experiments (Chappel and Hudson, 1978b; Hayes and Gessaman, 1980). Nevertheless, it is essential to incorporate computer modeling to predict the thermoregulatory requirement of the free-ranging animal because of the vast array of possible environment–animal interactions. The modeling efforts have ranged from simply assigning laboratory-determined heat loss measurements to the free-ranging animal relative to the prevailing meteorological conditions, to modeling the physical processes of heat flow from the body to the exterior as affected by the environment (Porter and Gates, 1969; Moen, 1973; Bakken, 1976; Campbell, 1977; Bakken, 1980). Although few biologists have the mathematical capability to model such complex biological and physical systems, detailed models may indicate useful simplifications that are based on knowledge rather than on unwarranted assumptions or ignorance (Bakken, 1976).

Once heat loss has been estimated, one of the major unresolved questions relative to thermoregulation is to what extent metabolic heat generated incidental to other efforts (such as during locomotion or in the work of food digestion and metabolism) can substitute for thermoregulatory expenditures below the lower critical temperature. If substitution occurs, then the lower critical temperature of the fed animal can be far lower than those estimated relative to basal metabolism (Fig. 7.14). In general, the heat increment of feeding probably does substitute for thermoregulatory costs below the lower critical temperature. However, the heat increment of feeding may be exceedingly transitory, lasting for only a short period of time after a feeding bout.

The question of whether heat generated during activity can substitute for thermoregulatory costs is a far more difficult question. The ecologist has asked the question since, if substitution occurs, an animal actively searching for food

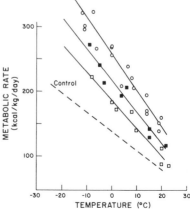

Fig. 7.16. Heat loss in black ducks as a function of contamination by 5 g (clear squares), 10 g (darkened squares), and 20 g (clear circles) of lubricating oil in comparison to the noncontaminated bird. (From Hartung, 1967, courtesy of the *Journal of Wildlife Management.*)

on a very cold day could conceivably be expending energy at the same rate as the inactive animal (i.e., basal metabolism and activity equaling basal metabolism and thermoregulation). Most studies have concluded, though, that very minimal or no substitution occurs (Hart, 1950, 1952, 1971; Hart and Jansky, 1963; Yousef *et al.*, 1973; Gates and Hudson, 1979; Schuchmann, 1979; the exception being Paladino, 1979). However, results generated in one of the major approaches to answering the question could easily be misinterpreted. For example, Schuchmann (1979) concluded that since the regressions between ambient air temperature (X) and energy expenditure (Y) of resting and flying hummingbirds were parallel, with the flying hummingbird always expending more energy than the resting hummingbird, substitution never occurred. However, it is seldom recognized that the resting and active animals invariably experience totally different thermal environments. In the preceding example, the flying hummingbird has a much greater surface area over which convective heat losses imposed by flying in cold temperatures would always be higher than for the resting bird. Thus, the comparison of energy expenditure solely as a function of air temperature is totally erroneous from a thermoregulatory perspective, since the effective thermal environment experienced by the active and nonactive animals can be quite different. Because of the possible importance of heat substitution in determining energy budgets and life strategies of temperate- and arctic-dwelling wildlife, far more work is needed to unequivocably determine the degree of thermogenic substitution from both the physiological and ecological perspective.

D. Daily and Seasonal Energy Expenditure for Maintenance

Energy expenditures over time in variable environments are rarely constant, since noncompensatory changes in basal metabolism, activity, or thermoregulation alter the daily energy budget for maintenance (Kendeigh, 1976; Wakeley, 1978). However, because of the predictability of specific energy expenditures as functions of body weight, very general relationships between body weight and energy expenditure for maintenance are expected. Such relationships should be used only as generalizations for interspecific comparisons and not as constants applicable to all members of the group at all times. The overzealous use of such approximations would ignore the rich diversity of life strategies and, consequently, energy expenditures between individuals and species over time.

1. Existence Metabolism and Average Daily Metabolic Rate of Captive Wildlife

Existence metabolism and average daily metabolic rates are the daily energy expenditures of captive animals. Both are composites of the energy expended for basal metabolism, activity, thermoregulation, if measured outside the zone of thermoneutrality, and the inefficiency of feed utilization, or heat increment

(Grodzinski and Wunder, 1975; French *et al.*, 1976; Kendeigh, 1976; Pimm, 1976). The term *existence metabolism* has been reserved for birds because of the recognition that the measurement represents a minimal level of energy expenditure that must be increased if it is applied to the free-ranging animal. Average daily metabolic rate (ADMR) measurements have been made primarily on captive rodents and insectivores.

The energy expenditure of captive birds and small mammals is strongly temperature dependent because of their very small thermal inertia. Existence metabolism measured at an ambient temperature of 30°C is 31% higher than the basal metabolic rate for passerines and 26% higher than the BMR for nonpasserines (Fig. 7.17). Since 30°C is within the zone of thermoneutrality for birds, the increased energy expenditure is due to the cost of very minimal activity and food processing. Existence metabolism is much higher when measured at 0°C and ranges from 3.20 times *BMR* in a 10-g passerine to 1.57 times *BMR* in a 5000-g nonpasserine.

Most ADMR measurements have been made at an ambient temperature of 20°C, which is close to the nest temperature of many rodents (French *et al.*, 1976). Average daily metabolic (*ADMR*) rates are expressed by the following equations:

$$\text{Rodents} \quad ADMR = 85.65W^{0.54}$$

$$\text{Insectivores} \quad ADMR = 66.85W^{0.43}$$

where ADMR is in kilocalories per animal per day and body weight (*W*) is in kilograms. The ADMR of other captive mammals ranging in size from weasels to elk is $140.2W^{0.75}$, or 2 times the interspecific *BMR* (Golley *et al.*, 1965; Jense, 1968; Ullrey *et al.*, 1970; Thompson *et al.*, 1973; Holter *et al.*, 1974; Moors, 1977; Harper *et al.*, 1978; Simpson *et al.*, 1978a; D. L. Baker *et al.*, 1979; Mould and Robbins, 1981).

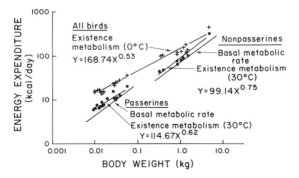

Fig. 7.17. Existence metabolism measured at various ambient temperatures relative to the basal metabolic rate in birds. (From Kendeigh, 1970, courtesy of the Cooper Ornithological Society.)

2. Daily Energy Expenditure of Free-Ranging Wildlife

Since the application of either existence metabolism or average daily metabolic rates to the free-ranging animal requires several assumptions and corrections, the energy budgets of free-ranging wildlife are being increasingly estimated using more direct approaches. Methods range in complexity from assigning energy equivalents determined by calorimetry to observed activity patterns or physiological parameters, to using radioactive isotopes whose biological decay or accumulation is proportional to energy expended. The two most popular methods currently used are time budget analyses and doubly labeled water. Utter and LeFebvre (1973) used both methods to determine the energy expenditure of free-flying purple martins, and their results are used here to illustrate time budget analyses. The martins rested for 9 hr during the night, or 37% of the 24-hr day (Table 7.7). Because of their size and diet, purple martins should become postabsorptive quite rapidly. Consequently, Utter and LeFebvre chose the interspecific basal metabolic rate as the nighttime resting cost. During the day, the resting cost was arbitrarily estimated as 1.5 times the interspecific BMR (Aschoff and Pohl, 1970a) because of the more intense daytime activity of the sitting bird, such as watching for prey, singing, preening, and interacting with conspecifics. Since flying costs 4.84 times the basal rate in martins, the longer flying time by the females increased their total daily energy expenditure by 14% relative to the more sedentary males.

Doubly labeled water refers to a method in which both deuterium and oxygen-18 isotopes are injected into the experimental animal. A blood sample is taken after the isotopes have equilibrated, which was 45 min for the martins. Several days later (1–2 days for the martins, although up to a week is possible), the same animal is recaptured and blood sampled. Since oxygen is lost from the body in both water and carbon dioxide, whereas hydrogen is lost mainly via water, the difference in the respective biological decays is proportional to the carbon dioxide produced and, therefore, energy expenditure (Lifson et al., 1955; LeFebvre, 1964; Lifson and McClintock, 1966; Utter and LeFebvre, 1970; Mullen, 1973; Utter and LeFebvre, 1973; Little and Lifson, 1975; Nagy and Milton, 1979). Although D_2O^{18} shows the most promise for elucidating energy expenditure of free-ranging animals because of minimal assumptions and disturbance to the animal, its use is currently limited to the smaller wildlife species because of the exceedingly high cost of the isotopes (Mullen, 1973; Nagy and Milton, 1979). Doubly labeled water should estimate a slightly higher daily energy expenditure than that estimated by activity budget analysis, since the energy expended in food processing (i.e., heat increment) is measured in the D_2O^{18} method but not in activity analysis.

Other methods have also been used to estimate daily energy expenditure or food intake and include pellet analyses of predatory birds (Graber, 1962), gas-

TABLE 7.7
The Time Budget and Estimated Energy Expenditure of Purple Martins during the Breeding Season[a]

Sex	Weight (kg)	Nighttime inactivity		Daytime activity				Total daily energy expenditure (kcal)
				Sitting		Flying		
		Hours	Percentage of 24 hrs	Hours	Percentage of 24 hrs	Hours	Percentage of 24 hrs	
Males	0.052	9.0	0.37	7.9	0.33	7.1	0.30	
Energy expenditure calculations		$(0.37)(114.8\ W^{0.726})$ $= 5.0$ kcal		$(0.33)(1.5)(140.9\ W^{0.704})$ $= 8.7$ kcal		$(0.30)(4.84)(140.9\ W^{0.704})$ $= 25.5$ kcal		39.2
Females	0.050	9.0	0.37	5.1	0.22	9.9	0.41	
Energy expenditure calculations		$(0.37)(114.8\ W^{0.726})$ $= 5.0$ kcal		$(0.22)(1.5)(140.9\ W^{0.704})$ $= 5.6$ kcal		$(0.41)(4.84)(140.9\ W^{0.704})$ $= 33.9$ kcal		44.5

[a] Adapted from Utter and LeFebvre, 1973.

135

trointestinal content measurements (Schmid, 1965; Collins and Smith, 1976), cardiac or respiratory rates (Owen, 1969, 1970; Berger et al., 1970; Morhardt and Morhardt, 1971; Gessaman, 1973; Holter et al., 1976; Lund and Folk, 1976; Flynn and Gessaman, 1979; Robbins et al., 1979; Pauls, 1980), and many other isotopes (Wagner, 1968; Baker et al., 1970; Chew, 1971; Sawby, 1973; Green, 1978; Green and Eberhard, 1979; Holleman and Stephenson, 1981). While feed intake measures do not actually quantify an energy requirement, they do provide necessary information for relating requirements to food resources.

Several authors have previously compiled daily energy expenditures for differing taxonomic groups (Table 7.8). Both birds and small mammals operate at an average rate of 2–3 times the basal rate. The most striking differences are for birds that search and feed during flight as compared to those in which flight is less important (Walsberg, 1980). Aerial feeders that do most of their prey searching from a perch are similar to ground feeders in daily energy expenditure (Ettinger and King, 1980).

Few detailed studies of larger mammals have thus far been accomplished. The estimates of daily energy expenditure for animals ranging in size from monkeys to elk have ranged from 85 to 190 kcal/kg$^{0.75}$/day (Moen, 1973; White et al., 1975; Zervanos and Day, 1977; Powell, 1979; Nagy and Milton, 1979).

Although the comparison of daily energy expenditures of differing taxonomic groups provides insight into the differing intensities of daily life, it does not provide the necessary understanding of the specific interactions between the animal and its environment determining individual, daily, or seasonal variation (Wakeley, 1978; Walsberg, 1980). For example, daily energy expenditures in birds can range from 1.2–2.1 times the basal metabolic rate when self-maintenance is the only requirement, to 2.4–4.0 times BMR during reproductive activities (Utter, 1971; Hails and Bryant, 1979).

One of the major areas of investigation initiated in recent years relative to understanding activity patterns, and therefore variation in daily energy budgets, has been the investigation of animal foraging strategies (Emlen, 1966; Mac-

TABLE 7.8

Estimates of Daily Energy Expenditures (kcal/day) for Several Taxonomic Groups as a Function of Body Weight (kg)

Groups	Equation	Reference
Rodents	DEE = 179.8 $W^{0.67}$	King (1974)
Rodents and lagomorphs	DEE = 211.4 $W^{0.69}$	Nagy et al. (1978)
All birds	DEE = 189.3 $W^{0.61}$	Walsberg (1980)
Birds who search and forage during flight	DEE = 318.1 $W^{0.66}$	Walsberg (1980)
Other birds	DEE = 195.1 $W^{0.65}$	Walsberg (1980)

Arthur and Pianka, 1966). The basic assumption in such studies is that activity, and therefore daily energy expenditure, is not a random event or process but is determined by a series of highly predictable interactions between the animal's requirements and its perception of the distribution of resources for meeting those requirements. Since many animals attempt to maximize their net rate of energy intake in optimizing the cost–benefit constraints of the biological system (reviewed by Pyke *et al.*, 1977), the energy expenditure for food-searching activities is determined by the cost of different search efforts, prey density, distribution, and energy content, and the efficiency and success rate of capture. Larger predators should use less energy-consuming search methods than should smaller predators if the efficiency of capture and prey size are similar. Similarly, small predators should be able to exploit more dispersed food sources than should larger predators when the preceding constraints apply (Norberg, 1977). Because of the higher total energy expenditure per unit of activity in the larger animal, the larger animal will reach a critical threshold of energy expenditure sooner relative to prey energy density than will the smaller animal (Robbins, 1973; Norberg, 1977). As our knowledge of the factors determining the activity patterns of free-ranging animals increases, our estimates and comparisons of daily energy expenditure will become more dependent on a multidimensional analysis. Rather than simply indicating the cost, we will be able to state why and to what extent changes in either the animal population or the environment affect the cost of daily life.

REFERENCES

Adam, I., Young, B. A., and Nicol, A. M. (1979). Energy cost to cattle of ingesting feed. *Agric. For. Bull., Univ. Alberta, Special Issue,* 38–39.

Alexander, R. McN., Jayes, A. S., and Ker, R. F. (1980). Estimates of energy cost for quadrupedal running gaits. *J. Zool.* **190**, 155–192.

Aschoff, J. (1981). Thermal conductance in mammals and birds, its dependence on body size and circadian phase. *Comp. Biochem. Physiol.* **69A**, 611–619.

Aschoff, J., and Pohl, H. (1970a). Rhythmic variations in energy metabolism. *Fed. Proc.* **29**, 1541–1552.

Aschoff, J., and Pohl, H. (1970b). Der Ruheumsatz von Vögeln als Funktion der Tageszeit und der Körpergrösse. *J. Ornithol.* **111**, 38–47.

Ashwell-Erickson, S., Elsner, R., and Wartzok, D. (1979). Metabolism and nutrition of Bering Sea harbor and spotted seals. *Proc. Alaska Sci. Conf.* **29**, 651–665.

Au, D., and Weihs, D. (1980). At high speeds dolphins save energy by leaping. *Nature* **284**, 548–550.

Baker, C. E., Dunaway, P. B., and Auerbach, S. I. (1970). Relationship of cesium-137 and iron-59 elimination rates to metabolic rates of small rodents. *Oak Ridge National Lab. Publ. ORNL-4568.*

Baker, D. L., Johnson, D. E., Carpenter, L. H., Wallmo, O. C., and Gill, R. B. (1979). Energy requirements of mule deer fawns in winter. *J. Wildl. Manage.* **43**, 162–169.

Bakken, G. S. (1976). A heat transfer analysis of animals: Unifying concepts and the application of metabolism chamber data to field ecology. *J. Theor. Biol.* **60**, 337–384.

Bakken, G. S. (1980). The use of standard operative temperature in the study of the thermal energetics of birds. *Physiol. Zool.* **53**, 108–119.

Bakker, R. T. (1972). Locomotor energetics of lizards and mammals compared. *Physiologist* **15**, 273.

Bartholomew, G. A., and Hudson, J. W. (1962). Hibernation, estivation, temperature regulation, evaporative water loss, and heart rate of the pygmy possum, *Cercaertus nanus. Physiol. Zool.* **35**, 94–107.

Barott, H. G., Fritz, J. C., Pringle, E. M., and Titus, H. W. (1938). Heat production and gaseous metabolism of young male chickens. *J. Nutr.* **15**, 145–167.

Baudinette, R. V., Nagle, K. A., and Scott, R. A. D. (1976). Locomotory energetics in dasyurid marsupials. *J. Comp. Physiol.* **109B**, 159–168.

Benedict, F. G. (1936). "The Physiology of the Elephant." Carnegie Inst. Publ. No. 474, Washington, D.C.

Benedict, F. G., and Fox, E. L. (1933). Der Grundumsatz von kleinen Vögeln (Spatzen, Kanarienvögeln and Sittichen). *Pfluger's Arch. ges. Physiol.* **232**, 357–388.

Berger, M., and Hart, J. S. (1974). Physiology and energetics of flight. *In* "Avian Biology," Vol. IV (D. S. Farner and J. R. King, eds.), pp. 415–477. Academic Press, New York.

Berger, M., Hart, J. S., and Roy, O. Z. (1970). Respiration, oxygen consumption and heart rate in some birds during rest and flight. *Z. Vergl. Physiol.* **66**, 201–214.

Birkebak, R. C. (1966). Heat transfer in biological systems. *Int. Rev. Gen. Exp. Zool.* **2**, 269–344.

Blaxter, K. L. (1962). "The Energy Metabolism of Ruminants." Thomas, Springfield, Illinois.

Brackenbury, J. H., and Avery, P. (1980). Energy consumption and ventilatory mechanisms in the exercising fowl. *Comp. Biochem. Physiol.* **66A**, 439–445.

Bradley, S. R., and Deavers, D. R. (1980). A re-examination of the relationship between thermal conductance and body weight in mammals. *Comp. Biochem. Physiol.* **65A**, 465–476.

Brenner, F. J., and Malin, W. F. (1965). Metabolism and survival time of the red-winged blackbird. *Wilson Bull.* **77**, 282–289.

Brockway, J. M., and Gessaman, J. A. (1977). The energy cost of locomotion on the level and on gradients for the red deer (*Cervus elaphus*). *Q. J. Physiol.* **62**, 333–339.

Brockway, J. M., and Maloiy, G. M. O. (1967). Energy metabolism of the red deer. *J. Physiol.* **194**, 22p–24p.

Brody, S. (1945). "Bioenergetics and Growth." Hafner, New York.

Brown, J. H., and Bartholomew, G. A. (1969). Periodicity and energetics of torpor in the kangaroo mouse, *Microdipodops pallidus. Ecology* **50**, 705–711.

Calder, W. A., and King, J. R. (1974). Thermal and caloric relations of birds. *In* "Avian Biology," Vol. IV (D. S. Farner and J. R. King, eds.), pp. 259–413. Academic Press, New York.

Campbell, G. S. (1977). "An Introduction to Environmental Biophysics." Springer-Verlag, New York.

Casey, T. M., Withers, P. C., and Casey, K. K. (1979). Metabolic and respiratory responses of arctic mammals to ambient temperature during the summer. *Comp. Biochem. Physiol.* **64A**, 331–341.

Cena, K., and Monteith, J. L. (1975). Transfer processes in animal coats. I. Radiative transfer. *Proc. R. Soc. Lond. B. Biol. Sci.* **188**, 377–393.

Chappel, R. W., and Hudson, R. J. (1978a). Energy cost of feeding in Rocky Mountain bighorn sheep. *Acta Theriol.* **23**, 359–363.

Chappel, R. W., and Hudson, R. J. (1978b). Winter bioenergetics of Rocky Mountain bighorn sheep. *Can. J. Zool.* **56**, 2388–2393.

Chappel, R. W., and Hudson, R. J. (1979). Energy cost of standing in Rocky Mountain bighorn sheep. *J. Wildl. Manage.* **43,** 261–263.

Chassin, P. S., Taylor, C. R., Heglund, N. C., and Seeherman, H. J. (1976). Locomotion in lions: Energetic cost and maximum aerobic capacity. *Physiol. Zool.* **49,** 1–10.

Chew, R. M. (1971). The excretion of ^{65}Zn and ^{54}Mn as indices of energy metabolism of *Peromyscus polionotus. J. Mammal.* **52,** 337–350.

Clapperton, J. L. (1964). The energy metabolism of sheep walking on the level and on gradients. *Brit. J. Nutr.* **18,** 47–54.

Clark, B. D., and Bemis, W. (1979). Kinematics of swimming of penguins at the Detroit Zoo. *J. Zool.* **188,** 411–428.

Clark, R. M., Holter, J. B., Colovos, N. F., and Hayes, H. H. (1972). Effect of postural position and position changes on energy expenditure in fasting dairy cattle. *J. Dairy Sci.* **55,** 257–260.

Cohen, Y., Robbins, C. T., and Davitt, B. B. (1978). Oxygen utilization by elk calves during horizontal and vertical locomotion compared to other species. *Comp. Biochem. Physiol.* **61A,** 43–48.

Collins, V. R., and Smith, M. H. (1976). Field determination of energy flow in a small nocturnal mammal. *J. Mammal.* **57,** 149–158.

Costello, R. R., and Whittow, G. C. (1975). Oxygen cost of swimming in a trained California sea lion. *Comp. Biochem. Physiol.* **50A,** 645–647.

Cranford, J. A. (1978). Hibernation in the western jumping mouse *(Zapus princeps). J. Mammal.* **59,** 496–509.

Crawford, H. S., and Whelan, J. B. (1973). Estimating food intake by observing mastications by tractable deer. *J. Range Manage.* **26,** 372–375.

Davis, L. B., and Birkebak, R. C. (1975). Convective energy transfer in fur. *In* "Perspectives of Biophysical Ecology" (D. M. Gates and R. B. Schmerl, eds.), pp. 525–548. Springer-Verlag, New York.

Dawson, T. J. (1976). Energetic cost of locomotion in Australian hopping mice. *Nature* **259,** 305–306.

Dawson, T. J., and Hulbert, A. J. (1970). Standard metabolism, body temperature and surface areas of Australian marsupials. *Am. J. Physiol.* **218,** 1233–1238.

Dawson, T. J., and Taylor, C. R. (1973). Energetic cost of locomotion in kangaroos. *Nature* **246,** 313–314.

Dawson, T. J., Grant, T. R., and Fanning, D. (1979). Standard metabolism of monotremes and the evolution of homeothermy. *Aust. J. Zool.* **27,** 511–515.

Dawson, W. R. Bennett, A. F., and Hudson, J. W. (1976). Metabolism and thermoregulation in hatchling ring-billed gulls. *Condor* **78,** 49–60.

Emlen, J. M. (1966). The role of time and energy in food preference. *Am. Nat.* **100,** 611–617.

Ettinger, A. O., and King, J. R. (1980). Time and energy budgets of the willow flycatcher *(Empidonax traillii)* during the breeding season. *Auk* **97,** 533–546.

Fedak, M. A., Pinshow, B., and Schmidt-Nielsen, K. (1974). Energy cost of bipedal running. *Am. J. Physiol.* **227,** 1038–1044.

Fedak, M. A., and Seeherman, H. J. (1979). Reappraisal of energetics of locomotion shows identical cost in bipeds and quadrupeds including ostrich and horse. *Nature* **282,** 713–716.

Flynn, R. K., and Gessaman, J. A. (1979). An evaluation of heart rate as a measure of daily metabolism in pigeons *(Columba livia). Comp. Biochem. Physiol.* **63A,** 511–514.

French, N. R., Grant, W. E., Grodzinski, W., and Swift, D. M. (1976). Small mammal energetics in grassland ecosystems. *Ecol. Monogr.* **46,** 201–220.

Gates, C. C., and Hudson, R. J. (1978). Energy costs of locomotion in wapiti. *Acta Theriol.* **23,** 365–370.

Gates, C. C., and Hudson, R. J. (1979). Effects of posture and activity on metabolic responses of wapiti to cold. *J. Wildl. Manage.* **43**, 564–567.

Gebczynski, M. (1971). Oxygen consumption in starving shrews. *Acta Theriol.* **16**, 288–292.

Gessaman, J. A. (Ed.). (1973). "Ecological Energetics of Homeotherms." Monogr. Ser. Vol. 20, Utah State Univ. Press, Logan.

Golley, F. B., Petrides, G. A., Rauber, E. L., and Jenkins, J. H. (1965). Food intake and assimilation by bobcats under laboratory conditions. *J. Wildl. Manage.* **29**, 442–447.

Graber, R. R. (1962). Food and oxygen consumption in three species of owls (Strigidae). *Condor* **64**, 473–487.

Graham, N. McC. (1964). Energy costs of feeding activities and energy expenditure of grazing sheep. *Aust. J. Agric. Res.* **15**, 969–973.

Green, B. (1978). Estimation of food consumption in the dingo, *Canis familiaris dingo*, by means of ^{22}Na turnover. *Ecology,* **59**, 207–210.

Green, B., and Eberhard, I. (1979). Energy requirements and sodium and water turnovers in two captive marsupial carnivores: The Tasmanian devil, *Sarcophilus harrissi,* and the native cat, *Dasyurus viverrinus. Aust. J. Zool.* **27**, 1–8.

Grodzinski, W., and Wunder, B. A. (1975). Ecological energetics of small mammals. *In* "Small Mammals: Their Productivity and Population Dynamics" (F. B. Golley, K. Petrusewicz, and L. Ryszkowski, eds.), pp. 173–204. Int. Biological Programme 5, Cambridge Univ. Press, London.

Hails, C. J. (1979). A comparison of flight energetics in Hirundines and other birds. *Comp. Biochem. Physiol.* **63A**, 581–585.

Hails, C. J., and Bryant, D. M. (1979). Reproductive energetics of a free-living bird. *J. Anim. Ecol.* **48**, 471–482.

Harper, R. B., Travis, H. F., and Glinsky, M. S. (1978). Metabolizable energy requirement for maintenance and body composition of growing farm-raised male pastel mink *(Mustela vison). J. Nutr.* **108**, 1937–1943.

Hart, J. S. (1950). Interrelations of daily metabolic cycle, activity and environmental temperature of mice. *Can. J. Res.* **28**, 293–307.

Hart, J. S. (1952). Effects of temperature and work on metabolism, body temperature, and insulation: Results with mice. *Can. J. Zool.* **30**, 90–98.

Hart, J. S. (1971). Rodents. *In* "Comparative Physiology of Thermoregulation" (G. C. Whittow, ed.), pp. 1–149. Academic Press, New York.

Hart, J. S., and Jansky, L. (1963). Thermogenesis due to exercise and cold in warm- and cold-acclimated rats. *Can. J. Biochem. Physiol.* **41**, 629–634.

Hartung, R. (1967). Energy metabolism of oil-covered ducks. *J. Wildl. Manage.* **31**, 798–804.

Hayes, S. R., and Gessaman, J. A. (1980). The combined effects of air temperature, wind and radiation on the resting metabolism of avian raptors. *J. Therm. Biol.* **5**, 119–125.

Hepburn, R. L. (1959). Effects of snow cover on mobility and distribution of deer in Algonquin Park. Master's thesis, Univ. of Toronto.

Hinds, D. S. (1973). Acclimatization of thermoregulation in the desert cottontail, *Sylvilagus auduboni. J. Mammal.* **54**, 708–725.

Holleman, D. F., and Stephenson, R. O. (1981). Prey selection and consumption by Alaskan wolves in winter. *J. Wildl. Manage.* **45**, 620–628.

Holleman, D. F., Luick, J. R., and White, R. G. (1979). Lichen intake estimates for reindeer and caribou during winter. *J. Wildl. Manage.* **43**, 192–201.

Holter, J. B., Tyler, G., and Walski, T. W. (1974). Nutrition of the snowshoe hare *(Lepus americanus). Can. J. Zool.* **52**, 1553–1558.

Holter, J. B., Urban, W. E., Jr., Hayes, H. H., and Silver, H. (1976). Predicting metabolic rate from telemetered heart rate in white-tailed deer. *J. Wildl. Manage.* **40**, 626–629.

Hoyt, D. F., and Taylor, C. R. (1981). Gait and the energetics of locomotion in horses. *Nature* **292**, 239–240.

Hudson, J. W., and Deavers, D. R. (1976). Thyroid function and basal metabolism in the ground squirrels *Ammospermophilus leucurus* and *Spermophilus* spp. *Physiol. Zool.* **49**, 425–444.

Hudson, J. W., Deavers, D. R., and Bradley, S. R. (1972). A comparative study of temperature regulation in ground squirrels with special reference to the desert species. *Symp. Zool. Soc. London* **31**, 191–213.

Irving, L. (1972). "Arctic Life of Birds and Mammals, Including Man." Springer-Verlag, New York.

Jacobsen, N. K. (1980). Differences in thermal properties of white-tailed deer pelage between seasons and body regions. *J. Therm. Biol.* **5**, 151–158.

Jense, G. K. (1968). Food habits and energy utilization of badgers. Master's thesis, South Dakota State Univ., Brookings.

Kayser, C. (1961). "The Physiology of Natural Hibernation." Pergamon, New York.

Kayser, C. (1964). Stoffwechsel und Winterschlaf. *Helgolander wiss. Meeresunters.* **9**, 155–186.

Kelsall, J. P., and Prescott, W. (1971). Moose and deer behavior in snow in Fundy National Park, New Brunswick. *Can. Wildl. Serv. Dept. Ser. No. 15.*

Kendeigh, S. C. (1970). Energy requirements for existence in relation to size of bird. *Condor* **72**, 60–65.

Kendeigh, S. C. (1976). Latitudinal trends in the metabolic adjustments of the house sparrow. *Ecology* **57**, 509–519.

Ketterson, E. D., and King, J. R. (1977). Metabolic and behavioral responses to fasting in the white-crowned sparrow *(Zonotrichia leucophrys gambelii)*. *Physiol. Zool.* **50**, 115–129.

King, J. R. (1974). Seasonal allocation of time and energy resources in birds. *In* "Avian Energetics" (R. A. Paynter, Jr., ed.), pp. 4–85. Nuttall Ornith. Club Publ. No. 15, Cambridge, Mass.

King, J. R., and Farner, D. S. (1961). Energy metabolism, thermoregulation and body temperature. *In* "Biology and Comparative Physiology of Birds," Vol. II (A. J. Marshall, ed.), pp. 215–288. Academic Press, New York.

Kleiber, M. (1947). Body size and metabolic rate. *Physiol. Rev.* **27**, 511–541.

Kleiber, M. (1961). "The Fire of Life." Wiley, New York.

Knight, M. H., and Skinner, J. D. (1981). Thermoregulatory, reproductive and behavioral adaptations of the big eared desert mouse, *Malacothrix typica*, to its environment. *J. Arid Envir.* **4**, 137–145.

Lasiewski, R. C. (1963). Oxygen consumption of torpid, resting, active, and flying hummingbirds. *Physiol. Zool.* **36**, 122–140.

Lasiewski, R. C., and Dawson, W. R. (1967). A re-examination of the relation between standard metabolic rate and body weight in birds. *Condor* **69**, 13–23.

LeFebvre, E. A. (1964). The use of D_2O^{18} for measuring energy metabolism in *Columba livia* at rest and in flight. *Auk* **81**, 403–416.

Le Maho, Y. (1977). The emperor penguin: A strategy to live and breed in the cold. *Am. Sci.* **65**, 680–693.

Lifson, N., and McClintock, R. (1966). Theory of use of the turnover rates of body water for measuring energy and material balance. *J. Theor. Biol.* **12**, 46–74.

Lifson, N., Gordon, G. B., and McClintock, R. M. (1955). Measurement of total carbon dioxide production by means of D_2O^{18}. *J. Appl. Physiol.* **7**, 704–710.

Little, W. S., and Lifson, N. (1975). Validation study of $D_2{}^{18}O$ method for determination of CO_2 output of the eastern chipmunk *(Tamias striatus)*. *Comp. Biochem. Physiol.* **50A**, 55–56.

Lund, G. F., and Folk, G. E., Jr. (1976). Simultaneous measurements of heart rate and oxygen consumption in black-tailed prairie dogs *(Cynomys ludovicianus)*. *Comp. Biochem. Physiol.* **55A**, 201–206.

MacArthur, R. H., and Pianka, E. R. (1966). On optimal use of a patchy environment. *Am. Nat.* **100,** 603–609.

McEwan, E. H. (1970). Energy metabolism of barren ground caribou *(Rangifer tarandus). Can. J. Zool.* **48,** 391–392.

McEwan, E. H., and Koelink, A. F. C. (1973). The heat production of oiled mallards and scaup. *Can. J. Zool.* **51,** 27–31.

McEwan, E. H., Aitchison, N., and Whitehead, P. E. (1974). Energy metabolism of oiled muskrats. *Can. J. Zool.* **52,** 1057–1061.

McNab, B. K. (1974). The energetics of endotherms. *Ohio J. Sci.* **74,** 370–380.

McNab, B. K. (1978a). The comparative energetics of neotropical marsupials. *J. Comp. Physiol.* **125,** 115–128.

McNab, B. K. (1978b). Energetics of arboreal folivores: Physiological problems and ecological consequences of feeding on an ubiquitous food supply. *In* "The Ecology of Arboreal Folivores" (G. G. Montgomery, ed.), pp. 153–162. Smithsonian Inst., Washington, D.C.

McNab, B. K. (1979). Climatic adaptation in the energetics of heteromyid rodents. *Comp. Biochem. Physiol.* **62A,** 813–820.

Malloy, L. G., and Herreid, C. F., II. (1979). Energetics of behavior in single and paired mice. *Am. Zool.* **19,** 991.

Marston, H. R. (1948). Energy transactions in the sheep. I. The basal heat production and heat increment. *Aust. J. Sci. Res. B.* **1,** 93–129.

Martin, R. A., Fiorentini, M., and Connors, F. (1980). Social facilitation of reduced oxygen consumption in *Mus musculus* and *Meriones unguiculatus. Comp. Biochem. Physiol.* **65A,** 519–522.

Mattfeld, G. F. (1974). The energetics of winter foraging by white-tailed deer: A perspective on winter concentration. Doctoral dissertation, SUNY College of Forestry, Syracuse, N.Y.

Mautz, W. W., and Fair, J. (1980). Energy expenditure and heart rate for activities of white-tailed deer. *J. Wildl. Manage.* **44,** 333–342.

Moen, A. N. (1973). "Wildlife Ecology, An Analytical Approach." Freeman, San Francisco.

Moen, A. N. (1978). Seasonal changes in heart rates, activity, metabolism, and forage intake of white-tailed deer. *J. Wildl. Manage.* **42,** 715–738.

Moors, P. J. (1977). Studies of the metabolism, food consumption and assimilation efficiency of a small carnivore, the weasel *(Mustela nivalis* L.). *Oecologia* **27,** 185–202.

Morhardt, J. E., and Morhardt, S. S. (1971). Correlations between heart rate and oxygen consumption in rodents. *Am. J. Physiol.* **221,** 1580–1586.

Morowitz, H. J. (1968). "Energy Flow in Biology: Biological Organization as a Problem in Thermal Physics." Academic Press, New York.

Morrison, P. (1960). Some interrelations between weight and hibernation function. *Bull. Harvard Mus. Comp. Zool.* **124,** 75–91.

Mould, E. D., and Robbins, C. T. (1981). Nitrogen metabolism in elk. *J. Wildl. Manage.* **45,** 323–334.

Mullen, R. K. (1973). The $D_2^{18}O$ method of measuring the energy metabolism of free-living animals. *In* "Ecological Energetics of Homeotherms" (J. A. Gessaman, ed.), pp. 32–43. Monogr. Ser. Vol. 20, Utah State Univ. Press, Logan.

Nagy, K. A., and Milton, K. (1979). Energy metabolism and food consumption by wild howler monkeys *(Alouatta palliata). Ecology* **60,** 475–480.

Nagy, K. A., Seymour, R. S., Lee, A. K., and Braithwaite, R. (1978). Energy and water budgets in free-living *Antechinus stuartii (Marsupialia: Dasyuridae). J. Mammal.* **59,** 60–68.

Newman, J. R., and Rudd, R. L. (1978). Minimum and maximum metabolic rates of *Sorex sinuosus. Acta Theriol.* **23,** 371–380.

Noll-Banholzer, U. (1979). Body temperature, oxygen consumption, evaporative water loss and heart rate in the fennec. *Comp. Biochem. Physiol.* **62A,** 585–592.

Norberg, R. A. (1977). An ecological theory on foraging time and energetics and choice of optimal food-searching method. *J. Anim. Ecol.* **46,** 511–529.

Osuji, P. O. (1974). The physiology of eating and the energy expenditure of the ruminant at pasture. *J. Range Manage.* **27,** 437–443.

Osuji, P. O., Gordon, J. G., and Webster, A. J. F. (1975). Energy exchange associated with eating and rumination in sheep given grass diets of different physical forms. *Brit. J. Nutr.* **34,** 59–71.

Owen, R. B., Jr. (1969). Heart rate, a measure of metabolism in blue-winged teal. *Comp. Biochem. Physiol.* **31,** 431–436.

Owen, R. B., Jr. (1970). The bioenergetics of captive blue-winged teal under controlled and outdoor conditions. *Condor* **72,** 153–163.

Paladino, F. V. (1979). Energetics of terrestrial locomotion in white-crowned sparrows *(Zonotrichia leucophorys gambelii).* Doctoral dissertation, Washington State Univ., Pullman.

Parker, K. A. (1981). The energetics of survival in deer and elk. Unpublished manuscript, Washington State University, Pullman.

Parsons, P. E., and Taylor, C. R. (1977). Energetics of brachiation versus walking: A comparison of a suspended and an inverted pendulum mechanism. *Physiol. Zool.* **50,** 182–188.

Passmore, R. (1971). The regulation of body weight in man. *Proc. Nutr. Soc.* **30,** 122–127.

Passmore, R., and Durnin, J. V. G. A. (1955). Human energy expenditure. *Physiol. Rev.* **35,** 801–840.

Pauls, R. W. (1980). Heart rate as an index of energy expenditure in red squirrels *(Tamiasciurus husonicus). Comp. Biochem. Physiol.* **67A,** 409–418.

Pennycuick, C. J. (1975). On the running of the gnu *(Connochaetus taurinus)* and other animals. *J. Exp. Biol.* **63,** 775–799.

Pimm, S. L. (1976). Existence metabolism. *Condor* **78,** 121–124.

Pinshow, B., Fedak, M. A., and Schmidt-Neilsen, K. (1977). Terrestrial locomotion in penguins: It costs more to waddle. *Science* **195,** 592–594.

Poczopko, P. (1971). Metabolic levels in adult homeotherms. *Acta Theriol.* **16,** 1–21.

Poczopko, P. (1979). Metabolic rate and body size relationships in adult and growing homeotherms. *Acta Theriol.* **24,** 125–136.

Porter, W. P., and Gates, D. M. (1969). Thermodynamic equilibria of animals with environment. *Ecol. Monogr.* **39,** 227–244.

Powell, R. A. (1979). Ecological energetics and foraging strategies of the fisher *(Martes pennanti). J. Anim. Ecol.* **48,** 195–212.

Prange, H. D., and Schmidt-Nielsen, K. (1970). The metabolic costs of swimming in ducks. *J. Exp. Biol.* **53,** 763–777.

Prinzinger, R., and Hänssler, I. (1980). Metabolism—weight relationship in some small non-passerine birds. *Experimentia* **36,** 1299–1300.

Pyke, G. H. (1981). Optimal travel speeds of animals. *Am. Nat.* **118,** 475–487.

Pyke, G. H., Pulliam, H. R., and Charnov, E. L. (1977). Optimal foraging: A selective review of theory and tests. *Q. Rev. Biol.* **52,** 137–154.

Raab, J. L., Eng, P., and Waschler, R. A. (1976). Metabolic cost of grade running in dogs. *J. Appl. Physiol.* **41,** 532–535.

Reese, J. B., and Haines, H. (1978). Effects of dehydration on metabolic rate and fluid distribution in the jackrabbit, *Lepus californicus. Physiol. Zool.* **51,** 155–165.

Reichman, O. J., and Aitchison, S. (1981). Mammal trials on mountain slopes: Optimal paths in relation to slope angle and body weight. *Am. Nat.* **117,** 416–420.

Renecker, L. A., Hudson, R. J., Christophersen, M. K., and Arelis, C. (1978). Effect of posture, feeding, low temperature, and wind on energy expenditures of moose calves. *Proc. 14th Ann. North Am. Moose Conf. and Workshop,* 126–140.

Ribeiro, J. M. de C. R., Brockway, J. M., and Webster, A. J. F. (1977). A note on the energy cost of walking in cattle. *Anim. Prod.* **25,** 107–110.

Ringberg, T. (1979). The Spitzbergen reindeer—a winter-dormant ungulate? *Acta Physiol. Scandinavica* **105,** 268–273.

Robbins, C. T. (unpublished). Basal metabolic rate in elk. Washington State Univ., Pullman.

Robbins, C. T. (1973). The biological basis for the determination of carrying capacity. Doctoral dissertation, Cornell Univ., Ithaca, N.Y.

Robbins, C. T., Cohen, Y., and Davitt, B. B. (1979). Energy expenditure by elk calves. *J. Wildl. Manage.* **43,** 445–453.

Rogerson, A. (1968). Energy utilization by the eland and wildebeest. *Symp. Zool. Soc. London* **21,** 153–161.

Romijn, C., and Lokhorst, W. (1961). Some aspects of energy metabolism in birds. *In* "Symposium on Energy Metabolism" (E. Brouwer and A. J. H. Van Es, eds.), pp. 49–59. European Association for Anim. Prod. Publ. No. 10.

Sawby, S. W. (1973). An evaluation of radioisotopic methods of measuring free-living metabolism. *In* "Ecological Energetics of Homeotherms" (J. A. Gessaman, ed.), pp. 86–93. Monogr. Ser. Vol. 20, Utah State Univ. Press, Logan.

Schmid, W. D. (1965). Energy intake of the mourning dove *Zenaidura macroura marginella. Science* **150,** 1171–1172.

Schmidt-Nielsen, K. (1970). Energy metabolism, body size and the problems of scaling. *Fed. Proc.* **29,** 1524–1532.

Schmidt-Nielsen, K. (1972). Locomotion: Energy cost of swimming, flying and running. *Science* **177,** 222–228.

Schnell, G. D., and Hellack, J. J. (1979). Bird flight speeds in nature: Optimized or a compromise. *Am. Nat.* **113,** 53–66.

Scholander, P. F. (1978). Rhapsody in science. *Ann. Rev. Physiol.* **40,** 1–17.

Scholander, P. F., Walter, V., Hock, R., and Irving, L. (1950a). Body insulation of some arctic and tropical mammals and birds. *Biol. Bull.* **99,** 225–236.

Scholander, P. F., Hock, R., Walters, V., Johnson, F., and Irving, L. (1950b). Heat regulation in some arctic and tropical mammals and birds. *Biol. Bull.* **99,** 237–258.

Scholander, P. F., Hock, R., Walters, V., and Irving, L. (1950c). Adaptations to cold in arctic and tropical mammals and birds in relation to body temperature, insulation, and basal metabolic rate. *Biol. Bull.* **99,** 259–271.

Schuchmann, K.-L. (1979). Metabolism of flying hummingbirds. *Ibis* **121,** 85–86.

Scott, I. M., Yousef, M. K., and Bradley, W. G. (1972). Body fat content and metabolic rate of rodents: Desert and mountain. *Proc. Soc. Exp. Biol. Med.* **141,** 818–821.

Silver, H. (1968). Deer nutrition studies. *In* "White-tailed Deer of New Hampshire" (H. R. Siegler, ed.), pp. 182–196. N. H. Fish and Game Dept., Concord.

Silver, H., Colovos, N. F., and Hayes, H. H. (1959). Basal metabolism of white-tailed deer—a pilot study. *J. Wildl. Manage.* **23,** 434–438.

Simpson, A. M., Webster, A. J. F., Smith, J. S., and Simpson, C. A. (1978a). Energy and nitrogen metabolism of red deer *(Cervus elaphus)* in cold environments: A comparison with cattle and sheep. *Comp. Biochem. Physiol.* **60A,** 251–256.

Simpson, A. M., Webster, A. J. F., Smith, J. S., and Simpson, C. A. (1978b). The efficiency of utilization of dietary energy for growth in sheep *(Ovis ovis)* and red deer *(Cervus elaphus). Comp. Biochem. Physiol.* **59A,** 95–99.

South, F. E., and House, W. A. (1967). Energy metabolism in hibernation. *In* "Mammalian

Hibernation,'' Vol. III (K. C. Fisher, A. R. Dawe, C. P. Lyman, E. Schönbaum, and F. E. South, Jr., eds.), pp. 305–324. Amer. Elsevier, New York.

Southwick, E. E. (1980). Seasonal thermoregulatory adjustments in white-crowned sparrows. *Auk* **97**, 76–85.

Stevenson, J. (1933). Experiments on the digestion of food by birds. *Wilson Bull.* **45**, 155–167.

Taylor, C. R. (1977). The energetics of terrestrial locomotion and body size in vertebrates. *In* "Scale Effects in Animal Locomotion" (T. J. Pedley, ed.), pp. 127–141. Academic Press, London.

Taylor, C. R., Schmidt-Nielsen, K., and Raab, J. L. (1970). Scaling of energetic cost of running to body size in mammals. *Am. J. Physiol.* **219**, 1104–1107.

Taylor, C. R., Caldwell, S. L., and Rowntree, V. J. (1972). Running up and down hills: Some consequences of size. *Science* **178**, 1096–1097.

Telfer, E. S. (1967). Comparison of moose and deer winter range in Nova Scotia. *J. Wildl. Manage.* **31**, 418–425.

Telfer, E. S., and Kelsall, J. P. (1979). Studies of morphological parameters affecting ungulate locomotion in snow. *Can. J. Zool.* **57**, 2153–2159.

Thing, H. (1977). Behavior, mechanics and energetics associated with winter cratering by caribou in northwestern Alaska. *Biol. Pap. Univ. Alaska No. 18.*

Thomas, S. P., and Suthers, R. A. (1972). The physiology and energetics of bat flight. *J. Exp. Biol.* **57**, 317–335.

Thompson, C. B., Holter, J. B., Hayes, H. H., Silver, H., and Urban, W. E., Jr. (1973). Nutrition of white-tailed deer. I. Energy requirements of fawns. *J. Wildl. Manage.* **37**, 301–311.

Thonney, M. L., Touchberry, R. W., Goodrich, R. D., and Meiske, J. C. (1976). Intraspecies relationship between fasting heat production and body weight: A reevaluation of $W^{.75}$. *J. Anim. Sci.* **43**, 692–704.

Torre-Bueno, J. R., and Larochelle, J. (1978). The metabolic cost of flight in unrestrained birds. *J. Exp. Biol.* **75**, 223–229.

Toutain, P. L., Toutain, C., Webster, A. J. F., and McDonald, J. D. (1977). Sleep and activity, age and fatness, and the energy expenditure of confined sheep. *Brit. J. Nutr.* **38**, 445–454.

Tucker, V. A. (1969). The energetics of bird flight. *Sci. Am.* **220**, 70–78.

Tucker, V. A. (1970). Energetic cost of locomotion in animals. *Comp. Biochem. Physiol.* **34A**, 841–846.

Ullrey, D. E., Youatt, W. G., Johnson, H. E., Fay, L. D., Schoepke, B. L., and Magee, W. T. (1970). Digestible and metabolizable energy requirements for winter maintenance of Michigan white-tailed does. *J. Wildl. Manage.* **34**, 863–869.

Utter, J. M. (1971). Daily energy expenditures of free-living purple martins (*Progne subis*) and mockingbirds (*Mimus polyglottos*) with a comparison to two northern populations of mockingbirds. Doctoral dissertation, Rutgers Univ., New Brunswick, N.J.

Utter, J. M., and LeFebvre, E. A. (1970). Energy expenditure for free flight by the purple martin (*Progne subis*). *Comp. Biochem. Physiol.* **35**, 713–719.

Utter, J. M., and LeFebvre, E. A. (1973). Daily energy expenditure of purple martins (*Progne subis*) during the breeding season: Estimates using D_2O^{18} and time budget methods. *Ecology* **54**, 597–604.

Vleck, D. (1979). The energy cost of burrowing by the pocket gopher, *Thomomys bottae*. *Physiol. Zool.* **52**, 122–136.

Vogt, F. D., and Robert, L. G. (1978). Effects of temperature and nesting on the energetics of daily torpor in mice. *Am. Zool.* **18**, 591.

Wagner, C. K. (1968). Relationship between oxygen consumption, ambient temperature, and excretion of 32-phosphorus in laboratory and field populations of cotton rats. Master's thesis, Univ. of Georgia, Athens.

Wakeley, J. S. (1978). Activity budgets, energy expenditures, and energy intakes of nesting Ferruginous hawks. *Auk* **95**, 667–676.

Walker, J. M., Graber, A., Berger, R. J., and Heller, H. C. (1979). Sleep and estivation (shallow torpor): Continuous processes of energy conservation. *Science* **204**, 1098–1100.

Wallgren, H. (1954). Energy metabolism of two species of the genus *Emberiza* as correlated with distribution and migration. *Acta Zool. Fennica* 84.

Walsberg, G. E. (1980). Energy expenditure in free-living birds: Patterns and diversity. Unpublished manuscript, Washington State Univ., Pullman.

Walsberg, G. E., and King, J. R. (1978). The relationship of external surface area of birds to skin surface area and body mass. *J. Exp. Biol.* **76**, 185–189.

Walsberg, G. E., Campbell, G. S., and King, J. R. (1978). Animal coat color and radiative heat gain: A re-evaluation. *J. Comp. Physiol.* **126B**, 211–222.

Wang, L. C. H. (1979). Time patterns and metabolic rates of natural torpor in the Richardson's ground squirrel. *Can. J. Zool.* **57**, 149–155.

Wang, L. C. H., Jones, D. L., MacArthur, R. A., and Fuller, W. A. (1973). Adaptation to cold: Energy metabolism in an atypical lagomorph, the arctic hare (*Lepus arcticus*). *Can. J. Zool.* **51**, 841–846.

Ward, J. M., Jr., and Armitage, K. B. (1978). Effects of water restriction and temperature on metabolism of the marmot. *Am. Zool.* **18**, 591.

Watts, P. D., Oritsland, N. A., Jonkel, C., and Ronald, K. (1981). Mammalian hibernation and the oxygen consumption of a denning black bear (*Ursus americanus*). *Comp. Biochem. Physiol.* **69A**, 121–123.

Webster, A. J. F., and Hayes, F. L. (1968). Effects of beta-adrenergic blockade on the heart rate and energy expenditure of sheep during feeding and during acute cold exposure. *Can. J. Physiol. Pharmacol.* **46**, 577–583.

Weiner, J. (1977). Energy metabolism of the roe deer. *Acta Theriol.* **22**, 3–24.

Wesley, D. E., Knox, K. L., and Nagy, J. G. (1970). Energy flux and water kinetics in young pronghorn antelope. *J. Wildl. Manage.*, **34**, 908–912.

Wesley, D. E., Knox, K. L., and Nagy, J. G. (1973). Energy metabolism of pronghorn antelopes. *J. Wildl. Manage.* **37**, 563–573.

West, G. C. (1972). Seasonal differences in resting metabolic rate of Alaskan ptarmigan. *Comp. Biochem. Physiol.* **42A**, 867–876.

Westerterp, K. (1977). How rats economize—energy loss in starvation. *Physiol. Zool.* **50**, 331–362.

White, R. G., and Yousef, M. K. (1977). Energy expenditure in reindeer walking on roads and on tundra. *Can. J. Zool.* **56**, 215–223.

White, R. G., Thomson, B. R., Skogland, T., Person, S. J., Russell, D. E., Holleman, D. F., and Luick, J. R. (1975). Ecology of caribou at Prudhoe Bay, Alaska. *Biol. Pap. Univ. Alaska, Special Rep. No. 2.*

Whittow, G. C., and Findlay, J. D. (1968). Oxygen cost of panting in the domestic ox. *Am. J. Physiol.* **214**, 94–99.

Wickstrom, M. (1981). Determination of foraging costs and efficiencies of elk and mule deer. Unpublished manuscript, Washington St. Univ., Pullman.

Wooley, J. B., Jr., and Owen, R. B., Jr. (1978). Energy costs of activity and daily energy expenditure in the black duck. *J. Wildl. Manage.* **42**, 739–745.

Wunder, B. A. (1975). A model for estimating metabolic rate of active or resting mammals. *J. Theor. Biol.* **49**, 345–354.

Wunder, B. A., and Morrison, P. R. (1974). Red squirrel metabolism during incline running. *Comp. Biochem. Physiol.* **48A**, 153–161.

Young, B. A. (1966). Energy expenditure and respiratory activity of sheep during feeding. *Aust. J. Agric. Res.* **17**, 355–362.

Young, B. A., and Degen, A. A. (1980). Ingestion of snow by cattle. *J. Anim. Sci.* **51,** 811–815.

Yousef, M. K., Dill, D. B., and Freeland, D. V. (1972). Energetic cost of grade walking in man and burro, *Equus asinus:* Desert and mountain. *J. Appl. Physiol.* **33,** 337–340.

Yousef, M., Robertson, W., Dill, D., and Johnson, H. (1973). Energetic cost of running in the antelope ground squirrel *Ammospermophilus leucurus. Physiol. Zool.* **46,** 139–147.

Zar, J. H. (1968). Standard metabolism comparisons between orders of birds. *Condor* **70,** 278.

Zar, J. H. (1969). The use of the allometric model for avian standard metabolism–body weight relationships. *Comp. Biochem. Physiol.* **29,** 227–234.

Zervanos, S. M. (1975). Seasonal effects of temperature on the respiratory metabolism of the collared peccary (*Tayassu tajacu*). *Comp. Biochem. Physiol.* **50A,** 365–371.

Zervanos, S. M., and Day, G. I. (1977). Water and energy requirements of captive and free-living collared peccaries. *J. Wildl. Manage.* **41,** 527–532.

8

Protein Requirements for Maintenance

I. DIETARY PROTEIN CONCENTRATION REQUIREMENTS FOR CAPTIVE WILDLIFE

Protein requirements for captive wildlife as a percentage of the diet decrease with increasing age (Fig. 8.1). The very rapid weight gain during early life requires more protein than does the constant replacement of proteinaceous tissues in the adult animal. For example, the requirement for Galliformes decreases from as much as 28% during very early growth to 12% for adult maintenance. The requirement for the adult bird increases during egg-laying and perhaps averages 19–20% dietary protein. Since these requirement estimates are a qualitative description of the amount of crude protein necessary to supply essential amino acids as well as other nonprotein nitrogen sources, the actual dietary protein requirement will fluctuate depending on dietary protein quality. For example, the

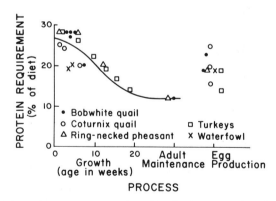

Fig. 8.1. Protein requirements as a percentage of the diet for several life processes in birds. (Data from Callenbach and Hiller, 1933; Norris, 1935; Norris *et al.*, 1936; Nestler *et al.*, 1942, 1944; M. L. Scott and Reynolds, 1949; Baldini *et al.*, 1950, 1953; Holm and Scott, 1954; M. L. Scott *et al.*, 1954; Weber and Reid, 1967; Johnson, 1971; Andrews *et al.*, 1973; Johri and Vohra, 1977; NRC, 1977;Woodard *et al.*, 1976; Woodard *et al.*, 1977; M. L. Scott, 1977–1978; Yamane *et al.*, 1979; Allen and Young, 1980; Schwartz and Allen, 1981.)

protein requirement for early growth in bobwhite quail could be reduced to 20% if the diet were supplemented with 1.3% lysine (Baldini *et al.*, 1953).

Dietary protein requirements for early growth in weanling mammals range from 33% for cats (P. P. Scott, 1968); 25% for mink, foxes, and guinea pigs (NRC, 1968, 1978a), 19% for the rhesus monkey (NRC, 1978b); to 13–20% for white-tailed deer (French *et al.*, 1956; McEwen *et al.*, 1957; Ullrey *et al.*, 1967; Smith *et al.*, 1975). Such estimates are not applicable to nursing animals totally dependent on their mother's milk since the protein content of milk dry matter often far exceeds these levels. The dietary protein requirements for maintenance of adult mammals range from 19–25% for adult carnivores to as low as 5.5–9% for wild ruminants (French *et al.*, 1956; McEwan *et al.*, 1957; Eisfeld, 1974a; Holter *et al.*, 1979). Other protein requirements, such as for gestation and hair and antler growth, will be intermediate to the requirements for rapid neonatal growth and adult maintenance.

II. FACTORIAL ESTIMATE OF NITROGEN REQUIREMENTS FOR MAINTENANCE

While such estimates are suitable for developing diets for captive wildlife, the nutritionist has a more complex and difficult problem in estimating protein requirements for free-ranging populations. For example, egg production in wild birds is never prolonged and maximized as in the game farm pheasant, quail, or turkey by continuous egg removal. While the cost per egg to either the wild or captive bird is the same, the distribution of that cost in time may be quite different. Egg protein synthesized by the game farm bird must come from immediate food sources if 90 eggs are to be laid in 100 days (Yamane *et al.*, 1979). However, the wild bird is able to rely on protein reserves accumulated prior to laying a far smaller number of eggs as it distributes the cost over both laying and nonlaying intervals (Raveling, 1979).

Similarly, one must question from an ecological perspective whether a maximum growth rate imparts greater fitness than does a slightly reduced rate, because many authors have observed identical weights at maturity for captive animals on protein intakes below those suggested as the requirement. Very high growth rates of captive animals fed concentrate diets are increasingly associated with bone and tendon abnormalities (Serafin, 1981). The ad libitum feed availability of most requirement studies using captive animals may exceed that available to free-ranging animals and require an interplay between food quality and quantity. Thus, the estimation of a protein requirement as a percentage of the diet using captive animals is woefully inadequate for understanding the nutritional complexities of free-ranging animals. Protein requirements can also be estimated by summing the expenditures for appropriate maintenance and productive processes.

Maintenance requirements for nitrogen are equal to the minimal losses in the feces (metabolic fecal nitrogen, MFN) and urine (endogenous urinary nitrogen, EUN), which must be balanced by intake for nitrogen equilibrium to occur. Very small amounts of nitrogen are also lost in sweat and sloughed skin cells. However, these losses have never been adequately quantified for wildlife and thus are assumed to be negligible. MFN is composed of nonabsorbed digestive enzymes, intestinal cellular debris, undigested bacteria or protozoa associated with fermentation, mucus, and any other nitrogenous product of immediate animal origin excreted in the feces. Theoretically, MFN and EUN could be most easily determined by feeding diets devoid of nitrogen but adequate in energy to prevent the unnecessary loss of catabolized tissue protein, and simply measuring fecal and urinary nitrogen excretion. When animals are placed on nitrogen-free diets, nitrogen excretion decreases over a number of days until a minimal constant level is attained. The number of days ranges from 6 days for rats, 8 for guinea pigs, to 15 days for rabbits (Smuts, 1935). However, since few animals will consume nitrogen-free diets, more indirect approaches have been used to estimate minimal nitrogen losses.

The most common indirect approach has been to feed animals different levels of the same ration or several rations of different protein content to obtain markedly different nitrogen intake levels. The regression between intake (X) and excretion (Y) when extrapolated to zero nitrogen intake estimates MFN and EUN because all nitrogen in the urine or feces at a hypothetical zero nitrogen intake must be of animal origin. The use of captive animals to estimate minimal nitrogen losses of free-ranging animals assumes that (*a*) internal nitrogen metabolism is not altered by confinement and (*b*) other dietary ingredients that the free-ranging animal may encounter have either no effect or at least a predictable effect on nitrogen loss.

A. Birds

Separate estimates of MFN and EUN in birds are virtually impossible in any simple way, since both urine and feces are eliminated together via the cloaca. Attempts have been made to physically separate urinary and fecal nitrogen in birds eliminating solid excreta by scraping the ureate crystals off the feces or by correcting the total nitrogen excretion for its chemically determined uric acid content (Billingsley and Arner, 1970). Such efforts are only partially successful, since urinary nitrogen in birds is not entirely uric acid and some mixing of urine and feces undoubtedly occurs prior to excretion. Uric acid content of turkey vulture urinary nitrogen ranged from 76.1 to 86.8% with most of the residue being ammonia and urea (McNabb *et al.*, 1980). Although the excretory ducts on

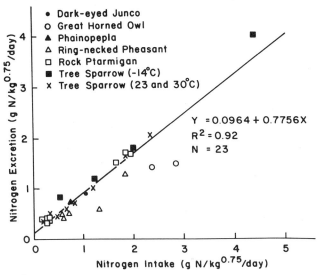

Fig. 8.2. Total fecal and urinary nitrogen excretion in birds as a function of nitrogen intake. (Data from Martin, 1968; Duke *et al.*, 1973; Labisky and Anderson, 1973; Moss and Parkinson, 1975; Walsberg, 1975; Parrish and Martin, 1977.) Data points for the great horned owl and tree sparrow ($-14°C$) are not included in the regression. (From C. T. Robbins, 1981, courtesy of Cooper Ornithological Society.)

experimental birds can be surgically repositioned (i.e., colostomized; Okumura *et al.*, 1981), emptying of urinary nitrogen into the distal intestinal tract, retrograde flow of urine into the large intestine and ceca (Skadhauge, 1976), and lower-tract fermentation in herbivorous birds could be an important nitrogen conservation mechanism (Mortensen and Tindall, 1981). If urinary nitrogen, whether as ammonia or amino acids, is reabsorbed from the avian lower intestine, endogenous losses on surgically manipulated birds would overestimate the actual values for the intact bird. The comparison of nitrogen excretion in colostomized and intact birds will be useful in determining the role of the avian cloaca in nitrogen conservation.

Consequently, the combined EUN–MFN losses in birds have been estimated by comparing nitrogen intake to total excretion on a per kilogram metabolic body weight basis (Fig. 8.2). The estimated nitrogen excretion at zero nitrogen intake is $0.096 g/kg^{0.75}/day$ (0.07 for nonpasserines and 0.13 for passerines), or approximately 1 mg per basal kilocalorie (Aschoff and Pohl, 1970). However, several observations are consistently different from the regression line. For example, the endogenous losses by tree sparrows housed at $-14°C$ is $0.25 g/kg^{0.75}/day$ compared with 0.12 for those housed at 30°C. Similarly, the two great horned owl observations are well below the regression, indicating either a very efficient retention of ingested nitrogen or lower endogenous losses. Nitrogen equilibrium—that is, the point at which intake and excretion are equal—is 0.43g/kg$^{0.75}$/day (C. T. Robbins, 1981).

B. Mammals

1. Metabolic Fecal Nitrogen

Mammalian MFN excretion increases from monogastrics with minimal lower-tract fermentation to adult ruminants and cecal digestors with extensive gastrointestinal microbial populations (Table 8.1, Fig. 8.3). The very young ruminant or cecal digestor that is consuming only milk and consequently has not developed a fermentation system will have an MFN excretion similar to the 1–2 g N/kg dry-matter intake characteristic of the monogastric (Table 8.1). The MFN excretion of elk calves and white-tailed deer fawns averages 26.5% of that occurring in the adults and indicates one of the costs of ingesting very fibrous feeds requiring microbial digestion. Since macropod marsupials have a ruminant-like microbial digestion system, their average MFN excretion (3.51±0.99) is higher than that of the monogastric but less than that of wild ruminants (4.97±1.34).

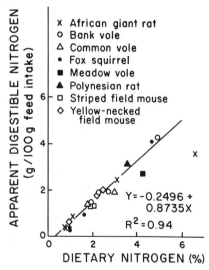

Fig. 8.3. Relationship between dietary nitrogen content and its apparent digestibility for small rodents. (Data from Baumgras, 1944; Cowan *et al.*, 1968; Drozdz, 1968; Garrison and Breidenstein, 1970; Garrison *et al.*, 1975; Tewe and Ajayi, 1979.) Values for animals consuming acorns and the disparate point for the African giant rat consuming groundnut cake have been omitted from the regression because of their suspected content of soluble digestion inhibitors. Metabolic fecal nitrogen (g/100 g dry-feed intake) is the negative Y intercept. Ultimately, separate values should be generated based on gastrointestinal morphology and diet.

Unfortunately, broad conclusions about cecal digestors are difficult, since their MFN excretion ranges from 3 to 9 g N/kg dry-matter intake (Table 8.1). One possible problem relative to protein metabolism that the cecal digestor faces is that the microbial protein produced during fermentation is not subjected to enzymatic digestion by the host prior to excretion. Cecal droppings are very high in protein content relative to fecal matter bypassing the cecum. Rodents and lagomorphs have solved the problem by excreting cecal pellets at regular intervals and consuming them as they emerge from the rectum (Haga, 1960; Slade and Robinson, 1970; Kenagy and Hoyt, 1980; Ouellette and Heisinger, 1980).

The meaning of MFN in species having significant gastrointestinal microbial populations, particularly ruminants, has always been questioned. Most of the MFN in ruminants (80–90%) consists of bacterial residues (Virtanen, 1966; Mason, 1969) that originate from either fermented plant nitrogen that is simply never absorbed or from endogenous animal sources. If the bacterial fecal nitrogen is simply plant nitrogen or very simple forms of endogenous nitrogen, such as urea, then the MFN loss is largely an excretory process. For example, the entry of urea nitrogen into the gastrointestinal tract of the rock hyrax and its subsequent utilization by microorganisms equals or exceeds the MFN losses (Hume et al., 1980). However, if the bacteria are fermenting complex endogenous substrates, such as sloughed gastrointestinal epithelial cells, mucus, or digestive enzymes, then the MFN represents a requirement in the true sense in which additional nitrogen must be ingested, absorbed, and synthesized into the replacement structures (Cheng and Costerton, 1980; Kennedy and Milligan, 1980). Although currently available estimates of MFN in ruminants must be regarded with some skepticism, all nitrogen losses must be balanced by intake if constant weight of both the host and symbiotic microflora is to be attained.

Since MFN could be affected by the amount of food ingested, its dry-matter digestibility, and the subsequent amount of nondigested fecal residues (Schneider, 1935; Bosshardt and Barnes, 1946; Williams and Senior, 1978), MFN might be more appropriately expressed as a function of fecal dry-matter excretion. Several hypotheses are suggested in using such an approach with ruminants and include:

1. The more rapid rumen or lower-tract fermentation in smaller species might increase bacterial protein and therefore MFN excretion.
2. Small animals have a larger gut surface area to volume ratio, which might increase MFN excretion in the smaller species if host nitrogen is an important component of MFN formation.
3. Slower rates of feed passage in larger ruminants might provide a longer time for greater bacterial protein synthesis or more complete dry-matter and fiber digestion, which would have the opposite effect of the first two hypotheses in increasing MFN in the larger species (Arman et al., 1975).

TABLE 8.1
Metabolic Fecal Nitrogen Excretion by Mammals

Species	Diet	Metabolic fecal nitrogen (g N/kg dry-matter intake)	Reference
Monogastrics and fecal digestors			
Black-tailed jackrabbits	Alfalfa and herbage	4.61	Data from Nagy et al. (1976)
Capybara		4.00	Van Soest (1981), citing unpublished data
Domestic horse	Pelleted diet and concentrates	4.80	Slade and Robinson (1970)
Domestic rabbits	Mixture of fish meal, almond hulls, and oat hay	8.00	Slade and Robinson (1970)
Elk calves	Milk	1.26	C. T. Robbins et al. (1981)
Guinea pig	Mixture of fish meal, almond hulls, and oat hay	9.44	Slade and Robinson (1970)
Rhesus monkey	Commercially formulated diet	1.04	Data from R. C. Robbins and Gavan (1966)
Rock hyrax	Concentrates—grains, fruits, beet pulp	3.69	Hume et al. (1980)
Small rodents	Herbage, nuts and seeds, commercial diets	2.50	Fig. 8.3
Snowshoe hares	Browse and pelleted diets	6.10	Holter et al. (1974)
		9.00	Combined data of Holter et al. (1974), Walski and Mautz (1977)
White-tailed deer fawns	Milk	2.37	Robbins, unpublished data
Macropod marsupials			
Brushtail possum	High fiber diet	3.36	Wellard and Hume (1981)
	Low fiber diet	1.79	

154

Euro	Hay and concentrates	2.70	Brown and Main (1967)
	Hays	3.80	Data from Hume (1974)
Red kangaroo	Hays and straw	4.73	Data from Foot and Romberg (1964), Hume (1974)
Tammar wallaby	Alfalfa hay and sucrose	4.10	Hume (1977)
	Straw and concentrates	4.11	Barker (1968)
Ruminants			
Bison	Hays	4.14	Data from Richmond et al. (1977)
Bush duiker	Pelleted diet	3.63	Arman and Hopcraft (1975)
Caribou		5.45	Data from McEwan and Whitehead (1970), Jacobsen and Skjenneberg (1972)
Eland	Pelleted diet	3.53	Arman and Hopcraft (1975)
Elk	Alfalfa and grass hay	5.58	Mould and Robbins (1981)
Hartebeest	Pelleted diet	4.16	Arman and Hopcraft (1975)
Red deer	Pelleted diet	6.98	Data from Maloiy (1968)
Thomson's gazelle	Pelleted diet	4.77	Arman and Hopcraft (1975)
White-tailed deer	Pelleted diet	3.75	Data from Smith et al. (1975), Holter et al. (1977)
White-tailed and mule deer	Browses and alfalfa	5.30	Holter et al. (1979)
Yak	Hays	7.81	C. T. Robbins et al. (1975)
		4.57	Data from Richmond et al. (1977)

The first two hypotheses are apparently more important since MFN as a percentage of fecal dry matter (Y) increases in the smaller species listed in Table 8.2 $(Y = 1.33 - 0.0018W, R^2 = 0.67$ where W is body weight in kilograms).

Use of fecal nitrogen content has also been suggested as an indirect field technique for evaluating the dietary quality of free-ranging wildlife (Arman *et al.*, 1975; Gates and Hudson, 1979). The suggestion is based on two very general observations that (*a*) fecal nitrogen is positively related to feed nitrogen and (*b*) dry-matter digestibility is positively related to feed nitrogen content. Nevertheless, many other factors are involved in the relationships and currently reduce the usefulness of the fecal nitrogen index. For example, fecal nitrogen is increased if the ingested feeds contain soluble phenolics, or tannins (Fig. 8.4). Tannins are a heterogeneous group of compounds that precipitate protein and are widely distributed in plants. Similarly, the relationship between feed nitrogen content and dry-matter digestibility is equally variable, because many other plant components are more responsible for dry-matter digestibility than is simply protein.

Thus, when one compares a limited number of closely related forages, as in the equation for grass and alfalfa diets fed to elk, the feed–fecal nitrogen relationship can be quite good (Fig. 8.4). However, when browses, grass, and alfalfa hays, and pelleted diets are compared, as in the deer examples, far more variability is evident (Table 8.2). Currently, fecal nitrogen content should be used only as a very general indicator of protein intake, such as qualitative seasonal changes, and not as an absolute predictor. As further studies quantify the effects of different plant constituents or animal characteristics on fecal nitrogen content, the technique will become increasingly useful.

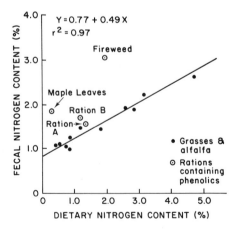

Fig. 8.4. Relationship between feed and fecal nitrogen concentrations for elk. (From Mould and Robbins, 1981, courtesy of the *Journal of Wildlife Management.*)

TABLE 8.2

Linear Regressions of Fecal Nitrogen Content (Y in %) versus Dietary Nitrogen Content (X in %) in Several Wild Ruminants[a]

Species	Animal weight (kg)	Equation	R^2
Duiker	7–11	$Y = 1.21 + 0.253X$	
Eland	130–300	$Y = 0.83 + 0.288X$	
Elk			
(Nontannin-containing plants)	200	$Y = 0.77 + 0.486X$	0.97
(Nontannin and tannin plants)	200	$Y = 1.02 + 0.443X$	0.80
Hartebeest	69–79	$Y = 1.23 + 0.224X$	
Mule deer	40	$Y = 1.36 + 0.336X$	0.24
Thomson's gazelle	9–21	$Y = 1.16 + 0.267X$	
White-tailed deer	40	$Y = 1.39 + 0.340X$	0.62

[a]Data from Silver and Colovos, 1957; Dietz *et al.*, 1962; Ullrey *et al.*, 1964; Short and Reagor, 1970; Ullrey *et al.*, 1971a; Ullrey *et al.*, 1971b; Mothershead *et al.*, 1972; Thompson *et al.*, 1973; Arman *et al.*, 1975; C. T. Robbins *et al.*, 1975; Mould and Robbins, 1981.

2. Endogenous Urinary Nitrogen

Endogenous urinary nitrogen studies were initiated by Folin (1905). He noticed that as dietary protein intake decreased, the relative distribution of urinary nitrogen constituents changed as the total urea excretion decreased and creatinin excretion remained constant. He suggested that ''we are forced to assume that protein katabolism is not all of one kind. . . . There must be at least two kinds . . . which are essentially independent and quite different [p. 122].'' He termed the constant form characterized by creatinin, tissue or endogenous metabolism, and the variable form, exogenous metabolism. The endogenous form in conjunction with MFN establishes the minimal nitrogen intake necessary for nitrogen equilibrium, whereas the variable form is an inefficiency in the utilization of dietary nitrogen. Although Folin stated that the two types of nitrogen metabolism formed ''an exceedingly interesting and important [p. 122]'' area of investigation, EUN studies are even fewer than those on MFN in wildlife. The few available EUN estimates were largely derived by comparing urinary nitrogen to nitrogen intake.

The average endogenous urinary nitrogen for marsupials is 63 ± 19 mg N/$kg^{0.75}$/day and for the placental mammals 105 ± 36 (Table 8.3). Since endogenous urinary nitrogen and basal metabolic rate represent the minimal internal idling rates of nitrogen and energy metabolism, EUN might have a constant relationship to basal metabolism (Smuts, 1935). Earlier studies suggested that mammals excreted 2 mg of EUN per kilocalorie of basal metabolism, or 140 mg N/$kg^{0.75}$/day (Smuts, 1935). However, this constant would overestimate the

TABLE 8.3

Endogenous Urinary Nitrogen (EUN) and the Ratio between EUN and Basal Metabolic (BMR) Rate[a] for Several Wildlife Species

Species or group	EUN (mg N/$W_{kg}^{0.75}$/day)	EUN:BMR (mg N/kcal)	Reference
Monogastrics and cecal digestors			
Black-tailed jackrabbit	128	1.83	Data of Nagy et al. (1976)
Marsupials			
Koala	77	1.58	Data of Harrop and Degabriele (1976)
Red kangaroo	54	1.11	Data of Foot and Romberg (1965), Hume (1974)
Red-necked pademelon	80	1.65	Hume (1977)
Tammar wallaby	39	0.80	Hume (1977)
Ruminants and camels			
Camel	60	0.86	Schmidt-Nielsen et al. (1957)
Elk	160	2.29	Mould and Robbins (1981)
Red deer	90	1.29	Data of Maloiy et al. (1970)
Roe deer	78	1.12	Eisfeld (1974b)
White-tailed deer	115	1.64	C. T. Robbins et al. (1975)

[a]Basal metabolism for eutherians = $70W_{kg}^{.75}$ and marsupials = $48.6W_{kg}^{.75}$ (Kleiber, 1947, Dawson and Hulbert, 1970).

EUN excretion of most wildlife (Moir, 1968), since the average ratio between EUN and basal metabolism for marsupials is 1.28 and for placental mammals 1.50 (Table 8.3). Because of the general relationship between EUN and BMR, the endogenous urinary nitrogen of young mammals is generally much higher per unit of body weight than that of adults. Similarly, EUN excretion in hibernating animals is one-tenth or even less than that in the active animal (Kayser, 1961). The hibernating bear does not even urinate during its 3–5 months of winter inactivity (Lundberg et al., 1976)!

C. Conclusions

One of the major difficulties in estimating the nitrogen requirement for maintenance is that MFN is dependent on dry-matter intake, whereas EUN is predicted from metabolic body weight. For the free-ranging animal, body weight is far easier to determine than actual dry-matter intake. Consequently, several authors have estimated the total maintenance nitrogen requirement as a function of metabolic body weight (Table 8.4). Since energy requirements for maintenance are a function of metabolic body weight, both dry-matter intake and therefore MFN should also be. However, the use of metabolic body weight to estimate the total nitrogen requirement for maintenance implies that an animal can meet its requirement at any dietary nitrogen concentration as long as intake is not limited. An animal cannot ingest enough nitrogen if the dietary nitrogen concentration drops below the critical level of MFN losses. For example, if a ruminant loses 5 g N/kg dry-matter intake via MFN, the feed must contain at least 3.13% crude protein (5 × 6.25 = 31.3 g/kg) before the maintenance requirement can begin to be met. Of course, the necessary dietary protein concentration will be even higher, since not all feed nitrogen is absorbed and the preceding example does not include EUN losses.

Maintenance nitrogen requirements can also be estimated directly from MFN and EUN losses if a very simple computer model is constructed (Fig. 8.5). Body weight and the concentration and availability of dietary nitrogen are the inputs. The EUN requirement is estimated from body weight and compared to dietary nitrogen content to estimate the necessary dry-matter intake to meet EUN. How-

TABLE 8.4
Average Nitrogen Intake at Which Nitrogen Balance Is Achieved in Several Wildlife Species

Species or group	Nitrogen intake (g N/kg$^{0.75}$/day)	Reference
Birds		
Passerines and nonpasserines	0.43	C. T. Robbins (1981)
Monogastrics and cecal digestors		
Black-tailed jackrabbits	0.95	Data of Nagy et al. (1976)
Rhesus monkeys	0.68	Data of R. C. Robbins and Gavan (1966)
Rock hyrax	0.31	Hume et al. (1980)
Marsupials		
Brushtail possum	0.20	Wellard and Hume (1981)
Euro	0.29	Brown and Main (1967)
Koala	0.28	Data of Harrop and Degabriele (1976)
Red kangaroo	0.84	Data of Foot and Romberg (1965), Hume (1974)
Red-necked pademelon	0.60	Hume (1977)
Tammar wallaby	0.24	Hume (1977)
	0.29	Barker (1968)
Ruminants		
Caribou	0.82	McEwan and Whitehead (1970)
Red deer	0.68	Bubenik (1959), cited by Eisfeld (1974a)
Roe deer	0.41	Eisfeld (1974a, 1974b)
White-tailed deer	0.77	Holter et al. (1979)

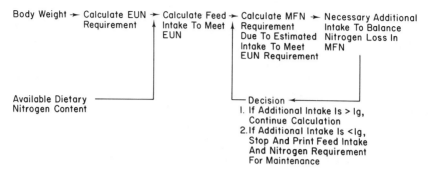

Body Weight ➤ Calculate EUN ➤ Calculate Feed ➤ Calculate MFN ➤ Necessary Additional
                 Requirement   Intake To Meet ↑ Requirement   Intake To Balance
                                 EUN   Due To Estimated Nitrogen Loss In
                                                  Intake To Meet   MFN
                                                  EUN Requirement

Available Dietary ─────────────────────────┐   ┌─ Decision ◄────────────────┘
Nitrogen Content   1. If Additional Intake Is > 1g,
                                                        Continue Calculation
                                                     2. If Additional Intake Is < 1g,
                                                        Stop And Print Feed Intake
                                                        And Nitrogen Requirement
                                                        For Maintenance

Fig. 8.5. Sequence of calculations necessary to estimate the total nitrogen requirement and necessary intake of an animal for maintenance.

ever, the feed ingested to meet the EUN requirement imposes an MFN loss. The iteration or loop procedure in the model is necessary since each unit of feed ingested to meet the preceding requirement necessitates a slightly increased intake. As long as the dietary nitrogen concentration is above the MFN loss, the additional intake will become so small that the calculations will stop when the necessary intake is within 1 g. When the program is used to estimate the necessary dry-matter and nitrogen intake for elk, intake increases curvilinearly as dietary protein content decreases (Fig. 8.6). Notice that the required maintenance nitrogen intake as a function of metabolic body weight covers the entire range of values in Table 8.4. Thus, the nitrogen requirement estimates of Table 8.4 are based on average rates of feed intake, dietary nitrogen concentrations, and en-

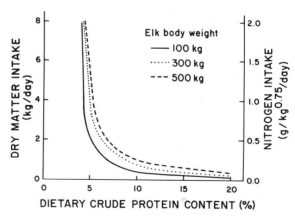

Fig. 8.6. Necessary dry-matter intake and the nitrogen requirement of elk at different body weights and dietary nitrogen concentrations. (From Mould and Robbins, 1981, courtesy of the *Journal of Wildlife Management*.)

dogenous nitrogen losses, and therefore do not define the requirement across the entire spectrum of possible interactions. In reality, dry-matter intake usually decreases at low dietary nitrogen concentrations, which minimizes MFN losses (Mould and Robbins, 1981).

REFERENCES

Allen, N. K., and Young, R. J. (1980). Studies on the amino acid and protein requirements of laying Japanese quail (*Coturnix coturnix japonica*). *Poult. Sci.* **59**, 2029–2037.

Andrews, T. L., Harms, R. H., and Wilson, H. R. (1973). Protein requirement of the bobwhite chick. *Poult. Sci.* **52**, 2199–2201.

Arman, P., and Hopcraft, D. (1975). Nutritional studies on East African herbivores. 1. Digestibilities of dry matter, crude fiber and crude protein in antelope, cattle and sheep. *Brit. J. Nutr.* **33**, 255–264.

Arman, P., Hopcraft, D., and McDonald, I. (1975). Nutritional studies on East African herbivores. 2. Losses of nitrogen in faeces. *Brit. J. Nutr.* **33**, 265–276.

Aschoff, J., and Pohl, H. (1970). Rhythmic variations in energy metabolism. *Fed. Proc.* **29**, 1541–1552.

Baldini, J. T., Roberts, R. E., and Kirkpatrick, C. M. (1950). A study of the protein requirements of bobwhite quail reared in confinement in battery brooders to eight weeks of age. *Poult. Sci.* **29**, 161–166.

Baldini, J. T., Roberts, R. E., and Kirkpatrick, C. M. (1953). Low protein rations for the bobwhite quail. *Poult. Sci.* **32**, 945–949.

Barker, S. (1968). Nitrogen balance and water intake in the Kangaroo Island wallaby, *Protemnodon eugenii* (Desmarest). *Aust. J. Exp. Biol. Med. Sci.* **46**, 17–32.

Baumgras, P. (1944). Experimental feeding of captive fox squirrels. *J. Wildl. Manage.* **8**, 296–300.

Billingsley, B. B., Jr., and Arner, D. H. (1970). The nutritive value and digestibility of some winter foods of the eastern wild turkey. *J. Wildl. Manage.* **34**, 176–182.

Bosshardt, D. K., and Barnes, R. H. (1946). The determination of metabolic fecal nitrogen and protein digestibility. *J. Nutr.* **31**, 13–21.

Brown, G. D., and Main, A. R. (1967). Studies on marsupial nutrition. V. The nitrogen requirements of the euro, *Macropus robustus*. *Aust. J. Zool.* **15**, 7–27.

Callenbach, E. W., and Hiller, C. A. (1933). The artificial propagation of ringnecked pheasants. *Penn. Agr. Exp. Sta. Bull.* **299**.

Cheng, K.-J., and Costerton, J. W. (1980). Adherent rumen bacteria—their role in the digestion of plant material, urea and epithelial cells. *In* "Digestive Physiology and Metabolism in Ruminants" (Y. Ruckebusch and P. Thivend, eds.), pp. 227–250. MTP Press, Lancaster, England.

Cowan, R. L., Long, T. L., and Jarrett, J. (1968). Digestive capacity of the meadow vole (*Microtus pennsylvanicus*). *J. Anim. Sci.* **27**, 1517.

Dawson, T. J., and Hulbert, A. J. (1970). Standard metabolism, body temperature and surface areas of Australian marsupials. *Am. J. Physiol.* **218**, 1233–1238.

Dietz, D. R., Udall, R. H., and Yeager, L. E. (1962). Chemical composition and digestibility by mule deer of selected forage species, Cache La Poudre Range, Colorado. *Colo. Game Fish Dept. Tech. Publ.* **14**.

Drozdz, A. (1968). Digestibility and assimilation of natural foods in small rodents. *Acta Theriol.* **13**, 367–389.

Duke, G. E., Ciganek, J. G., and Evanson, O. A. (1973). Food consumption and energy, water, and nitrogen budgets in captive great-horned owls (*Bubo virginianus*). *Comp. Biochem. Physiol.* **44A**, 283–292.

Eisfeld, D. (1974a). Protein requirements of roe deer (*Capreolus capreolus* L.) for maintenance. *Int. Congr. Game Biol.* **11,** 133–138.

Eisfeld, D. (1974b). Der Proteinbedarf des Rehes (*Capreolus capreolus* L.) zur Erhaltung. *Z. Jagdwiss.* **20,** 43–48.

Folin, O. (1905). A theory of protein metabolism. *Am. J. Physiol.* **13,** 117–138.

Foot, J. Z., and Romberg, B. (1965). The utilization of roughage by sheep and the red kangaroo, *Macropus rufus* (Desmarest). *Aust. J. Agric. Res.* **16,** 429–435.

French, C. E., McEwen, L. C., Magruder, N. D., Ingram, R. H., and Swift, R. W. (1956). Nutrient requirements for growth and antler development in the white-tailed deer. *J. Wildl. Manage.* **20,** 221–232.

Garrison, M. V., and Breidenstein, C. P. (1970). Digestion of sugarcane by the polynesian rat. *J. Wildl. Manage.* **34,** 520–522.

Garrison, M. V., Reid, R. L., Fawley, P., and Breidenstein, C. P. (1975). Comparative digestibility of acid detergent fiber by laboratory albino and wild polynesian rats. *J. Nur.* **108,** 191–196.

Gates, C. C., and Hudson, R. J. (1979). Weight dynamics of free-ranging elk. *Agric. For. Bull., Special Issue, Univ. Alberta, Canada,* 80–82.

Haga, R. (1960). Observations of the ecology of the Japanese pika. *J. Mammal.* **41,** 200–212.

Harrop, C. J. F., and Degabriele, R. (1976). Digestion and nitrogen metabolism in the koala (*Phascolactos cinereus*). *Austr. J. Zool.* **24,** 201–215.

Holm, E. R., and Scott, M. L. (1954). Studies on the nutrition of wild waterfowl. *New York Fish Game J.* **1,** 171–187.

Holter, J. B., Tyler, G., and Walski, T. W. (1974). Nutrition of the snowshoe hare (*Lepus americanus*). *Can. J. Zool.* **52,** 1553–1558.

Holter, J. B., Urban, W. E. Jr., and Hayes, H. H. (1977). Nutrition of northern white-tailed deer throughout the year. *J. Anim. Sci.* **45,** 365–376.

Holter, J. B., Hayes, H. H., and Smith, S. H. (1979). Protein requirement of yearling white-tailed deer. *J. Wildl. Manage.* **43,** 872–879.

Hume, I. D. (1974). Nitrogen and sulphur retention and fibre digestion by euros, red kangaroos and sheep. *Aust. J. Zool.* **22,** 13–23.

Hume, I. D. (1977). Maintenance nitrogen requirements of the macropod marsupials *Thylogale thetis,* red-necked pademelon, and *Macropus eugenii,* tammar wallaby. *Aust. J. Zool.* **25,** 407–417.

Hume, I. D., Rubsamen, K., and Engelhardt, W. v. (1980). Nitrogen metabolism and urea kinetics in the rock hyrax (*Procavia habessinica*). *J. Comp. Physiol.* **138,** 307–314.

Jacobsen, E., and Skjenneberg, S. (1972). Some results from feeding experiments with reindeer. *Proc. Int. Reindeer and Caribou Symp.* **1,** 95–107.

Johnson, N. F. (1971). Effects of levels of dietary protein on wood duck growth. *J. Wildl. Manage.* **35,** 798–802.

Johri, T. S., and Vohra, P. (1977). Protein requirements of *Coturnix coturnix japonica* for reproduction using purified diets. *Poult. Sci.* **56,** 350–353.

Kayser, C. (1961). "The Physiology of Natural Hibernation." Pergamon Press, New York.

Kenagy, G. J., and Hoyt, D. F. (1980). Reingestion of feces in rodents and its daily rhythmicity. *Oecologia* **44,** 403–409.

Kennedy, P. M., and Milligan, L. P. (1980). Input of endogenous protein into the forestomachs of sheep. *Can. J. Anim. Sci.* **60,** 1029–1032.

Kleiber, M. (1947). Body size and metabolic rate. *Physiol. Rev.* **27,** 511–541.

Labisky, R. F., and Anderson, W. L. (1973). Nutritional responses of pheasants to corn, with special reference to high lysine corn. *Ill. Nat. Hist. Surv. Bull.* **31,** 87–112.

Lundberg, D. A., Nelson, R. A., Wahner, H. W., and Jones, J. D. (1976). Protein metabolism in the black bear before and during hibernation. *Mayo Clin. Proc.* **51,** 716–722.

McEwan, E. H., and Whitehead, P. E. (1970). Seasonal changes in the energy and nitrogen intake in reindeer and caribou. *Can. J. Zool.* **48,** 905–913.

McEwen, L. C., French, C. E., Magruder, N. D., Swift, R. W., and Ingram, R. H. (1957). Nutrient requirements of the white-tailed deer. *Trans. North Am. Wildl. Conf.* **22,** 119–132.

McNabb, F. M. A., McNabb, R. A., Prather, I. D., Conner, R. N., and Adkisson, C. S. (1980). Nitrogen excretion by turkey vultures. *Condor* **82,** 219–223.

Maloiy, G. M. (1968). The physiology of digestion and metabolism in the red deer (*Cervus elaphus* L.). Doctoral dissertation, Univ. Aberdeen, Scotland.

Maloiy, G. M. O., Kay, R. N. B., and Goodall, E. D. (1970). Digestion and nitrogen metabolism in sheep and red deer given large or small amounts of water and protein. *Brit. J. Nutr.* **24,** 843–854.

Martin. E. W. (1968). The effects of dietary protein on the energy and nitrogen balance of the tree sparrow (*Spizella arborea arborea*). *Physiol. Zool.* **41,** 313–331.

Mason, V. C. (1969). Some observations on the distribution and origin of nitrogen in sheep feces. *J. Agric. Sci.* **73,** 99–111.

Moir, R. J. (1968). Ruminant digestion and evolution. *In* "Handbook of Physiology: Alimentary canal," Vol. 5 (C. F. Code, ed.), pp. 2673–2694. Am. Physiol. Soc., Washington, D.C.

Moss, R., and Parkinson, J. A. (1975). The digestion of bulbils (*Polygonum viviparum* L.) and berries (*Vaccinium myrtillus* L. and *Empetrum* sp.) by captive ptarmigan (*Lagopus mutus*). *Brit. J. Nutr.* **33,** 197–206.

Mortensen, A., and Tindall, A. R. (1981). Caecal decomposition of uric acid in captive and free ranging willow ptarmigan (*Lagopus lagopus lagopus*). *Acta Physiol. Scandinavica* **111,** 129–133.

Mothershead, C. L., Cowan, R. L., and Ammann, A. P. (1972). Variations in determinations of digestive capacity of white-tailed deer. *J. Wildl. Manage.* **36,** 1052–1060.

Mould, E. D., and Robbins, C. T. (1981). Nitrogen metabolism in elk. *J. Wildl. Manage.* **45,** 323–334.

Nagy, K. A., Shoemaker, V. H., and Costa, W. R. (1976). Water, electrolyte, and nitrogen budgets of jackrabbits (*Lepus californicus*) in the Mojave desert. *Physiol. Zool.* **49,** 351–363.

National Research Council (NRC). (1968). Nutrient requirements of mink and foxes. Publ. No. 1676, Nat. Acad. Sci., Washington, D.C.

National Research Council (NRC). (1977). Nutrient requirements of poultry. Publ. No. 2725, Nat. Acad. Sci., Washington, D.C.

National Research Council (NRC). (1978a). Nutrient requirements of laboratory animals. Publ. No. 2767, Nat. Acad. Sci., Washington, D.C.

National Research Council (NRC). (1978b). Nutrient requirements of nonhuman primates. Publ. No. 2786, Nat. Acad. Sci., Washington, D.C.

Nestler, R. B., Bailey, W. W., and McClure, H. E. (1942). Protein requirements of bobwhite quail chicks for survival, growth, and efficiency of feed utilization. *J. Wildl. Manage.* **6,** 185–193.

Nestler, R. B., Bailey, W. W., Llewellyn, L. M., and Rensberger, M. J. (1944). Winter protein requirements of bobwhite quail. *J. Wildl. Manage.* **8,** 218–222.

Norris, L. C. (1935). Nutrition of game birds. *New York Cons. Dept. Ann. Rep.* (1934) **24,** 289–291.

Norris, L. C., Elmore, L. J., Ringrose, R. C., and Bump, G. (1936). The protein requirements of ring-necked pheasant chicks. *Poult. Sci.* **15,** 454–459.

Okumura, J., Isshiki, Y., and Nakahiro, Y. (1981). Some factors affecting urinary and faecal nitrogen loss by chickens fed on a protein-free diet. *Brit. Poult. Sci.* **22,** 1–7.

Ouellette, D. E., and Heisinger, J. F. (1980). Reingestion of feces by *Microtus pennsylvanicus*. *J. Mammal.* **61,** 366–368.

Parrish, J. W., Jr., and Martin, E. W. (1977). The effect of dietary lysine level on the energy and nitrogen balance of the dark-eyed junco. *Condor* **79,** 24–30.

Raveling, D. G. (1979). The annual cycle of body composition of Canada geese with special reference to control of reproduction. *Auk* **96**, 234–252.

Richmond, R. J., Hudson, R. J., and Christopherson, R. J. (1977). Comparison of forage intake and digestibility by American bison, yak, and cattle. *Acta Theriol.* **22**, 225–230.

Robbins, C. T. (1981). Estimation of the relative protein cost of reproduction in birds. *Condor* **83**, 177–179.

Robbins, C. T., Van Soest, P. J., Mautz, W. W., and Moen, A. N. (1975). Feed analyses and digestion with reference to white-tailed deer. *J. Wildl. Manage.* **39**, 67–79.

Robbins, C. T., Podbielancik-Norman, R. S., Wilson, D. L., and Mould, E. D. (1981). Growth and nutrient consumption of elk calves compared to other ungulate species. *J. Wildl. Manage.* **45**, 172–186.

Robbins, R. C., and Gavan, J. A. (1966). Utilization of energy and protein of a commercial diet by rhesus monkeys (*Macaca mulatta*). *Lab. Anim. Care* **16**, 286–291.

Schmidt-Nielsen, B., Schmidt-Nielsen, K., Houpt, T. R., and Jarnum, S. A. (1957). Urea excretion in the camel. *Am. J. Physiol.* **188**, 477–484.

Schneider, B. H. (1935). The subdivision of the metabolic nitrogen in the feces of the rat, swine, and man. *J. Biol. Chem.* **109**, 249–278.

Schwartz, R. W., and Allen, N. K. (1981). Effect of aging on the protein requirements of mature female Japanese quail for egg production. *Poult. Sci.* **60**, 342–348.

Scott, M. L. (1977–1978). Twenty-five years of research in game bird nutrition. *World Pheasant Assoc. J.* **3**, 31–45.

Scott, M. L., and Reynolds, R. E. (1949). Studies on the nutrition of pheasant chicks. *Poult. Sci.* **28**, 392–397.

Scott, M. L., Holm, E. R., and Reynolds, R. E. (1954). Studies on pheasant nutrition. 2. Protein and fiber levels in diets for young pheasants. *Poult. Sci.* **33**, 1237–1244.

Scott, P. P. (1968). The special features of nutrition of cats, with observations on wild felidae nutrition in the London Zoo. *Symp. Zool. Soc. London* **21**, 21–36.

Serafin, J. A. (1981). The influence of diets differing in nutrient composition upon growth and development of sandhill cranes. Unpublished manuscript, Patuxent Wildl. Res. Center, Laurel, Maryland.

Short, H. L., and Reagor, J. C. (1970). Cell wall digestibility affects forage value of woody twigs. *J. Wildl. Manage.* **34**, 964–967.

Silver, H., and Colovos, N. F. (1957). Nutritive evaluation of some forage rations of deer. *New Hampshire Fish Game Dept. Tech. Circ.* **15**.

Skadhauge, E. (1976). Cloacal absorption of urine in birds. *Comp. Biochem. Physiol.* **55A**, 93–98.

Slade, L. M., and Robinson, D. W. (1970). Nitrogen metabolism in nonruminant herbivores. II. Comparative aspects of protein digestion. *J. Anim. Sci.* **30**, 761–763.

Smith, S. H., Holter, J. B., Hayes, H. H., and Silver, H. (1975). Protein requirement of white-tailed deer fawns. *J. Wildl. Manage.* **39**, 582–589.

Smuts, D. B. (1935). The relationship between the basal metabolism and the endogenous nitrogen metabolism, with particular reference to the estimation of the maintenance requirement of protein. *J. Nutr.* **9**, 403–433.

Tewe, O. O., and Ajayi, S. S. (1979). Utilization of some common tropical foodstuffs by the African giant rat (*Cricetomys gambianus*, Waterhouse). *Afr. J. Ecol.* **17**, 165–173.

Thompson, C. B., Holter, J. B., Hayes, H. H., Silver, H., and Urban, W. E., Jr. (1973). Nutrition of white-tailed deer. I. Energy requirements of fawns. *J. Wildl. Manage.* **37**, 301–311.

Ullrey, D. E., Youatt, W. G., Johnson, H. E., Ku, P. K., and Fay, L. D. (1964). Digestibility of cedar and aspen browse for the white-tailed deer. *J. Wildl. Manage.* **28**, 791–797.

Ullrey, D. E., Youatt, W. G., Johnson, H. E., Fay, L. D., and Bradley, B. L. (1967). Protein requirement of white-tailed deer fawns. *J. Wildl. Manage.* **31**, 679–685.

Ullrey, D. E., Johnson, H. E., Youatt, W. G., Fay, L. D., Schoepke, B. L., and Magee, W. T. (1971a). A basal diet for deer nutrition research. *J. Wildl. Manage.* **35**, 57–62.

Ullrey, D. E., Youatt, W. G., Johnson, H. E., Fay, L. D., Pursur, D. B., Schoepke, B. L., and Magee, W. T. (1971b). Limitations of winter aspen browse for the white-tailed deer. *J. Wildl. Manage.* **35**, 732–743.

Van Soest, P. J. (1981). "Nutritional Ecology of the Ruminant." O and B Books, Inc., Corvallis, Oreg.

Virtanen, A. I. (1966). Milk production of cows on protein-free feed. *Science* **153**, 1603–1614.

Walsberg, G. E. (1975). Digestive adaptations of *Phainopepla nitens* associated with the eating of mistletoe berries. *Condor* **77**, 169–174.

Walski, T. W., and Mautz, W. W. (1977). Nutritional evaluation of three winter browse species of snowshoe hares. *J. Wildl. Manage.* **41**, 144–147.

Weber, C. W., and Reid, B. L. (1967). Protein requirements of coturnix quail to five weeks of age. *Poult. Sci.* **46**, 1190–1194.

Wellard, G. A., and Hume, I. D. (1981). Nitrogen metabolism and nitrogen requirements of the brushtail possum, *Trichosurus vulpecula* (Kerr). *Aust. J. Zool.* **29**, 147–156.

Williams, V. J., and Senior, W. (1978). The effect of semi-starvation on the digestibility of food in young adult female rats. *Aust. J. Biol. Sci.* **31**, 593–599.

Woodard, A. E., Vohra, P., and Snyder, R. L. (1976). Protein level and growth in the pheasant. *Poult. Sci.* **55**, 2108.

Woodard, A. E., Vohra, P., and Snyder, R. L. (1977). Effect of protein levels in the diet on the growth of pheasants. *Poult. Sci.* **56**, 1492–1500.

Yamane, T., Ono, K., and Tanaka, T. (1979). Protein requirement of laying Japanese quail. *Brit. Poult. Sci.* **20**, 379–383.

9

Reproductive Costs

Wildlife reproductive characteristics and success are closely linked to food resources (Lack, 1968; Cody, 1971; Jones and Ward, 1976, 1979; Thorne *et al.*, 1976; Cengel *et al.*, 1978; Fogden and Fogden, 1979; Pattee and Beasom, 1979; Raveling, 1979; Ricklefs, 1980; Stenseth *et al.*, 1980). Reproductive requirements include the energy and matter deposited in the avian egg, that retained by the gravid uterus and enlarged mammary gland, that secreted as milk during lactation, and any additional heat or material wastes produced during incubation or fetal and neonatal maintenance. While the reproductive requirements are in addition to normal maintenance processes, maternal tissue can be an important source of either energy or matter in meeting the above requirements (Verme, 1967; Jones and Ward, 1976; Wypkema and Ankney, 1979; Ankney and Scott, 1980; Drobney, 1980). However, tissue mobilization does not lessen the requirement, since deficits must eventually be balanced, but it does provide a mechanism to distribute the very high cost over a much longer period of time.

Many other costs or consequences are involved in the total reproductive strategy of an animal. For example, many birds and mammals do not molt while reproductive demands are high (King, 1973). Increased time and energy expenditures during food searching or territorial defense may predispose reproducing adults to higher mortality (Madison, 1978; Bryant, 1979). Territorial establishment and defense, nest or burrow construction, and courtship are often additional energetic costs. Thus, the ecologist must recognize the far-reaching implications of reproductive demands.

I. BIRDS

A. Gamete Synthesis

Reproductive requirements for tissue synthesis can be estimated by determining energy and matter accumulated on a daily basis.* Maximum testicle growth

---

*Productive requirement estimates are net energy or protein costs (i.e., the amounts actually retained) and do not include the efficiencies of dietary energy and matter utilization unless otherwise stated.

TABLE 9.1
Estimated Energy Requirements of Semen Production in Birds

Species	Body weight (g)	Semen volume[a] (ml/day)	Energy content[b] (kcal/day)	Percentage BMR	Reference
Domestic chicken	2600	0.6	1.20	0.80	Ricklefs (1974)
Prairie falcon	625	0.1	0.20	0.38	Boyd (1978)
Prairie × peregrine falcon	625	0.3	0.60	1.15	Boyd (1978)

[a]Semen volumes for many other birds can be found in Lake (1978) and Gee and Temple (1978).
[b]Energy content estimated as 2 kcal/ml (Ricklefs, 1974).

and semen production increase the average energy requirement of the breeding male as a percentage of the basal metabolic rate 0.72 and 0.78%, respectively (Ricklefs, 1974) (Table 9.1). The daily protein requirement was increased 3.5% above maintenance for maximum testicle growth (Robbins, 1981). Thus, gamete synthesis in the breeding male produces very small increases in energy and protein requirements.

Requirements for egg production are dependent on growth of the ovary and oviduct preparatory to egg synthesis, the number of eggs laid, their composition, and the temporal sequence of yolk and albumin synthesis. Ovary–oviduct growth increases the energy requirement by 4.2% of the basal metabolic rate and protein by 27.7% of maintenance (Ricklefs, 1974; Drobney, 1980; Robbins, 1981). The average egg size in interspecific comparisons is generally inversely proportional to body weight—that is, smaller species lay larger eggs relative to body weight than do larger birds (Lack, 1968; Rahn et al., 1975). Although the number and size of eggs laid are characteristic of a species, both do vary with age of the female, time of laying, and food availability (Cody, 1971; Batt and Prince, 1979).

Egg composition varies with the mode of development of the embryo and chick (Table 9.2). For example, the eggs of precocial species average 37% yolk, 53% albumin, and 10% shell, as compared to the eggs of altricial species having 22% yolk, 70% albumin, and 8% shell (Ar and Yom Tov, 1978). The larger yolk of precocial species assures a relatively advanced, functional chick at hatching that is capable of searching for its own food. The extremes of precocial evolution are exemplified by the megapod and kiwi, whose egg contents contain 61 and 62% yolk, respectively. The megapod female abandons the newly laid egg and relies on rotting vegetation to provide the necessary thermal conditions for incubation, whereas the large yolk of the kiwi provides the necessary reserves for the developing embryo in one of the longest known incubation periods from which the chick hatches fully feathered with an adult-type plumage (Calder, 1979; Calder et al., 1978).

TABLE 9.2
Composition of the Avian Egg[a]

Species	Weight of egg (g)	Albumin (%)	Yolk (%)	Shell (%)	Energy content/gram fresh weight (kcal)	Lipid content (%)	Protein content (%)
Precocial or semiprecocial							
Ostrich (*Struthio camelus*)	1400	53.4	32.5	14.1			
Emu (*Dromiceus novae-hollandiae*)	710	52.2	35.0	12.8			
Kiwi (*Apteryx* sp.)	350	37.1	57.0	5.9	2.74		
Cackling Canada goose (*Branta canadensis minima*)						12.70	14.80
Mallard duck (*Anas platyrhynchos*)	80	51.0	35.9	10.9	1.90	14.59	11.50
Wood duck (*Aix sponsa*)	43	53.5	35.8	10.7	2.05	14.10	13.50
Laughing gull (*Larus atricilla*)	42	57.7	33.2	9.1	1.62	10.63	
Guinea fowl (*Numida meleagris*)	40	49.8	32.5	12.6	1.64	10.75	10.98
Ring-necked pheasant (*Phasianus colchicus*)	30	53.1	36.3	10.6			
Blue-winged teal (*Anas discors*)	28	48.4	42.7	8.9	1.87	12.84	11.42
Ruffed grouse (*Bonasa umbellus*)	18			8.3			
Partridge	18	50.8	37.0	12.2			
Plover	15	50.7	40.8	8.5			
California quail (*Lophortyx californicus*)	10	59.8	30.6	6.9	1.48		
Coturnix quail (*Coturnix coturnix*)	10	54.4	31.3	15.2	1.63	10.37	

(continued)

[a]Data from Asmundson *et al.*, 1943; Bump *et al.*, 1947; Romanoff and Romanoff, 1949; Reid, 1971; Case and Robel, 1974; Lawrence and Schreiber, 1974; Ricklefs, 1974, 1977; Calder *et al.*, 1978; Jones, 1979; Raveling, 1979; Drobney, 1980; Rohwer 1980. A more extensive compilation is available in Carey *et al.* (1980).

The energy content (kilocalories per gram) of the various eggs increases as the relative concentrations of yolk and lipid increase (Fig. 9.1) and is therefore lowest in passerines, hawks, and owls and highest in the kiwi, with intermediate levels in the remaining groups (King, 1973; Ricklefs, 1974). The egg protein content is not significantly different between altricial and precocial species and averages 12.1±1.8% (Table 9.2). The daily laying requirement can be estimated by proportioning the energy and protein content of the egg over the number of days required for its synthesis and deposition. For example, yolk deposition occurs over a 4- to 26-day period, whereas the albumen must be deposited the day of laying if one egg is laid every day (King, 1973; Ricklefs, 1974; Roudybush *et al.*, 1979; Drobney, 1980, Hirsch and Grau, 1981).

TABLE 9.2   *Continued*

Species	Weight of egg (g)	Albumin (%)	Yolk (%)	Shell (%)	Energy content/gram fresh weight (kcal)	Lipid content (%)	Protein content (%)
Bobwhite quail (*Colinus virginianus*)	8				1.87		
Altricial							
African great white pelican (*Pelecanus onocrotalus*)	182		17.1		1.06		
Golden eagle (*Aquila chrysaetos*)	140	78.6	12.0	9.4			
Brown pelican (*Pelecanus occidentalis*)	92	66.4	26.0	12.2	1.37	6.99	14.10
Buzzard	60	76.8	14.0	9.2			
Pigeon (*Columba livia*)	17	74.0	17.9	8.1			
Jay	8.5	68.1	26.6	5.3			
Starling (*Sturnus vulgaris*)	7.2	70.6	17.0	12.5	1.04	5.93	9.58
Mourning dove (*Zenaidura macroura*)	6.4	55.8	29.3	14.9	1.24	7.52	11.12
Brewer's blackbird (*Euphagus cyanocephalus*)	4.6	72.7	19.9	6.8			
Mockingbird (*Mimus polyglottos*)	4.1	74.4	17.8	7.1			
Robin (*Turdus migratorius*)	2.5	70.3	24.2	5.5			
European tree sparrow (*Passer montanus*)	2.2				1.02		
Barn swallow (*Hirundo rustica*)	1.9		27.4	7.9			
Long-billed marsh wren (*Telmatodytes palustris*)	1.2				1.11		
Golden-crested wren	1.0	71.0	24.1	4.9			
Hummingbird	0.5	69.7	25.3	5.0			

The daily energy requirement for egg laying ranges from 29% of the basal metabolic rate in hawks and owls, which lay one egg approximately every third day, to 135% in waterfowl, which lay a relatively large, high-energy content egg each day until the clutch is complete (Table 9.3). Similarly, the estimated daily protein requirement increases from 86% above the maintenance requirement in hawks and owls to 230% in waterfowl, gulls, and terns. Since the protein increment necessary for gamete synthesis in birds is far greater than the energy increment, one may question the almost exclusive emphasis of many avian biologists on understanding the energy parameters. Many observations of females switching from diets of seeds or vegetable matter to high-protein insects during laying, the generally higher consumption of insects by laying females

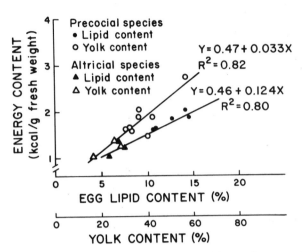

Fig. 9.1. Energy content of the egg as a function of either lipid or yolk content. Data points are from Table 9.2, while the equations are based on the more extensive data of Carey et al. (1980).

than by males, and the preferential mobilization of body protein further suggests that the protein requirements for egg laying are physiologically more important in many birds (Krapu, 1974; Swanson et al., 1974; Jones and Ward, 1976; Fogden and Fogden, 1979; Raveling, 1979; Ankney and Scott, 1980; Drobney, 1980).

TABLE 9.3
Energy[a] and Protein Costs of Egg Production[b]

| | | | Egg content | | Laying interval | | Protein requirement | |
Group	Egg weight (g)	Body weight (g)	Energy (kcal)	Protein (g)[c]	correction factor	Energy requirement[d] (% BMR)	Percentage endogenous loss[e]	Percentage maintenance[f]
Passerines	2.6	25.2	2.7	0.28	1.00	34	736	165
Galliformes	16.2	273.1	27.1	1.78	1.00	95	782	175
Raptors	32.8	502.2	34.5	3.61	0.38	29	392	86
Ducks	53.4	947.0	96.1	5.87	1.00	135	1015	228
Shorebirds	13.2	77.0	22.5	1.45	0.56	112	921	207
Gulls and Terns	34.4	220.5	58.6	3.79	0.53	128	1037	232

[a]Ricklefs, 1974.
[b]From Robbins (1981), courtesy of Cooper Ornithological Society.
[c]Average egg protein content estimated as 11% (Ricklefs, 1974).
[d]Energy requirements from Ricklefs but uncorrected for the efficiency of energy retention.
[e]Endogenous loss estimated as 0.60 g protein/$W_{kg}^{0.75}$/day.
[f]Maintenance costs estimated as 2.68 g protein/$W_{kg}^{0.75}$/day.

B. Incubation

The cost of incubation is any excess heat generated to initiate and continue embryonic development within the egg above what would normally be produced by the nonincubating bird. Energy expenditures for incubation in free-ranging birds have been difficult to measure directly. The cavity nester is essentially the only experimental situation that lends itself to calorimetric measurements (Gessaman and Findell, 1979). Consequently, the energy cost of incubation is poorly understood and is currently one of the most controversial areas of avian bioenergetics.

Ricklefs (1974) has estimated the incubation costs for passerines and nonpasserines as a function of the temperature gradient across the eggs between the bird and ambient air and the thermal conductance of the clutch. As the temperature gradient increases below the bird's lower critical temperature, the amount of heat transferred from the incubating bird to the egg must increase (Fig. 9.2). The cost increases curvilinearly from the larger to the smaller bird, since the smaller egg loses heat faster because of its higher surface area to volume ratio.

Estimates of the amount of heat transferred to the eggs in small, free-ranging passerines have ranged from 10 to 30% of the adult basal metabolic rate (King, 1973). If the brood patch covers a similar proportion (i.e., 10–30%) of the bird's surface area or if the bird is able to preferentially shunt heat to the eggs (C. M. Vleck, 1981), the continuous heat loss in the resting state, when combined with the additional insulation of the nest, would be adequate for incubation when the temperature gradient is not extreme. Incubation may, therefore, not necessitate an additional cost. As incubation progresses, the developing embryo also becomes an increasingly important source of heat (Drent, 1970; Hoyt et al., 1978; Gessaman and Findell, 1979; C. M. Vleck et al., 1979, 1980). Thermal model-

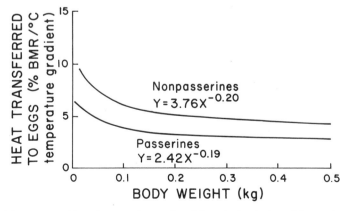

Fig. 9.2. Amount of heat that must be transferred from the incubating adult to a clutch of eggs equal to adult body weight. (From Ricklefs, 1974.)

ing estimates and measurements on cavity nesters have tended to point out that, while incubation does impose an additional cost below the thermoneutral environment relative to the bird sitting on a nest without eggs (i.e., the eggs essentially being poorly insulated extensions of the bird's body), the incubating bird can exist at the same energy expenditure rate as the nonnesting bird (Mertens, 1977; C. E. Vleck, 1977; Walsberg and King, 1978; Gessaman and Findell, 1979; C. M. Vleck, 1981). The most significant cost of incubation may occur during parental return to the nest when rewarming of cold eggs is necessary (C. E. Vleck, 1977).

Ecologically, one of the main costs of incubation for many birds is the reduced time available for feeding. For example, arctic nesting female Canada geese lost 42% of their peak spring weight between arrival on the nesting ground and hatching of their eggs. Since the females do all the incubation and therefore can do minimal feeding, the emaciated females feed almost continuously after the eggs have hatched (Raveling, 1979).

II. MAMMALS

A. Gestation

The most significant cost of fetal production in mammals is the direct cost to the female of embryonic development. While the cost of testicle growth and sperm production in males and ovary–oviduct growth in females has not been studied, the costs are undoubtedly quite small in both the absolute and ecological sense. The mammalian oviduct, in contrast to the avian oviduct, which enlarges to produce egg albumin and shell, remains quite small. Thus, female gestation requirements will be largely dependent on the amount of energy and matter retained by the gravid uterus and enlarging mammary gland and the increased maintenance costs of these structures.

Fetal growth during gestation is a curvilinear function in which most of the increase in mass occurs after 50–60% of the gestation period has elapsed (Fig. 9.3). Mammalian gestation periods and birth weights increase as power functions of maternal weight (Table 9.4). Since the average exponent of the relationships between birth weight and maternal weight in mammals is 0.76 ± 0.06 (Table 9.4), birth weight is a reflection of metabolic body weight in which the larger species produce a neonate that represents an increasingly smaller proportion of adult weight. When birth weight is compared to gestation length and those species experiencing delayed implantation or fertilization are excluded in an effort to compare the average rate of fetal growth, primate fetuses grow at approximately 24% of the rate observed in other mammals (Fig. 9.4).

However, comparisons of mass or crude growth rates ignore the important differences in embryonic composition during gestation and between neonates of

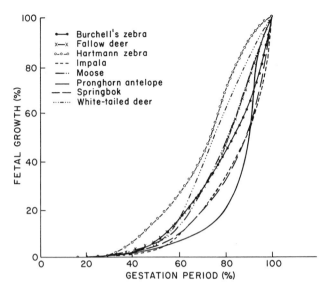

Fig. 9.3. Relative fetal growth (percentage of birth weight) in several ungulate species. (Reprinted from C. T. Robbins and B. L. Robbins, Fetal and neonatal growth patterns and maternal reproductive effort in ungulates and subungulates, *American Naturalist,* Vol. 114, 1979, by permission of The University of Chicago Press. Copyright © 1979 by University of Chicago Press.)

different species. For example, the water content of the developing fetus decreases while the fat, protein, and mineral contents increase during gestation (Robbins and Moen, 1975a; Havera, 1978). The mammalian neonate averages $12.0 \pm 1.9\%$ protein and $2.7 \pm 0.8\%$ ash at birth (Table 9.5). Neonatal fat, water, and energy contents do vary between species. The neonatal grey seal, guinea pig, human, and, perhaps, other primates contain 4–8 times more fat than do the other mammals, whose body fat content averages $2.0 \pm 0.9\%$. The fat reserves of the guinea pig are catabolized during the first few days postpartum (Widdowson and McCance, 1955; Adolph and Heggeness, 1971), whereas those of the seal may be necessary because of the cold environment and extremely short lactation (Stewart and Lavigne, 1980). The very low fat content of most neonatal mammals occurs irrespective of whether they are precocial or altricial at birth. For example, the white-tailed deer, which is born with a well-developed hair coat and follows its mother shortly after birth, has the same fat content as the very altricial mouse or vole (Table 9.5).

The caloric costs of gestation can be estimated using the birth weight to maternal weight relationships, the maternal weight to gestation period regressions, the composition and growth characteristics of the gravid uterus, and an

TABLE 9.4

Relationships between Adult Weight (g), Total Birth Weight (g), and Gestation Period (Days) in Mammals

Group	Birth weight (Y): adult weight (X)	Gestation period (Y) adult weight (X)	Reference
Placental mammals	$Y = 0.54X^{0.83}$		Leitch *et al.* (1959)
	$Y = 0.34X^{0.83}$		Leutenegger (1976)
	$Y = 0.20X^{0.71}$		Millar (1977)
		$Y = 12.0X^{0.25}$	Blaxter (1964)
		$Y = 11.6X^{0.24}$	Zeveloff and Boyce (1980)
Carnivores			
Fissipeds	$Y = 0.39X^{0.66}$		Case (1978)
Pinnipeds	$Y = 5.14X^{0.63}$		Case (1978)
		$Y = 16.2X^{0.16}$	Kihlstrom (1972)
Cetaceans		$Y = 201.5X^{0.04}$	Kihlstrom (1972)
Chiropterans	$Y = 0.40X^{0.76}$	$Y = 39.6X^{0.17}$	Case (1978), Kihlstrom (1972)
Insectivores	$Y = 0.12X^{0.80}$	$Y = 16.5X^{0.14}$	Case (1978), Kihlstrom (1972)
Lagomorphs	$Y = 0.29X^{0.71}$		Millar (1977)
Primates	$Y = 0.41X^{0.78}$	$Y = 31.3X^{0.16}$	Leutenegger (1973), Kihlstrom (1972)
Rodents	$Y = 0.13X^{0.82}$	$Y = 10.9X^{0.18}$	Case (1978), Kihlstrom (1972)
Ungulates and sub-ungulates	$Y = 0.89X^{0.79}$	$Y = 29.6X^{0.19}$	Robbins and Robbins (1979), Kihlstrom (1972)

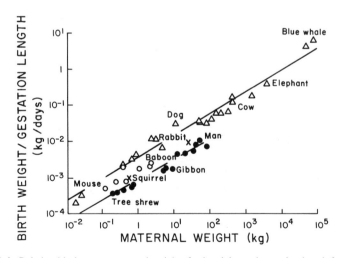

Fig. 9.4. Relationship between maternal weight, fetal weight, and gestation length for primates ($Y = 0.00084X^{0.56}$) and other mammals ($Y = 0.0035X^{0.60}$). (From Payne and Wheeler, 1967, courtesy of Cambridge University Press.)

estimate of the oxygen consumption of the gravid uterus per kilogram of tissue. Such values are currently available for ungulates and include:

1. The regressions relating maternal weight to birth weight and gestation period (Table 9.4).
2. Composition and growth characteristics of the gravid uterus in white-tailed deer and domestic ungulates as summarized by Robbins (1973) and Robbins and Robbins (1979).
3. The oxygen consumption of the gravid uterus, which includes the fetus, becomes asymptotic at 69.6 kcal/kg of tissue per day after approximately 50% of the gestation period has elapsed (Huckabee *et al.*, 1961; Crenshaw *et al.*, 1968; Clapp *et al.*, 1971). This caloric cost, which is virtually identical to determinations on smaller mammals (Moll *et al.*, 1970), includes the heat increment of tissue synthesis. Productive costs are defined as the net energy retained by the gravid uterus—that is, the gross energies of all accumulated tissue. The maintenance cost is all heat produced by the gravid uterus. The amniotic and allantoic fluids are assumed to be metabolically inactive.

For ungulates weighing more than 10 kg, the energy expenditure for maintenance of the gravid uterus becomes relatively asymptotic at 85% of the total cost (Figs. 9.5 and 9.6). This estimate agrees with the experimentally determined

TABLE 9.5
Average Weight and Composition of Mammalian Neonates or Full-Term Fetuses

Species	Average weight (g)	Percentage composition				Energy[a] content kcal/g wet weight	Reference
		Water	Protein	Fat	Ash		
White-tailed deer	3574	77.7	14.7	2.2	4.0	1.03	Robbins and Moen (1975a)
Human	3564	69.1	11.9	16.1	2.9	2.11	Widdowson (1950)
Grey seal			9.0				Widdowson (1950)
Domestic pig	1460	84.1	11.3	1.1	3.5	0.71	Widdowson (1950)
Domestic cat	118.0	80.7	14.9	1.8	2.6	0.97	Widdowson (1950)
Guinea pig	80.1	70.9	14.9	10.1	4.1	1.73	Widdowson (1950)
Domestic rabbit	54.0	84.6	11.1	2.0	2.3	0.78	Widdowson (1950)
Fox squirrel	22.0	82.4	9.8			0.86	Havera (1978)
Laboratory rat	5.85	86.0	10.8	1.0	2.2	0.68	Widdowson (1950)
Bank vole	2.13	84.5	9.8	3.8	1.6	0.88	Sawicka-Kapusta (1974)
Common vole	1.88	83.5	10.6	2.7	2.1	0.82	Sawicka-Kapusta (1970)
Old-field mouse	1.59	82.6	12.2	1.5	2.0	0.82	Kaufman and Kaufman (1977)
Laboratory mouse	1.55	83.3	12.5	2.1	2.1	0.87	Widdowson (1950)

[a]Energy content estimated as 5.43 kcal/g protein and 9.11 kcal/g fat if not experimentally determined.

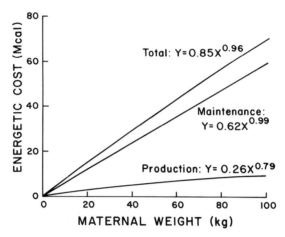

Fig. 9.5. Total energy cost of the developing gravid uterus in different sized female ungulates. (See text for sources of information and assumptions.)

average of 86% for domestic sheep (Robinson *et al.*, 1980). If one can extend these observations to other species, the maintenance requirement of the gravid uterus is the major cost of gestation for virtually all mammals. Supporting evidence for this conclusion relative to smaller species comes from Battaglia and Meschia (1981), who, using a similar approach, have estimated that 60% of the energy metabolized by the gravid uterus of a 1.2-kg guinea pig is lost as heat. The exception to this rule may occur in very small mammals, such as rodents and

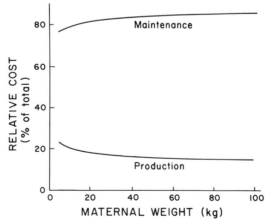

Fig. 9.6. Relative distribution of maintenance to productive energy costs for the gravid uterus in different sized female ungulates.

insectivores, which have an extremely short gestation period and relatively large litters.

It has been previously suggested that gestation is far less costly and stressful for the larger animal because the relatively smaller neonate of the larger species is produced over a far longer period of time (Blaxter, 1964; Payne and Wheeler, 1967, 1968). For example, the exponents (0.56 and 0.60, Fig. 9.4) of the Payne and Wheeler (1968) equations indicate that the productive cost of offspring synthesis as an increment of basal metabolic rate is less in the larger species. This is also true of ungulates, since the average fetal weight and gestation periods are functions of maternal weight raised to the 0.79 and 0.19 powers (Table 9.4), respectively. Consequently, the crude rate of offspring synthesis in ungulates is a function of maternal weight to the 0.60 power ($X^{0.79}/X^{0.19} = X^{0.60}$).

However, when one compares the total cost (maintenance and production) of the gravid uterus to the gestation period, the cost is indeed proportional to maternal metabolic body weight ($X^{0.96}/X^{0.19} = X^{0.77}$, Fig. 9.5). The relatively lower daily productive costs of the larger species are counterbalanced by the increasing maintenance costs during the longer gestation periods. Because of this proportionality, the time-dependent cost of gestation as a percentage of the basal metabolic rate peaks just prior to parturition at 44% and is generally the same for ungulates of different weights (Fig. 9.7). These relationships suggest some very fundamental energetic constraints to the evolution of fetal size and gestation periods. For example, Geist (1971) and Bailey (1980) have suggested that an ungulate living in a harsh seasonal environment should have a relatively long gestation period in which fewer offspring are produced. The supporting argument is that the longer gestation period reduces the daily maternal demand by

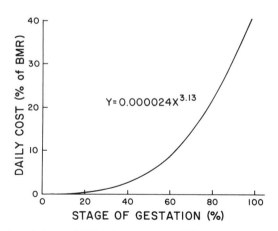

Fig. 9.7. Relative caloric cost (BMR in kcal/day as $70W_{kg}^{0.75}$) of uterine development in ungulates throughout the gestation period.

distributing the total cost over a longer period of time. While the production of fewer offspring of similar size obviously reduces the requirement, lengthening the gestation period would be counterproductive from an energetic perspective since any savings in productive energy is small compared to the increased maintenance cost. However, lengthening the gestation period relative to the demands for protein or other nutrients will indeed reduce their daily cost if maintenance is a relatively small proportion of that cost.

Protein accumulates curvilinearly in the gravid uterus and mammary gland during gestation. For example, the protein content of the uterus in white-tailed deer increases from 15.3 g at conception to 1430 g at term (Robbins and Moen, 1975a). The total protein accumulated in the gravid uterus averages 0.186 g/g of fetus produced. Consequently, protein accumulated in the gravid uterus of ungulates can be predicted from the following equation:

$$Y = 0.186(0.89W^{0.79})$$

or

$$Y = 0.17W^{0.79},$$

where Y is in grams, W is maternal weight in grams, and $0.89W^{0.79}$ is the relationship between maternal weight and fetal weight (Table 9.4). Because of the longer gestation period in larger species, protein is accumulated at a decreasing rate per kilogram of fetus produced (Fig. 9.8).

The mammary gland begins enlarging after approximately two-thirds of the gestation period has elapsed and ultimately accumulates 1700 g protein in the domestic cow (38 g protein per kilogram of fetus) and 106 g in the human (32 g protein per kilogram of fetus) (Garry and Stiven, 1936, Jakobsen, 1957). If one assumes an average mammary gland protein accumulation of 35 g/kg of fetus, the total protein cost of mammary gland development in ungulates during pregnancy is predicted by the equation:

$$Y = 0.035(0.89W^{0.79})$$

or

$$Y = 0.03W^{0.79},$$

where Y is in grams protein and W is maternal weight in grams. Thus, the mammary gland increases the total productive requirement by approximately 18% ($0.03/0.17 \times 100$). Therefore, the total interspecific productive protein requirement for pregnancy in ungulates is equal to $0.20W_{g}^{0.79}$.

Estimation of the maintenance cost of the gravid uterus for protein can only be speculative at this time. However, if EUN losses attributable to the gravid uterus equal 1.5 mg N/kcal of maintenance, the total protein cost (g) of maintaining the gravid uterus during gestation in ungulates is equal to $0.0060W^{0.99}$ where W is

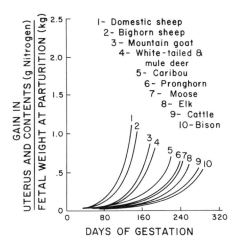

Fig. 9.8. Rate at which protein is accumulated on a daily basis in the gravid uterus of several wild ungulates. (From Robbins and Moen, 1975, courtesy of the *Journal of Wildlife Management.*)

maternal weight in grams. When the maintenance and productive protein costs of the uterus and mammary gland are summed ($Y = 0.1501W^{0.84}$), the larger animal with the disproportionately longer gestation period ($W^{0.84}/W^{0.19} = W^{0.65}$) does reduce the cost per unit of gestation. Similarly, the reproductive protein cost increases the total maternal endogenous requirement (i.e., as a percentage of endogenous urinary nitrogen excretion) from 55% in a 10-kg female ungulate to 35% in a 1000-kg animal.

Mineral requirements also increase curvilinearly during gestation as fetal mass enlarges and bones become ossified. However, unlike protein and energy, a fetal maintenance requirement for minerals does not exist, since minerals can be continuously recycled from maternal to fetal circulations. As long as the mineral retention abilities of the kidney and gastrointestinal tract remain unchanged by pregnancy, the total maintenance requirement of the mother–fetus complex will be identical to the nonpregnant female. Since the minerals retained in productive processes are the only additional cost of gestation, the larger female that produces a relatively smaller fetus during a longer gestation period will experience a relatively smaller increased daily mineral requirement than will the smaller species.

Thus, the estimation of costs for gestation and evaluation of the evolutionary consequences of changes in fetal size or gestation period length are far more complex than simply comparing relative growth rates. The pregnant female can benefit by lengthening the gestation period if protein or minerals are limiting, but gains nothing by prolonging the gestation period if energy is limiting. Many females, when provided with food ad libitum, ingest nutrients in excess of the actual cost of gestation (Blaxter, 1964; Randolph *et al.*, 1977). Since lactation is more costly than gestation, the female can accumulate reserves during gestation, if food is available, that can be mobilized later for milk synthesis.

B. Lactation

The production of milk by the postpartum mammalian female provides a dietary buffer between the passive, totally dependent fetus and the weaned, nutritionally independent individual often required to process possibly toxic, less digestible, or difficult-to-acquire plant or animal tissue. Lactation often enables the neonate to continue growth in an almost embryonic manner anatomically separated from the female, thus freeing her from the locomotory, nutritional, and anatomical constraints of transporting and bearing an excessively large fetus(es) (Pond, 1977). However, nutrient transfer and corresponding neonatal growth rates during lactation are much higher than rates occurring *in utero* (Blaxter, 1964; Pond, 1977).

Although few exceptions exist, mammals are not unique in nourishing their offspring with parental secretions. Pigeons and doves (Vandeputte-Poma, 1980), penguins (Prévost and Vilter, 1963), and flamingoes (Lang, 1963) feed esophageal or crop secretions to the hatchling. Because of the analogous role of these avian secretions, the pigeon is said to produce crop milk during lactation. Pigeon milk is produced by both parents and consists of sloughed, fat-loaded, epithelial cells of the crop wall. The fat content generally decreases during the lactation period (Desmeth and Vandeputte-Poma, 1980). Crop milk is the only nourishment for the young pigeon during the first 3 days. Subsequently, the milk is increasingly mixed with grain until milk production stops at 3–4 weeks.

1. Milk Composition

"The present status of knowledge of milk composition is virtually over-whelmed by numerous and detailed studies on seven [domestic] species, miscellaneous and scattered data on about 150 more and no data on nearly 4000 species [Jenness, 1974, p. 3]." Many of the milk samples collected from wildlife have either been the first milk, particularly from zoo animals whose young have died shortly after birth, or milk from very late in the lactation cycle, such as from free-ranging animals killed during fall and winter hunting seasons. Since milk composition changes quite markedly during a lactation cycle, one must be careful in evaluating the meaning of a single milk analysis reported for a particular species.

The first milk produced during each lactation cycle is called colostrum. Since the neonate is born into a very septic environment relative to the sterile uterus, development of an immune system is essential. Colostrum is often noted for its high concentration of maternal antibodies, or immunoglobulins, and active phagocytic cells (Cockson and McNeice, 1980). While the phagocytic cells are important to all animals in countering infection, the significance of colostral IgG immunoglobulins is species specific. For example, neonatal humans, monkeys, and rabbits acquire all their circulating maternal immunoglobulins *in utero.* Since *in utero* transmission of immunoglobulins does not occur in ungulates, marsupials, and mink, colostrum is the sole source of a passive immune system. Intermediate to these groups are rats, mice, cats, and dogs, which acquire maternal immunoglobulins both *in utero* and in colostrum (Yadav, 1971; Butler, 1974).

The secretion of colostral IgG immunoglobulins is practical only as long as the neonatal gut remains permeable to their absorption and upper-tract digestion is minimal. The intestine of ungulates remains permeable to the intact immunoglobulin for only 24–36 hr after birth (Lecce and Morgan, 1962; McCoy *et al.,* 1970), but continues in mink for 8 days (Hanson and Johansson, 1970), in mice and rats for 16–20 days (Halliday, 1955; Abrahamson *et al.,* 1979), and in macropod marsupials for 170–200 days (Yadav, 1971). The very prolonged absorption capabilities in marsupials corresponds to the time that the young resides in the pouch. Other types of immunoglobulins, such as IgA, continue to occur in milks after absorption of intact molecules has ceased and are important in the protection of the neonatal gut from infection (Hanson and Johansson, 1970; Butler, 1974).

The major constituents of milk are water, minerals, protein (such as casein and whey), fat, and carbohydrates, ranging from monosaccharides to oligosaccharides (Table 9.6). Protein concentration ranges from under 10 g/liter in human beings and certain other primates to nearly 200 g/liter in the snowshoe hare and white-tailed jackrabbit. Fat varies from traces in the milk of rhinoceroses and horses to 500 g/liter in many seals and whales. The principal carbohydrate of placental mammal milk is the disaccharide lactose, a polymer of glucose and galactose. Lactose content ranges from traces in the milk of certain seals, walruses, and sea lions of the Pacific Ocean, manatees, and marsupials to more than 100 g/liter in some primates. Marsupial milk, nevertheless, can be very high in carbohydrates, but the sugars are primarily oligo- and polysaccharides rich in galactose (Jenness, 1970, 1974; Kretchmer, 1972; Messer and Mossop, 1977; Bachman and Irvine, 1979; Messer and Green, 1979; Reidman and Ortiz, 1979; Walcott and Messer, 1980). Saccharide content is generally inversely proportional to the mineral content, since both are important in regulating the osmotic potential of milk in the mammary gland.

Most aquatic mammals produce highly concentrated milks (Jenness and Sloan, 1970; Jenness, 1974) (Table 9.6). The reduction in milk water content in aquatic animals provides a high-energy, low-bulk diet that is useful in offsetting neonatal heat loss in cold environments and conserves water in those mothers, such as the northern elephant seal, that abstain entirely from food and water during a rela-

TABLE 9.6
Mid-lactation Milk Composition (%) and Energy Content (kcal/g)[a]

Group or species	Water	Fat	Protein	Sugar	Ash	Energy
Marsupials						
Brush-tailed possum (*Trichosurus vulpecula*)	75.6	6.1	9.2	3.2	1.6	1.21
Red kangaroo (*Magaleia rufa*)	77.2	4.9	6.7	2.0	1.4	0.90
Primates						
Baboons (*Papio* sp.)	86.0	4.6	1.5	7.7	0.3	0.80
Lemurs (*Lemur* sp.)	—	2.3	1.9	6.7	0.3	0.57
Talapoin monkey (*Ceropithecus talapoin*)	87.7	3.0	2.1	7.2	0.3	0.67
Lagomorphs						
Domestic rabbit (*Oryctolagus cuniculus*)	68.8	15.2	10.3	1.8	1.8	2.04
Eastern cottontail (*Sylvilagus floridanus*)	64.8	14.4	15.8	2.7	2.1	2.33
Rodents						
Chinchilla (*Chinchilla chinchilla*)	—	11.2	7.3	1.7	1.0	1.50
Common rat (*Rattus norvegicus*)	77.9	8.8	8.1	3.8	1.2	1.43
European beaver (*Castor fiber*)	65.9	19.0	11.2	1.7	1.1	2.46
Golden hamster (*Mesocricetus auratus*)	77.4	4.9	9.4	4.9	1.4	1.20
Guinea pig (*Cavia porcellus*)	—	5.3	6.1	4.6	0.8	1.01
House mouse (*Mus musculus*)	70.7	13.1	9.0	3.0	1.5	1.85
Carnivores—fissipeds						
Arctic fox (*Alopea lagopus*)	71.4	13.5	11.1	3.0	1.0	1.98
Brown bear (*Ursus arctos*)	66.4	18.5	8.5	2.3	1.5	2.28
Domestic cat (*Felis catus*)	—	10.8	10.6	3.7	1.0	1.74
Domestic dog (*Canis familiaris*)	78.0	11.1	7.7	3.3	1.1	1.58
Mink (*Mustela vison*)	74.9	8.6	8.7	5.5	1.0	1.46
Racoon dog (*Nyctereutes procyonides*)	81.4	3.4	7.8	—	1.1	—
Red fox (*Vulpes vulpes*)	81.9	5.8	6.7	4.6	0.9	1.09
Carnivores—pinnipeds						
Harp seal (*Pagophilus groenlandicus*)	48.3	42.2	8.7	0.1	0.7	4.34
Northern elephant seal (*Mirounga angustirostris*)	35.5	48.8	7.6	0.3	—	4.88
Northern fur seal (*Callorhinus ursinus*)	39.0	49.4	10.2	0.1	0.5	5.09
Southern elephant seal (*Mirounga leonina*)	51.2	39.0	9.0	—	—	4.07

(continued)

[a]From Oftedal (1981), courtesy of Olav T. Oftedal, National Zoological Park, Washington, D.C.

TABLE 9.6 *Continued*

Group or species	Water	Fat	Protein	Sugar	Ash	Energy
Weddell seal (*Leptonychotes weddelli*)	42.8	42.1	15.8	1.0	—	4.08
Subungulates						
African elephant (*Loxodonta africana*)	82.7	5.0	4.0	5.3	0.7	0.88
Asian elephant (*Elephas maximus*)	82.3	7.3	4.5	5.2	0.6	1.12
Ungulates—perissodactyls						
Ass (*Equus asinus*)	91.5	0.6	1.4	6.1	0.4	0.38
Black rhinoceros (*Diceros bicornis*)	91.2	0.2	1.4	6.6	0.3	0.35
Domestic horse (*Equus caballus*)	89.1	1.6	2.2	6.4	0.4	0.52
Ungulates—artiodactyls						
Black-tailed deer (*Odocoileus hemionus*)	—	12.6	7.2	4.8	1.4	1.76
Dall sheep (*Ovis dalli*)	77.1	9.5	7.2	5.3	0.9	1.48
Domestic cow (*Bos taurus*)	87.6	3.7	3.2	4.6	0.7	0.71
Domestic goat (*Capra hircus*)	88.0	3.8	2.9	4.7	0.8	0.69
Domestic sheep (*Ovis aries*)	81.8	7.1	5.0	4.9	0.8	1.12
Eland (*Taurotragus oryx*)	78.1	9.9	6.3	4.4	1.1	1.43
Elk (*Cervus elaphus*)	81.0	6.7	5.7	4.2	1.3	1.10
Giraffe (*Giraffa camelopardalis*)	85.5	4.8	4.0	4.9	0.8	0.86
Moose (*Alces alces*)	78.5	10.0	8.4	3.0	1.5	1.51
Red deer (*Cervus elaphus*)	78.9	8.5	7.1	4.5	1.4	1.37
Reindeer (*Rangifer tarandus*)	73.7	10.9	9.5	3.4	1.3	1.66
Thar (*Hemitragus jemlahicus*)	—	9.8	5.8	3.3	—	1.35
Water buffalo (*Bubalis bubalis*)	83.2	6.5	4.3	4.9	0.8	1.02
White-tailed deer (*Odocoileus virginianus*)	77.5	7.7	8.2	4.6	1.5	1.35

tively short but intense lactation period. Similarly, some desert mammals produce concentrated milks as a water conservation mechanism (Jenness and Sloan, 1970; Baverstock *et al.*, 1976). The reduction of milk water content also enables the female to minimize udder size and weight. Species in which the neonate nurses on demand, such as marsupials, primates, perissodactyls, and some artiodactyls in which the infant stays with the mother and has relatively free access to the mammary gland, tend to produce very dilute milks. Intermediate in milk concentration are terrestrial species that nurse on a scheduled basis, such as lagomorphs, most rodents, most carnivores, and many artiodactyls (Ben Shaul, 1962b; Jenness, 1974). Milk energy and protein concentrations in related groups of animals are usually directly proportional to each other (Fig. 9.9). For exam-

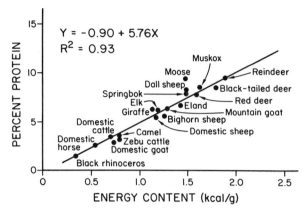

Fig. 9.9 The energy to protein content of wild and domestic ungulate milks during mid-lactation. (Data from Tener, 1956; Chen et al., 1965; Gregory et al., 1965; Treus and Kravchenko, 1968; Jenness and Sloan, 1970; Van Zyl and Wehmeyer, 1970; Arman et al., 1974; Luick et al., 1974; Franzmann et al., 1975; Hall-Martin et al., 1977; Mueller and Sadleir, 1977; Krzywinski et al., 1980; Robbins et al., 1981, Robbins and Stevens, unpublished.)

ple, the energy and protein content of the milk from many species of ungulates are linearly related even though the dry-matter content varies widely. Such a proportionality may reflect the requirements of the neonate. However, convincing proof that milk composition directly reflects neonatal requirements is lacking (Jenness, 1974).

2. Milk Intake

Estimates of milk intake by maternal-raised neonates, when combined with milk composition data, provide a basis for estimating maternal production requirements and for understanding the processes and requirements of neonatal growth. Available methods for determining milk intake include weighing the neonate before and after a controlled nursing period (Griffiths, 1965; Martin, 1966; Linzell, 1972; Arman et al., 1974; Sadleir, 1980a, 1980b), hand milking the lactating female after a controlled period of neonate removal (Yazan and Knorre, 1964; Treus and Kravchenko, 1968; Arman et al., 1974), and using isotopes, primarily deuterium or tritium oxide, to quantify the dilution of the neonate's body water pool by ingested milk (Macfarlane et al., 1969; Buss and Voss, 1971; McEwan and Whitehead, 1971; Holleman et al., 1975; Baverstock and Elhay, 1979; Robbins et al., 1981). Other isotopes, such as cesium and sodium, whose transfer from the adult to the neonate or dilution in the neonate is proportional to milk intake, have also been used (Holleman et al., 1975; Green and Newgrain, 1979). Two other procedures for estimating milk intake that have had limited use include (a) feeding trial experiments that attempt to duplicate the growth rates of maternal-nursed neonates with those of bottle-fed infants (Buss and Voss, 1971; Robbins and Moen, 1975b) and (*b*) the hypothetical or experimental estimation of necessary milk intake to meet the neonate's maintenance and productive requirements (Brody and Nisbet, 1938; Moen, 1973).

Unfortunately, all the experimental methods, except the use of isotopes, are appropriate only for captive animals. Each method entails various assumptions or experimental and interpretational difficulties. The major difficulty with hand milking is equating observed female production to actual neonatal intake. The interval between milkings, if quite short, may reflect physiological milk production capabilities and overestimate neonatal intake, while if the interval is too long, milk accumulation in the udder can reduce production and therefore underestimate intake (Linzell, 1972). Excitable wild animals often cannot be milked without injections of immobilizers, tranquilizers, or oxytocin (Arman *et al.*, 1974; Mueller and Sadleir, 1977; Arman, 1979).

Investigators weighing the neonate or mother before and after a nursing need to correct for fecal and urine excretion in those species in which the mother stimulates the nursing neonate to void. Red deer calves wore "nappies" to collect the urine and feces when stimulated by the mother or were stimulated by the investigator before rejoining the mother (Arman *et al.*, 1974). As in hand milking, the interval in which the infant is removed from its mother must be very carefully evaluated. For some species—such as the tree shrew, which nourishes its young only once in 48 hr (Martin, 1966), and rabbits and hares, which suckle once in 24 hr (Linzell, 1972; Broekhuizen and Maaskamp, 1980)—the normal frequency can be easily used.

Isotopes provide the greatest promise for estimating milk intake of both captive and free-ranging wildlife. Prior to the consumption of free water and preformed water in other feeds, milk is the only external water source. Consequently, milk intake can be directly determined from measurements of neonatal water kinetics and milk composition. As other sources of water are ingested, either the mother can be injected with a different water isotope than the neonate to distinguish between the various water sources or, as is often possible with captive animals, consumption of other sources of water can be directly measured (Holleman *et al.*, 1975; Robbins *et al.*, 1981). The major assumptions or requisite determinations of the water isotope methods are that cross-nursing between different neonate–mother combinations does not occur, water cycling between littermates and between neonate and mother is either negligible or measureable if it occurs, and water isotope kinetics reflect body water kinetics (Macfarlane *et al.*, 1969; Holleman *et al.*, 1975; Dove and Freer, 1979). Isotope cycling between neonate, mother, and other littermates is particularly significant in the smaller altricial mammals in which the litter is confined to a relatively small space or the mother consumes all urine and feces (Baverstock and Elhay, 1979; Baverstock *et al.*, 1979). When this occurs in litter-bearing species, one neonate per litter can be left uninjected to serve as a control to correct for isotope cycling.

Milk production generally rises to a peak during early lactation, then begins falling when the infant becomes more independent as it ingests other foods. Because of the changing nature of the lactation curve and the desire of many investigators to understand the maximum stress on the lactating female, peak milk yields of well-fed females have been the principal basis of comparison. Since peak yields are a power function of maternal body weight (Table 9.7), milk production and mammary gland size decrease relative to maternal size in the larger species. For example, predicted peak daily milk production as a percentage of maternal weight decreases from 28% in the pygmy shrew to 1.25% in an elephant (Hanwell and Peaker, 1977).

The decline in milk production once the peak has been reached can last for as little as 5 days in mice to many months in large ungulates (J. M. White, 1975; Robbins *et al.*, 1981). The decline is generally longer and less steep in larger species because their offspring have a slower relative growth rate and longer maternal dependence (Fig. 9.10). Exceptions to a general trend between maternal size and length of lactation do occur, as indicated by the extremely short, intense 9-day lactation in harp seals (Stewart and Lavigne, 1980). Poorly nourished females and those nursing larger litters often reduce milk production faster than do well-nourished females. The decline in milk intake relative to infant requirements stimulates the young animal to ingest food characteristic of the adult diet. The point at which the offspring is weaned is not solely due to nutritional considerations. For example, ungulate females continue very low levels of milk production for many weeks after the young could be weaned from a nutritional perspective (Robbins and Moen, 1975b; Sadleir, 1980b; Robbins *et al.*, 1981). Maintenance of the mother–infant bond in those species in which adult investment per offspring is already considerable is apparently of survival benefit—perhaps in predator avoidance or defense, in development of migration patterns, and in social interactions.

TABLE 9.7

Interspecific Comparisons of Milk Yield (Y in kcal/day) at the Peak of Lactation as a Function of Maternal Body Weight (W in kg)

Group	Equation	Reference
Mammals, including dairy breeds	$Y = 144W^{0.71}$	Brody (1945)
	$Y = 124W^{0.75}$	Payne and Wheeler (1968)
Mammals, excluding dairy breeds	$Y = 145W^{0.69}$	Linzell (1972)
	$Y = 127W^{0.69}$	Hanwell and Peaker (1977)
Ungulates and subungulates nursing a single neonate, excluding dairy breeds	$Y = 236W^{0.52}$	Robbins and Robbins (1979)

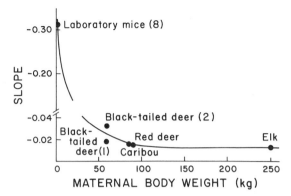

Fig. 9.10. Slope (*b*) of the declining portion of the lactation curve ($Y = ae^{-bx}$) for several wildlife species. Sources of information are Arman *et al.* (1974), J. M. White (1975), R. G. White *et al.* (1975), Sadleir (1980a), and Robbins *et al.* (1981).

The disproportionately longer lactation period in larger species distributes the total cost over a far longer period of time. Since the larger species is producing relatively low levels of milk both at the peak and during the prolonged tail end of the lactation cycle, the average daily cost as a function of the basal metabolic rate is lower in larger species (Fig. 9.11). Consequently, the lactation cost is high but of relatively short duration in the mouse as compared to the relatively lower cost but longer duration in the elk (Blaxter, 1971).

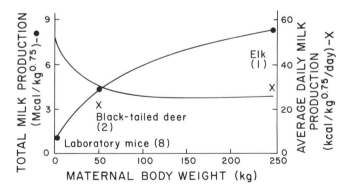

Fig. 9.11. Total and average daily milk production during lactation in mice, black-tailed deer, and elk as functions of metabolic body weight. Lactation periods were estimated as 21 days in mice, 205 days in deer, and 280 days in elk. Numbers in parentheses are number of young nursed. Sources of information are J. M. White (1975), Sadleir (1980a), and Robbins *et al.* (1981).

C. Care and Feeding of Orphaned Neonates

Numerous attempts have been made to raise orphaned wildlife (Table 9.8), but the successful care of very young animals is still largely an art requiring dedication and perseverance. For example, many infants do not defecate or urinate unless manually stimulated by rubbing or washing the anal and genital areas. Carnivores (Hoff, 1960; Ben Shaul, 1962a; Freiheit and Crotty, 1969; Weber, 1969; Gray, 1970; Baudy, 1971; Jantschke, 1973; Armstrong, 1975; Johnstone-Scott, 1975; Barnes, 1976; Weiher, 1976; Hughes, 1977; Johnstone, 1977; Toweill and Toweill, 1978), rodents (Yamamoto, 1962), artiodactyls (Zellmer, 1960; Ben Shaul, 1962; Stroman and Slaughter, 1972; Muller-Schwarze and Muller-Schwarze, 1973), and marsupials (Nakazato et al., 1971) are examples of only a few categories of neonates that must be stimulated. Since feces produced from the digestion of milk are normally quite soft and aromatic, their consumption by the lactating female as she licks the infant ensures cleanliness when the animals are confined to a nest or den, and possibly avoids attracting predators (Fig. 9.12). A similar response was observed in newly hatched western grebes, which would only defecate when swimming in water (Ratti, personal communication).

Fig. 9.12. A female white-tailed deer stimulating her nursing offspring to defecate and urinate. This act must often be simulated when human beings bottle raise wildlife. (Courtesy of Nadine Jacobsen, University of California, Davis.)

TABLE 9.8
Sources of Information on Raising Neonatal Mammals[a]

Animal	Reference
Marsupials	
Agile wallaby (*Protemnodon agilis*)	Nakazato *et al.* (1971)
Dama wallaby (*Protemnodon eugenii*)	Wilson (1971)
Kangaroos (in general)	Ben Shaul (1962a), Stephens (1975)
Silver-grey phalanger (*Trichosurus vulpecula*)	Ben Shaul (1962a)
Spotted cuscus (*Phalanger maculatus*)	Menzies (1972)
Insectivores	
Common tenrec (*Tenrec ecaudatus*)	Eisenberg and Muckenhirn (1968)
Giant hedgehog tenrec (*Setifer setosus*)	Eisenberg and Muckenhirn (1968)
Primates	
Baboons (*Papio* sp.)	Ben Shaul (1962a), Buss *et al.* (1970), Moore and Cummins (1979)
Black-faced spider monkey (*Ateles paniscus*)	Gensch (1965)
Chimpanzee (*Pan troglodytes*)	Ben Shaul (1962a), Yamamoto (1967)
Crowned lemur (*Lemur mongoz coronatus*)	Hick (1976)
Demidoff's bush baby (*Galago demidoffi*)	Ben Shaul (1962a)
Gelada baboon (*Theropithecus gelada*)	Ben Shaul (1962a)
Gibbon (*Hylobates lar*)	Sasaki (1962), Ben Shaul (1962a), Breznock *et al.* (1979)
Gorilla (*Gorilla gorilla*)	Carmichael *et al.* (1961), Ben Shaul (1962a), Kirchshofer *et al.* (1967), Fontaine (1968), Frueh (1968), Kirchshofer *et al.* (1968), Lotshaw (1971), Mallinson *et al.* (1976)
Mangabey (*Cercocebus* sp.)	Ben Shaul (1962a)
Marmosets (*Saguinus* and *Callithrix*)	Ogden (1979), Cicmanec *et al.* (1979)
Monkeys (in general)	Ben Shaul (1962a)
Patas monkey (*Erythrocebus patas*)	Ben Shaul (1962a)
Pig-tailed macaque (*Macaca nemestrina*)	Ben Shaul (1962a)
Rhesus monkey (*Macaca mulatta*)	Boelkins (1962), Ruppenthal (1979)
Ring-tailed lemur (*Lemur catta*)	Hick (1976)
Senegal galago (*Galago senegalensis*)	Ben Shaul (1962a)
Spider monkey (*Ateles paniscus* or *geoffroyi*)	Ben Shaul (1962a), Miles (1967)
Vervet monkey (*Cercopithecus aethiops*)	Ben Shaul (1962a)
Edentates	
Giant anteater (*Myrmecophaga tridactyla*)	Ben Shaul (1962a), Bickel *et al.* (1976), Hardin (1976)
Seven-banded armadillo (*Dasypus septemcinctus*)	Block (1974)
Six-banded armadillo (*Euphractus sexcinctus*)	Gucwinska (1971)
Lagomorphs	
European hares (*Lepus europaeus*)	Cooper (1970)
Rabbit (*Oryctolagus cuniculus*)	Hills and MacDonald (1956)

(*continued*)

TABLE 9.8 *Continued*

Animal	Reference
Rodents	
Cane rat (*Thryonomys* sp.)	Ben Shaul (1962a)
Flying squirrel (*Petaurista philippensis*)	Ben Shaul (1962a)
Japanese squirrel (*Scriurus vulgaris lis*)	Yamamato (1962)
Indian porcupine (*Hystrix indica*)	Ben Shaul (1962a)
North American porcupine (*Erethizon dorsatum*)	Tryon (1947)
Patagonian cavies (*Dolichotis patagonum*)	Rosenthal (1974)
Sierra Leone striped squirrels (*Funisciurus pyrrhopus*)	Mallinson (1975)
Small green squirrel (*Heliosciurus poensis*)	Ben Shaul (1962a)
Springhaas (*Pedetes capensis*)	Rosenthal and Meritt (1973)
Squirrels (in general)	Ben Shaul (1962a)
Carnivores—Pinnipeds	
California sea lion (*Zalophus californianus*)	Ben Shaul (1962a)
Elephant seal (*Mirounga* sp.)	Ben Shaul (1962a)
Harbour or common seal (*Phoca vitulina*)	Bossanyi (1948), Ben Shaul (1962a), Reineck (1962), Cansdale (1970), Cansdale and Yeadon (1975)
South American fur seal (*Arctocephalus pusillus*)	Ben Shaul (1962a)
Carnivores—Fissipeds	
Badger (*Meles meles*)	Ben Shaul (1962a)
Black-backed jackel (*Canio mesomelas*)	Ben Shaul (1962a)
Black-footed cats (*Felis nigripes*)	Armstrong (1975)
Black bear (*Ursus americanus*)	Ben Shaul (1962a), Hulley (1976)
Bush dog (*Speothos venaticus*)	S. L. Kitchener (1971), Collier and Emerson (1973), Jantschke (1973)
Ringtails (*Bassariscus astutus flavus*)	Toweill and Toweill (1978)
Cape hunting dog (*Lycaon pictus*)	Dathe (1962), Encke (1962), Ben Shaul (1962a)
Cats (in general)	Baudy (1971)
Cheetah (*Acinonyx jubatus*)	Encke (1960)
Coati mundi (*Nasua nasua*)	Ben Shaul (1962a)
Coyote (*Canis latrans*)	Barnum *et al.* (1979)
Fanaloka (*Fossa fossa*)	Wemmer (1971)
Fennec (*Fennecus zerda*)	Weiher (1976)
Grizzly and Kodiak bear (*Ursus arctos*)	Ben Shaul (1962a), Quick (1969), Freiheit and Crotty (1969), Fransen and Emerson (1973)
Honey badger (*Mellivora capensis*)	Johnstone-Scott (1975)
Jaguar (*Panthera onca*)	Hunt (1967)
Kinkajou (*Potos flavus*)	Clift (1967), Bhatia and Desai (1972)
Lion (*Panthera leo*)	Ben Shaul (1962a)
Leopard (*Panthera pardus*)	Hoff (1960), Ben Shaul (1962a)
Leopard cat (*Felis bengalensis borneoensis*)	Birkenmeier and Birkenmeier (1971)
Malayan sun bear (*Helarctos malayanus*)	Weber (1969)

(continued)

TABLE 9.8 *Continued*

Animal	Reference
Maned wolves (*Chrysocyon brachyurus*)	Acosta (1972)
Marbled cat (*Felis marmorata*)	Barnes (1976)
Marsh mongoose (*Atilax paludinosus*)	Ben Shaul (1962a)
Mink (*Mustela vison*)	Ben Shaul (1962a)
Ocelot (*Felis pardalis*)	Dunn (1974)
Otters (*Aonyx capensis*)	Ben Shaul (1962a)
Polar bear (*Thalarctos maritimus*)	Ben Shaul (1962a), Michalowski (1971, 1972), Hess (1971), Wortman and Larue (1974)
Puma (*Felis concolor*)	Hoff (1960), Ben Shaul (1962a)
Raccoon (*Procyon lotar*)	Ben Shaul (1962a)
Red Fox (*Vulpes vulpes*)	Ben Shaul (1962a)
Red panda (*Ailurus fulgens*)	Gray (1970), Vogt *et al.* (1980)
Servals (*Felis serval*)	Johnstone (1977)
Small-spotted genet (*Genetta genetta*)	Flint (1975)
Spotted hyaena (*Crocuta crocuta*)	Ben Shaul (1962a)
Tiger (*Panthera tigris*)	Ben Shaul (1962a), Hoff (1960), Hughes (1977)
Subungulates	
African elephant (*Loxodonta africana*)	Ben Shaul (1962a), Bellinge and Woodley (1964), Bolwig *et al.* (1965), Reuther (1969)
Indian elephant (*Elephas maximus*)	H. J. Kitchener (1960), Ben Shaul (1962a), Reuther (1969)
Aardvarks	
Aardvark (*Orycteropus afer*)	Sampsell (1969)
Ungulates—Perissodactyls	
American tapir (*Tapirus terrestris*)	Young (1961)
Great Indian rhinoceros (*Rhinoceros unicornis*)	Hagenbeck (1969)
White rhinoceros (*Diceros s. simus*)	Wallach (1969)
Zebra (*Equus grevyi x E. asinus*)	Ben Shaul (1962a)
Ungulates—Artiodactyls	
Artiodactyls (in general)	Ben Shaul (1962a), Hutchison (1970)
Bactrian camel (*Camelus bactrianus*)	Encke (1970)
Bighorn sheep (*Ovis canadensis*)	Deming (1955)
Bison (*Bison bison*)	Ben Shaul (1962a)
Bohor reedbuck (*Redunca redunca*)	Ben Shaul (1962a)
Duikers (in general)	Ben Shaul (1962a), Hopkins (1966)
Eland (*Taurotragus oryx*)	Treus and Kravchenko (1968)
Elk (*Cervus canadensis*)	Anderson (1951), Hobbs and Baker (1979), Robbins *et al.* (1981)
Fallow deer (*Dama dama*)	Klopfer and Klopfer (1962), Wayre (1967), Gilbert (1974)

(*continued*)

TABLE 9.8 *Continued*

Animal	Reference
Giraffe (*Giraffa camelopardalis*)	Zellmer (1960), Clevenger (1981)
Grant's gazelle (*Gazella granti*)	Rosenthal and Meritt (1971)
Greater kudu (*Tragelaphus strepsiceros*)	Vice and Olin (1969)
Grysbok (*Raphicerus melanotis*)	Ben Shaul (1962a)
Hippopotamus (*Hippopotamus amphibius*)	Weber (1970), Vanforeest *et al.* (1978), Cook (1981), Tobias (1981)
Llama (*Lama glama*)	Ben Shaul (1962a)
Moose (*Alces alces*)	Dodds (1959), Yazan and Knorre (1964), Markgren (1966)
Mule deer (*Odocoileus hemionus*)	Halford and Alldredge (1978)
Musk-ox (*Ovibos moschatus*)	Banks (1978)
Pronghorn antelope (*Antilocapra americana*)	Ben Shaul (1962a), Muller-Schwarze and Muller-Schwarze (1973)
Pudu (*Pudu pudu*)	Hick (1969)
Pygmy hippopotamus (*Choeropsis liberiensis*)	Stroman and Slaughter (1972)
Red deer (*Cervus elaphus*)	Youngson (1970), Blaxter *et al.* (1974), Krzywinski *et al.* (1980)
Roe deer (*Capreolus capreolus*)	Pinter (1962), Wayre (1967)
Royal antelope (*Neotragus pygmaeus*)	Ben Shaul (1962a)
Saiga antelope (*Saiga tatarica*)	Orbell and Orbell (1976)
Steinbok (*Raphicerus campestris*)	Ben Shaul (1962a)
Thomson's gazelle (*Gazella thomsoni*)	Ben Shaul (1962), Rosenthal and Meritt (1971)
Wart hog (*Phacochoerus aethiopicus*)	Ben Shaul (1962a)
White-tailed deer (*Odocoileus virginianus*)	Murphy (1960), Silver (1961), Robbins and Moen (1975b), Buckland *et al.* (1975)
Wild pig (*Sus scrofa*)	Ben Shaul (1962a), Clift (1967)

[a]Most articles have information on milk formulas, intake frequency and methods of feeding, methods of weaning, behavior, and other pertinent information. However, many of the authors had varying degrees of success. Consequently, many of the articles are better examples of failures than successes.

Milk formulation is one of the major concerns in raising orphans. Excesses of milk sugars or fats can cause diarrhea and digestive upsets (Reineck, 1962; Reuther, 1969). Lactose is a particular problem when using cow's milk or lactose-containing milk replacers to raise neonates of species, such as seals, in which the natural milk is largely devoid of this disaccharide. When lactose is fed to neonates in which the enzyme lactase is inadequate for its digestion, severe diarrhea occurs due to the osmotic effect of undigested disaccharides in the intestine. Cataracts can also be produced when lactase-deficient species absorb minimal amounts of galactose if cellular enzyme systems are inadequate for its metabolism. Residual galactose is converted to an opaque sugar alcohol and retained in the lens of the eye (Stephens, 1975). Many other deficiencies, particularly amino acids in very young mammals, and pathogens can cause cataracts (Vainisi *et al.*, 1980). Disaccharide intolerance and the debilitating consequences also can be produced by feeding other sugars. For example, sucrase, the enzyme necessary to digest common table sugar or corn syrup, is very low or absent in many mammalian neonates (Kerry, 1969; Walcott and Messer, 1980). Several milk substitutes that are lactose and galactose free are available as a formula base for lactose-intolerant species.

Added fats that produce severe diarrhea have been primarily vegetable fats, such as common cooking oils. If fat is to be added, butterfat is usually preferred and can be obtained in pure form. Other fats, such as animal tallow or egg yolk, can also be used. When significant levels of fat are added, the milk should be homogenized to ensure maximum fat dispersion, minimal fat droplet size, and, therefore, maximum digestibility. Likewise, one should avoid feeding milks higher in fat content than the species normally produces.

Although neonates have been successfully fed diets very dissimilar from maternal milk composition (Fransen and Emerson, 1973; Hardin, 1976), most neonates do better when fed milks similar to maternal milk composition (Hills and MacDonald, 1956; Wayre, 1967; Buss *et al.*, 1970; Livers, 1973). The common statement that the gross composition of maternal milk could not be duplicated because of its radical difference from either cow's milk or other common milk replacers is patently incorrect. Since the principal ingredients of milk are available in pure form, milk replacers of any composition can be formulated. However, even far greater sophistication in milk formulation will become necessary as we recognize the importance of the fatty acid and amino acid spectra of milk replacers and the normal compositional changes that occur throughout lactation (Green *et al.*, 1980). For example, the difference in protein structure between the caseins of domestic cow's milk and rabbit's milk produced a hard (cow) or soft (rabbit) stomach coagulum when ingested by young rabbits. Because of this, young rabbits could not be raised on cow's milk (Fonty *et al.*, 1979).

It has been difficult to raise baby elephants on cow's milk. Because chronic diarrhea ending in death has been common, the fat of cow's milk was often suspected as the problem. Although elephant milk contains 2–3 times as much fat as does cow's milk, the fat droplets in elephant milk are one-half the size of those in raw cow's milk. Similarly, the fatty acids of elephant milk are 82% capric and lauric acids (10 and 12 carbons in length), whereas cow's milk fat contains only 5% of these acids and 52% palmitic and oleic acids (16 and 18 carbons). The fat of elephant milk is actually closer to coconut oil in composition than to the fat of cow's milk. The intolerance of baby elephants to cow's milk is likely due to the differences in fat composition and droplet size, which would increase the difficulty of their absorption and thus lead to digestive upsets when being bottle-raised (Garton, 1963; McCullagh *et al.,* 1969; McCullagh and Widdowson, 1970).

After the milk formula has been prepared, the nutritionist must determine the amount to be fed, the method of feeding, and the feeding schedule. The amount fed should be based on a knowledge of neonatal milk intake control, development, requirements, and efficiencies of nutrient utilization. Many infants have been fed ad libitum quantities of milk, with the assumption that the neonate would correctly control its intake. However, rarely are neonates capable of correctly regulating their intake. Milk is often ingested in excess of physiological capacities and results in diarrhea, vomiting, listlessness, potbellies, labored breathing, anorexia, and death (Pinter, 1962; Wayre, 1967; Fransen and Emerson, 1973; Wortman and Larue, 1974; Hick, 1976; Mallinson *et al.,* 1976). Conversely, because of the association between diarrhea and overfeeding, neonates often are purposefully underfed, with the only criteria of success being neonatal survival. Both approaches should be rejected and replaced by a thorough understanding of the entire lactation process.

The behaviorist observing nursing interactions has often noted whether the young or the mother behaviorally terminates a specific nursing bout. For example, well-fed wild ungulate mothers during the first week of lactation often allow the offspring seemingly unlimited access to the udder, with the young terminating all nursing bouts by simply moving away voluntarily from the udder. Later, the mother increasingly rebukes advances and behaviorally terminates all nursing bouts by moving away or forcefully rejecting the young before it is willing to relinquish the teat. These observations suggest that the young ungulate is quite capable of controlling its own intake from the very beginning, with the mother balancing her needs and resources with those of the neonate once weaning has begun. However, when the same neonatal wild ungulates are bottle-raised, they are often incapable of correctly controlling their intake and, if initially given free access to milk, will drastically overeat and develop acute diarrhea. Thus, if the mother-raised neonatal ungulate is indeed controlling its intake, the bottle-raising efforts are not providing the same cues upon which intake can be controlled. For example, the larger hole of most artificial nipples often delivers milk far faster than the maternal teat and may not provide the time cues necessary for the neonate to sense or judge its state of fill.

These paradoxes can be judged in light of milk intake control in young rats. The neonatal rat pup is often attached to the mother's teat for the majority of the day. However, attachment is not synonymous with milk intake and intake control. The 1-day-old pup is indeed incapable of correctly controlling its intake. Neonatal milk intake is controlled by the mother, which only intermittently lets milk down into the nipple for the pup's consumption. The healthy pup will eagerly and almost reflexively consume all milk that the female provides. If the pup is given access to more milk during a sucking bout, such as by cannulating the mother's teat or pup's mouth and artificially providing more milk, milk is consumed until the physical capacities of the stomach force milk into the intestinal tract as far as the large intestine. Similarly, milk will back up into the nose and mouth, and the pup will simply choke until it has cleared its respiratory tract of milk. Only the 15- to 20-day-old pup is able to control its intake correctly (Hall and Rosenblatt, 1977). Far more studies on this important question are needed if we are to successfully and scientifically raise many wildlife neonates.

Although the milk intakes of only a few mother-raised wildlife neonates have been determined, existing information can be used initially to estimate necessary milk intake for bottle-raised infants. For example, milk energy intake during the first month in ungulate neonates can be estimated from the expected rate of gain and average body weight (Fig. 9.13). Similarly, more meaningful success criteria than simply survival must be developed. Suitable criteria may include the comparisons of growth rates, longevity, general health, and eventual reproductive success between bottle- and maternal-raised infants.

Rarely have feeding schedules of the natural mother–infant interaction been used in bottle-raising programs. In general, the feeding schedules are based on personnel time commitments and the milk intake capabilities of the neonate. Milk intake per meal can frequently be far higher for the bottle-raised infant than is observed for the maternal-nursed infant. Captive, maternal-nursed elk calves during the first 3 months consumed an average of 400 g per nursing, even though 1200–1300 g could be consumed by the 4-week-old bottle-raised calf. However, the serum of bottle-raised calves consuming such high levels always showed a very pronounced postmeal lipemia that was never observed in maternal-nursed

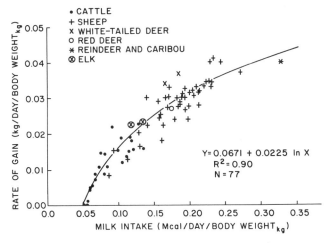

Fig. 9.13. Milk intake compared to growth rates during the first month of life in several ungulates. (From Robbins *et al.*, 1981, courtesy of the *Journal of Wildlife Management.*)

calves, which were apparently able to maintain a more constant nutrient supply at the tissue level rather than the surges produced by bottle feeding.

Unfortunately, as an indication of our lack of knowledge about lactation and neonatal metabolism, most bottle-raising efforts encounter diarrhea. While some of the diarrhea is primarily pathogenic in origin and totally independent of diet (Kramer *et al.*, 1971), many cases are due to the use of improper milk replacers and feeding schedules that predispose the neonate to gastrointestinal infection (Fonty *et al.*, 1979). Although antibiotics are often used in treating diarrhea, correction of the dietary cause of these diarrheas will be more productive. The pervasive trial-and-error approach to raising orphaned wildlife must be replaced by well-planned scientific efforts. While the preceding discussion concerns only nutritional considerations, problems in behavior and psychological development in bottle-raised infants are often equally important (Breznock *et al.*, 1979).

REFERENCES

Abrahamson, D. R., Powers, A., and Rodewald, R. (1979). Intestinal absorption of immune complexes by neonatal rats: A route of antigen transfer from mother to young. *Science* **206**, 567–569.

Acosta, A. L. (1972). Hand-rearing a litter of maned wolves (*Chrysocyon brachyurus*) at Los Angeles Zoo. *Int. Zoo. Ybk.* **12**, 170–174.

Adolph, E. F., and Heggeness, F. W. (1971). Age changes in body water and fat in fetal and infant mammals. *Growth* **35**, 55–63.

Anderson, C. C. (1951). Experimental feeding of calf elk. *Wyo. Wildl.* **15**, 24–27.

Ankney, C. D., and Scott, D. M. (1980). Changes in nutrient reserves and diet of breeding brown-headed cowbirds. *Auk* **97**, 684–696.

Ar, A., and Yom Tov, Y. (1978). The evolution of parental care in birds. *Evolution* **32**, 655–669.

Arman, P. (1979). Milk from semi-domesticated ruminants. *World Rev. Nutr. Diet.* **33**, 198–227.

Arman, P., Kay, R. N. B., Goodall, E. D., and Sharman, G. A. M. (1974). The composition and yield of milk from captive red deer (*Cervus elaphus* L.). *J. Reprod. Fert.* **37**, 67–84.

Armstrong, J. (1975). Hand-rearing black-footed cats (*Felis nigripes*) at the National Zoological Park, Washington. *Int. Zoo Ybk.* **15**, 245–249.

Asmundson, V. S., Baker, G. A., and Emlen, J. T. (1943). Certain relations between the parts of birds' eggs. *Auk* **60**, 34–44.

Bachman, K. C., and Irvine, A. B. (1979). Composition of milk from the Florida manatee, *Trichechus manatus latirostris*. *Comp. Biochem. Physiol.* **62A**, 873–878.

Bailey, J. A. (1980). Desert bighorn, forage competition, and zoogeography. *Wildl. Soc. Bull.* **8**, 208–216.

Banks, D. R. (1978). Hand-rearing a musk-ox (*Ovibos moschatus*) at Calgary Zoo. *Int. Zoo Ybk.* **18**, 213–215.

Barnes, R. G. (1976). Breeding and hand-rearing of the marbled cat (*Felis marmorata*) at the Los Angeles Zoo. *Int. Zoo Ybk.* **16**, 205–208.

Barnum, D. A., Green, J. S., Flinders, J. T., and Gates, N. L. (1979). Nutritional levels and growth rates of hand-reared coyote pups. *J. Mammal.* **60**, 820–823.

Batt, B. D., and Prince, H. H. (1979). Laying dates, clutch size and egg weight of captive mallards. *Condor* **81**, 35–41.

Battaglia, F. C., and Meschia, G. (1981). Foetal and placental metabolisms: Their interrelationship and impact upon maternal metabolism. *Proc. Nutr. Soc.* **40**, 99–113.

Baudy, R. E. (1971). Notes on breeding felids at the rare feline breeding center. *Int. Zoo Ybk.* **11**, 121–123.

Baverstock, P. R., and Elhay, S. (1979). Water-balance of small lactating rodents. IV. Rates of milk production in Australian rodents and the guinea pig. *Comp. Biochem. Physiol.* **63A**, 241–246.

Baverstock, P. R., Spencer, L., and Pollard, C. (1976). Water balance of small lactating rodents. II. Concentration and composition of milk of females on *ad libitum* and restricted water intakes. *Comp. Biochem. Physiol.* **53A**, 47–52.

Baverstock, P. R., Watts, C. H. S., and Spencer, L. (1979). Water-balance of small lactating rodents. V. The total water-balance picture of the mother-young unit. *Comp. Biochem. Physiol.* **63A**, 247–252.

Bellinge, W. H. S., and Woodley, F. W. (1964). Some notes on the rearing of young African elephants. *E. Afr. Wildl. J.* **2**, 71–74.

Ben Shaul, D. M. (compiler). (1962a). Notes on hand-rearing various species of mammals. *Int. Zoo Ybk.* **4**, 300–332.

Ben Shaul, D. M. (1962b). The composition of the milk of wild animals. *Int. Zoo Ybk.* **4**, 333–342.

Bhatia, C. L., and Desai, J. H. (1972). Growth and development of hand-reared kinkajou (*Potos flavus*) at Delhi Zoo. *Int. Zoo Ybk.* **12**, 176–177.

Bickel, C. L., Murdock, G. K., and Smith, M. L. (1976). Hand-rearing a giant anteater (*Myrmecophaga tridactyla*) at Denver Zoo. *Int. Zoo Ybk.* **16**, 195–198.

Birkenmeier, E., and Birkenmeier, E. (1971). Hand-rearing the leopard cat (*Felis bengalensis borneoensis*). *Int. Zoo Ybk.* **11**, 118–121.

Blaxter, K. L. (1964). Protein metabolism and requirements in pregnancy and lactation. *In* "Mammalian Protein Metabolism," Vol II (H. N. Munro and J. B. Allison, eds.), pp. 173–223. Academic Press, London.

Blaxter, K. L. (1971). The comparative biology of lactation. *In* "Lactation" (I. R. Falconer, ed.), pp. 51–69. Butterworth, London.

Blaxter, K. L., Kay, R. N. B., Sharman, G. A. M., Cunningham, J. M. M., and Hamilton, W. J. (1974). Farming the red deer. Dept. Agric. Fish. Scotland, Edinburgh.

Block, J. A. (1974). Hand-rearing seven-banded armadillos (*Dasypus septemcinctus*) at the National Zoological Park, Washington. *Int. Zoo Ybk.* **14**, 210–214.

Boelkins, R. C. (1962). Large scale rearing of infant Rhesus monkey (*Macaca mulatta*) in the laboratory. *Int. Zoo Ybk.* **4**, 285–289.

Bolwig, N., Hill, D. H., and Philpott, M. (1965). Hand-rearing an African elephant (*Loxodonta africana*). *Int. Zoo Ybk.* **5**, 152–154.

Bossanyi, J. (1948). Notes on the keeping of a baby common seal (*Phoca vitulina*). *J. Zool.* **117**, 791–792.

Boyd, L. L. (1978). Artifical insemination of falcons. *Symp. Zool. Soc. London* **43**, 73–80.

Breznock, A. W., Porter, S., Harrold, J. B., and Kawakami, T. G. (1979). Hand-rearing infant gibbons. *In* "Nursery Care of Non-human Primates" (G. C. Ruppenthal, ed.), pp. 287–298. Plenum, New York.

Brody, S. (1945). "Bioenergetics and Growth." Reinhold, New York.

Brody, S., and Nisbet, R. (1938). Growth and development. XLVII. A comparison of the amounts and energetic efficiencies of milk production in rat and dairy cow. *Missouri Agric. Exp. Sta. Res. Bull.* **285**, 1–30.

Broekhuizen, S., and Maaskamp, F. (1980). Behaviour of does and leverets of the European hare (*Lepus europaeus*) whilst nursing. *J. Zool.* **191**, 487–501.

Bryant, D. M. (1979). Reproductive costs in the house martin (*Delichon urbica*). *J. Anim. Ecol.* **48**, 655–675.

Buckland, D. E., Abler, W. A., Kirkpatrick, R. L., and Whelan, J. B. (1975). Improved husbandry system for rearing fawns in captivity. *J. Wildl. Manage.* **39**, 211–214.

Bump, G., Darrow, R. W., Edminster, F. C., and Crissey, W. F. (1947). "The Ruffed Grouse." New York State Cons. Dept., Albany.

Buss, D. H., and Voss, W. R. (1971). Evaluation of four methods for estimating the milk yield of baboons. *J. Nutr.* **101**, 901–909.

Buss, D. H., Voss, W. R., and Nora, A. H. (1970). A brief note on a modified formula for hand-rearing infant baboons (*Papio* sp.). *Int. Zoo Ybk.* **10**, 133–134.

Butler, J. E. (1974). Immunoglobulins of the mammary secretions. *In* "Lactation," Vol. III (B. L. Larson and V. R. Smith, eds.), pp. 217–255. Academic Press, New York.

Calder, W. A., III. (1979). The kiwi and egg design: Evolution as a package deal. *BioScience* **29**, 461–467.

Calder, W. A., III. Parr, C. R., and Karl, D. P. (1978). Energy content of eggs of the brown kiwi *Apteryx australis:* An extreme in avian evolution. *Comp. Biochem. Physiol.* **60**, 177–179.

Cansdale, G. (1970). Hand-rearing common seals (*Phoca vitulina*) at Skegness Natureland. *Int. Zoo Ybk.* **10**, 146–147.

Cansdale, G. S., and Yeadon, J. K. B. (1975). A further note on hand-rearing common seals (*Phoca vitulina*) at Natureland Marine Zoo, Skegness. *Int. Zoo Ybk.* **15**, 250–251.

Carey, C., Rahn, H., and Parisi, P. (1980). Calories, water, lipid and yolk in avian eggs. *Condor* **82**, 335–343.

Carmichael, L., Kraus, M. B., and Reed, T. (1961). The Washington National Zoological Park gorilla infant, Tomoko. *Int. Zoo Ybk.* **3**, 88–93.

Case, R. M., and Robel, R. J. (1974). Bioenergetics of the bobwhite. *J. Wildl. Manage.* **38**, 638–652.

Case, T. J. (1978). On the evolution and adaptive significance of postnatal growth rates in the terrestrial vertebrates. *Q. Rev. Biol.* **53**, 243–282.

Cengel, D. J., Estep, J. E., and Kirkpatrick, R. L. (1978). Pine vole reproduction in relation to food habits and body fat. *J. Wildl. Manage.* **42**, 822–833.

Chen, E. C. H., Blood, D. A., and Baker, B. E. (1965). Rocky Mountain bighorn sheep (*Ovis canadensis canadensis*) milk. *Can. J. Zool.* **43**, 885–888.

Cicmanec, J. L., Hernandez, D. M., Jenkins, S. R., Campbell, A. K., and Smith, J. A. (1979). Hand-rearing infant callitrichids (*Saguinus* spp. and *Callithrix jacchus*), owl monkeys (*Aotus trivirgatus*), and capuchins (*Cebus albifrons*). *In* "Nursery Care of Non-human Primates" (G. C. Ruppenthal, ed.), pp 307–312. Plenum, New York.

Clapp, J. F., Abrams, R. M., Caton, D., Cotter, J. R., and Barron, D. H. (1971). Fetal oxygen consumption in late gestation. *Q. J. Exp. Physiol.* **56**, 137–146.

Clevenger, M. A. (1981). Hand-rearing and development of a Rothschild's giraffe *Giraffa camelopardalis rothschildi* at the Oklahoma City Zoo. *Anim. Keepers' Forum* **8**, 30–35.

Clift, C. E. (1967). Notes on breeding and rearing a kinkajou (*Potos flavus*) at Syracuse Zoo. *Int. Zoo Ybk.* **7**, 126–127.

Cockson, A., and McNeice, R. (1980). Survival in the pouch: The role of macrophages and maternal milk cells. *Comp. Biochem. Physiol.* **66A**, 221–225.

Cody, M. L. (1971). Ecological aspects of reproduction. *In* "Avian Biology," Vol. I (D. S. Farner and J. R. King, eds.), pp. 461–512. Academic Press, New York.

Collier, C., and S. Emerson. 1973. Hand-raising bush dogs (*Speothos venaticus*) at the Los Angeles Zoo. *Int. Zoo Ybk.* **13**, 139–140.

Cook, M. (1981). Hand-rearing two baby hippopotamuses *Hippopotamus amphibius* at the Kansas City Zoo. *Anim. Keepers' Forum* **8**, 35–41.

Crenshaw, C., Huckabee, W. E., Curet, L. B., Mann, L., and Barron, D. H. (1968). A method for the estimation of the umbilical blood flow in unstressed sheep and goats with some results of its application. *Q. J. Exp. Physiol.* **53**, 65–75.

Dathe, H. (1962). Hand-rearing of a Cape hunting dog (*Lycaon pictus*). *Int. Zoo Ybk.* **4**, 291–292.

Deming, O. V. (1955). Rearing bighorn lambs in captivity. *Calif. Fish Game* **41**, 131–143.

Desmeth, M., and Vandeputte-Poma, J. (1980). Lipid composition of pigeon cropmilk—I. Total lipids and lipid classes. *Comp. Biochem. Physiol.* **66B**, 129–133.

Dodds, D. F. (1959). Feeding and growth of a captive moose calf. *J. Wildl. Manage.* **23**, 231–232.

Dove, H., and Freer, M. (1979). The accuracy of tritiated water turnover rate as an estimate of milk intake in lambs. *Aust. J. Agric. Res.* **30**, 725–739.

Drent, R. (1970). Functional aspects of incubation in the herring gull. *Behaviour Suppl.* **17**, 1–132.

Drobney, R. D. (1980). Reproductive bioenergetics of wood ducks. *Auk* **97**, 480–490.

Dunn, G. L. (1974). Use of a domestic cat as foster mother for an ocelot (*Felis pardalis*). *Int. Zoo Ybk.* **14**, 218–219.

Eisenberg, J., and Muckenhirn, N. (1968). The reproduction and rearing of tenrecoid insectivores in captivity. *Int. Zoo Ybk.* **8**, 106–110.

Encke, W. (1960). Birth and rearing of cheetahs at Krefeld Zoo. *Int. Zoo Ybk.* **2**, 85–86.

Encke, W. (1962). Hand-rearing Cape hunting dogs (*Lycaon pictus*) at the Krefeld Zoo. *Int. Zoo Ybk.* **4**, 292–293.

Encke, W. (1970). Birth and hand-rearing of a bactrian camel (*Camelus bactrianus*) at Krefeld Zoo. *Int. Zoo Ybk.* **10**, 90.

Flint, M. (1975). Hand-rearing the small-spotted genet (*Genetta genetta*) at Randolph Park Zoo, Tucson. *Int. Zoo Ybk.* **15**, 244–245.

Fogden, M. P. L., and Fogden, P. M. (1979). The role of fat and protein reserves in the annual cycle of the grey-backed camaroptera in Uganda (Aves: Sylvidae). *J. Zool.* **189**, 233–258.

Fontaine, P. A. (1968). Birth of four species of apes at Dallas Zoo. *Int. Zoo Ybk.* **8**, 115–118.

Fonty, G., Gouet, Ph., and Riou, Y. (1979). Effect of milk composition on the gastrointestinal microflora of artificially reared young rabbits. *Ann. Biol. Anim. Bioch. Biophys.* **19**, 567–571.

Fransen, D. R., and Emerson, S. B. (1973). Further notes on hand raising bears (*Ursus arctos horribilis* and *U. a. middendorffi*) at Los Angeles Zoo. *Int. Zoo Ybk.* **13**, 143–145.

Franzmann, A., Arneson, P. D., and Ullrey, D. E. (1975). Composition of milk from Alaskan moose in relation to other North American wild ruminants. *J. Zoo. Anim. Med.* **6**, 12–14.

Freiheit, C. F., and Crotty, M. J. (1969). Hand-rearing Kodiak bears (*Ursus arctos middendorffi*) at Buffalo Zoo. *Int. Zoo Ybk.* **9**, 158–160.

Frueh, R. J. (1968). A captive-born gorilla (*Gorilla g. gorilla*) at St. Louis Zoo. *Int. Zoo Ybk.* **8**, 128–131.

Garry, R. C., and Stiven, D. (1936). A review of recent work on dietary requirements in pregnancy and lactation, with an attempt to assess human requirements. *Nutr. Abst. Rev.* **5**, 855–887.

Garton, G. A. (1963). The composition and biosynthesis of milk lipids. *J. Lipid Res.* **4**, 237–254.

Gee, G. F., and Temple, S. A. (1978). Artificial insemination for breeding non-domestic birds. *Symp. Zool. Soc. London* **43**, 51–72.

Geist, V. (1971). "Mountain Sheep: A Study in Behavior and Evolution." Univ. Chicago Press, Chicago.

Gensch, W. (1965). Birth and rearing of a black-faced spider monkey (*Ateles paniscus chamek* Humboldt) at Dresden Zoo. *Int. Zoo Ybk.* **5**, 110.

Gessaman, J. A., and Findell, P. R. (1979). Energy cost of incubation in the American kestrel. *Comp. Biochem. Physiol.* **63A**, 57–62.

Gilbert, B. K. (1974). The influence of foster rearing on adult social behavior in fallow deer (*Dama dama*). *In* "The Behaviour of Ungulates and Its Relation to Management," pp. 247–273. IUCN Publ. New Series 24, Vol. I.

Gray, B. J. (1970). Care and development of a hand-reared red panda (*Ailurus fulgens*). *Int. Zoo Ybk.* **10**, 139–142.

Green, B., and Newgrain, K. (1979). Estimation of milk intake of sucklings by means of ^{22}Na. *J. Mammal.* **60**, 556–559.

Green, B., Newgrain, K., and Merchant, J. (1980). Changes in milk composition during lactation in the tammar wallaby (*Macropus eugenii*). *Aust. J. Biol. Sci.* **33**, 35–42.

Gregory, M. E., Rowland, S. J., and Thompson, S. Y. (1965). Changes during lactation in the composition of the milk of the African black rhinoceros (*Diceros bicornis*). *Proc. Zool. Soc. London* **145**, 327–333.

Griffiths, M. (1965). Rate of growth and intake of milk in a suckling echidna. *Comp. Biochem. Physiol.* **16A**, 383–392.

Gucwinska, H. (1971). Development of six-banded armadillos (*Euphractus sexcinctus*) at Wroclaw Zoo. *Int. Zoo Ybk.* **11**, 88–89.

Hagenbeck, C. H. (1969). Notes on the artificial rearing of a Great Indian rhinoceros (*Rhinoceros unicornis*) at Hamburg Zoo. *Int. Zoo Ybk.* **9**, 99–101.

Halford, D. K., and Alldredge, A. W. (1978). A method for artificially raising mule deer fawns. *Am. Midl. Nat.* **100**, 493–498.

Hall, G. H., and Rosenblatt, J. S. (1977). Suckling behavior and intake control in the developing rat pup. *J. Comp. Physiol. Psychol.* **91**, 1232–1247.

Hall-Martin, A. J., Skinner, J. D., and Smith, A. (1977). Observations on lactation and milk composition of the giraffe *Giraffa camelopardalis*. *S. Afr. J. Wildl. Res.* **7**, 67–71.

Halliday, R. (1955). The absorption of antibodies from immune sera by the gut of the young rat. *Proc. Roy. Soc. London* **143B**, 408–413.

Hanson, L. A., and Johansson, B. G. (1970). Immunological studies of milk. *In* "Milk Proteins: Chemistry and Molecular Biology," Vol. I (H. A. McKenzie, ed.), pp. 45–123. Academic Press, New York.

Hanwell, A., and Peaker, M. (1977). Physiological effects of lactation on the mother. *Symp. Zool. Soc. London* **41**, 297–312.

Hardin, C. J. (1976). Hand-rearing a giant anteater (*Myrmecophaga tridactyla*) at Toledo Zoo. *Int. Zoo Ybk.* **16**, 199–200.

Havera, S. P. (1978). Nutrition, supplemental feeding, and body composition of the fox squirrel, *Sciurus niger,* in central Illinois. Doctoral dissertation, Univ. of Illinois.

Hess, J. K. (1971). Hand-rearing polar bear cubs (*Thalarctos maritimus*) at St. Paul Zoo. *Int. Zoo Ybk.* **11**, 102–107.

Hick, U. (1969). Successful raising of a pudu at Cologne Zoo. *Int. Zoo Ybk.* **9**, 110–112.

Hick, U. (1976). Hand-rearing a ring-tailed lemur (*Lemur catta*) and a crowned lemur (*Lemur mongoz coronatus*) at Cologne Zoo. *Int. Zoo Ybk.* **16**, 187–189.

Hills, D. M., and MacDonald, I. (1956). Hand-rearing of rabbits. *Nature* **178**, 704–705.

Hirsch, K. V., and Grau, C. R. (1981). Yolk formation and oviposition in captive emus. *Condor* **83**, 381–382.

Hobbs, N. T., and Baker, D. L. (1979). Rearing and training elk calves for use in food habits studies. *J. Wildl. Manage.* **43**, 568–570.

Hoff, W. (1960). Hand raising baby cats at Lincoln Park Zoo, Chicago. *Int. Zoo Ybk.* **2**, 86–89.

Holleman, D. F., White, R. G., and Luick, J. R. (1975). New isotope methods for estimating milk intake and yield. *J. Dairy Sci.* **58**, 1814–1821.

Hopkins, J. M. (1966). Observations on the rearing and behavior of young duikers in captivity. *Nigerian Field* **31**, 118–131.

Hoyt, D. F., Vleck, D., and Vleck, C. M. (1978). Metabolism of avian embryos: Ontogeny and temperature effects in the ostrich. *Condor* **80**, 265–271.

Huckabee, W. E., Metcalfe, J., Prystowsky, H., and Barron, D. H. (1961). Blood flow and oxygen consumption of the pregnant uterus. *Am. J. Physiol.* **200**, 274–278.

Hughes, F. (1977). Hand-rearing a Sumatran tiger (*Panthera tigris sumatrae*) at Whipsnade Park. *Int. Zoo Ybk.* **17**, 219–221.

Hulley, J. T. (1976). Hand-rearing American black bear cubs (*Ursus americanus*) at Toronto Zoo. *Int. Zoo Ybk.* **16**, 202–205.

Hunt, H. (1967). Growth rate of a new-born, hand-reared jaguar (*Panthera onca*) at Topeka Zoo. *Int. Zoo Ybk.* **7**, 147–148.

Hutchison, M. (1970). Artificial rearing of some East African antelopes. *J. Zool.* **161**, 437–442.

Jakobsen, P. E. (1957). Proteinbehov og proteinsyntese ved fosterdannelse hos drøvtyggere. 299 Beretning fra Forsøgslaboratoriet, Copenhagen.

Jantschke, F. (1973). On the breeding and rearing of bush dogs (*Speothos venaticus*) at Frankfurt Zoo. *Int. Zoo Ybk.* **13**, 141–143.

Jenness, R. (1970). Protein composition of milk. *In* "Protein Composition of Milk" (H. A. McKenzie, ed.), pp. 17–43. Academic Press, New York.

Jenness, R. (1974). The composition of milk. *In* "Lactation," Vol. III (B. L. Larson and V. R. Smith, eds.), pp. 3–107. Academic Press, New York.

Jenness, R. (1979). Comparative aspects of milk proteins. *J. Dairy Res.* **46**, 197–210.

Jenness, R., and Odell, D. K. (1978). Composition of milk in the pygmy sperm whale (*Kogia breviceps*). *Comp. Biochem. Physiol.* **61A**, 383–386.

Jenness, R., and Sloan, R. E. (1970). The composition of milks of various species: A review. *Dairy Sci. Abstr.* **32**, 599–612.

Johnstone, P. (1977). Hand-rearing serval (*Felis serval*) at Mole Hall Wildlife Park. *Int. Zoo Ybk.* **17**, 218–219.

Johnstone-Scott, R. (1975). Hand-rearing a honey badger (*Mellivora capensis*) at Howletts Zoo Park, Bekesbourne. *Int. Zoo Ybk.* **15**, 241–244.

Jones, P. J. (1979). Variability of egg size and composition in the great white pelican (*Pelecanus onocrotalus*). *Auk* **96**, 407–408.

Jones, P. J., and Ward, P. (1976). The level of reserve protein as the proximate factor controlling the timing of breeding and clutch-size in the red-billed quelea, *Quelea quelea*. *Ibis* **118**, 547–574.

Jones, P. J., and Ward, P. (1979). A physiological basis for colony desertion by red-billed queleas (*Quelea quelea*). *J. Zool.* **189**, 1–19.

Kaufman, D. W., and Kaufman, G. A. (1977). Body composition of the old-field mouse (*Peromyscus polionotus*). *J. Mammal.* **58**, 429–434.

Kerry, K. R. (1969). Intestinal disaccharidase activity in a monotreme and eight species of marsupials (with an added note on the disaccharidases of five species of sea birds). *Comp. Biochem. Physiol.* **29**, 1015–1022.

Kihlstrom, J. E. (1972). Period of gestation and body weight in some placental mammals. *Comp. Biochem. Physiol.* **43A**, 673–679.

King, J. R. (1973). Energetics of reproduction in birds. *In* "Breeding Biology of Birds" (D. S. Farner, ed.), pp. 78–107. Nat. Acad. Sci., Washington, D.C.

Kirchshofer, R., Fradrich, H., Podolczak, D., and Podolczak, G. (1967). An account of the physical and behavioural development of the hand-reared gorilla infant (*Gorilla g. gorilla*) born at Frankfurt Zoo. *Int. Zoo Ybk.* **7**, 108–113.

Kirchshofer, R., Weisse, K., Berenz, K., Klose, H., and Klose, I. (1968). A preliminary account of the physical and behavioural development during the first 10 weeks of the hand-reared gorilla twins (*Gorilla g. gorilla*) born at Frankfurt Zoo. *Int. Zoo Ybk.* **8**, 121–128.

Kitchener, H. J. (1960). Elephant calves-their feeding and care. *Malay Nat. J.* **14**, 215–220.

Kitchener, S. L. (1971). Observations on the breeding of the bush dog (*Speothos venaticus*) at Lincoln Park Zoo, Chicago. *Int. Zoo Ybk.* **11**, 99–101.

Klopfer, P. H., and Klopfer, M. S. (1962). Notes on hand-rearing fallow deer (*Dama dama*). *Int. Zoo Ybk.* **4**, 295–296.

Kramer, T. T., Nagy, J. G., and Barber, T. A. (1971). Diarrhea in captive mule deer fawns attributed to *Escherichia coli*. *J. Wildl. Manage.* **35**, 205–209.

Krapu, G. L. (1974). Foods of breeding pintails in North Dakota. *J. Wildl. Manage.* **38**, 408–417.

Kretchmer, N. (1972). Lactose and lactase. *Sci. Am.* **227**, 70–78.

Krzywinski, A., Krzywinska, K., Kisza, J., Roskosz, A., and Kruk, A. (1980). Milk composition, lactation and the artificial rearing of red deer. *Acta Theriol.* **25**, 341–347.

Lack, D. (1968). "Ecological Adaptations for Breeding Birds." Methuen, London.

Lake, P. E. (1978). The principles and practice of semen collection and preservation in birds. *Symp. Zool. Soc. London* **43**, 31–49.

Lang, E. M. (1963). Flamingoes raise their young on a liquid containing blood. *Experientia* **19**, 532–533.

Lawrence, J. M., and Schreiber, R. W. (1974). Organic material and calories in the egg of the brown pelican, *Pelecanus occidentalis*. *Comp. Biochem. Physiol.* **47A**, 435–440.

Lecce, J. G., and Morgan, D. O. (1962). Effects of dietary regimen on cessation of intestinal absorption of large molecules (closure) in the neonatal pig and lamb. *J. Nutr.* **78**, 263–268.

Leitch, I., Hytten, F. E., and Billewicz, W. Z. (1959). The maternal and neonatal weights of some mammalia. *Proc. Zool. Soc. London* **133**, 11–28.

Leutenegger, W. (1973). Maternal–fetal weight relationships in primates. *Folia Primat.* **20**, 280–293.

Leutenegger, W. (1976). Allometry of neonatal size in eutherian mammals. *Nature* **263**, 229–230.

Linzell, J. L. (1972). Milk yield, energy loss in milk, and mammary gland weight in different species. *Dairy Sci. Abstr.* **34**, 351–360.

Livers, T. H. (1973). The use of milk replacers for hand-rearing carnivores. *Int. Zoo Ybk.* **13**, 211–213.

Lotshaw, R. (1971). Births of two lowland gorillas (*Gorilla g. gorilla*) at Cincinnati Zoo. *Int. Zoo Ybk.* **11**, 84–87.

Luick, J. R., White, R. G., Gau, A. M., and Jenness, R. (1974). Compositional changes in the milk secreted by grazing reindeer. I. Gross composition and ash. *J. Dairy Sci.* **57**, 1325–1333.

McCoy, G. C., Reneau, J. K., Hunter, A. G., and Williams, J. B. (1970). Effects of diet and time on blood serum proteins in the newborn calf. *J. Dairy Sci.* **53**, 358–362.

McCullagh, K. G., and Widdowson, E. M. (1970). The milk of the African elephant. *Brit. J. Nutr.* **24**, 109–117.

McCullagh, K. G., Lincoln, H. B., and Southgate, D. A. T. (1969). Fatty acid composition of milk fat of the African elephant. *Nature* **222**, 493–494.

McEwan, E. H., and Whitehead, P. E. (1971). Measurement of the milk intake of reindeer and caribou calves using tritiated water. *Can. J. Zool.* **49**, 443–447.

Macfarlane, W. V., Howard, B., and Siebert, B. D. (1969). Tritiated water in the measurement of milk intake and tissue growth of ruminants in the field. *Nature* **221**, 578–579.

Madison, D. M. (1978). Behavioral and sociochemical susceptibility of meadow voles (*Microtus pennsylvanicus*) to snake predation. *Am. Midl. Nat.* **100**, 23–28.

Mallinson, J. J. C. (1975). Notes on a breeding group of Sierra Leone striped squirrels (*Funisciurus pyrrhopus leonis*) at the Jersey Zoo. *Int. Zoo Ybk.* **15**, 237–240.

Mallinson, J. J. C., Coffey, P., and Usher-Smith, J. (1976). Breeding and hand-rearing lowland gorillas (*Gorilla g. gorilla*) at the Jersey Zoo. *Int. Zoo Ybk.* **16**, 189–194.

Markgren, G. (1966). A study of hand-reared moose calves. *Viltrevy* **4**, 1–42.

Martin, R. D. (1966). Tree shrews: Unique reproductive mechanism of systematic importance. *Science* **152**, 1402–1404.

Menzies, J. I. (1972). Notes on a hand-reared spotted cuscus (*Phalanger maculatus*). *Int. Zoo Ybk.* **12,** 97–98.

Mertens, J. A. L. (1977). The energy requirements for incubation in great tits, *Parus major* L. *Ardea* **65,** 184–196.

Messer, M., and Green, B. (1979). Milk carbohydrates of marsupials. II. Quantitative and qualitative changes in milk carbohydrates during lactation in the tammar wallaby (*Macropus eugenii*). *Aust. J. Biol. Sci.* **32,** 519–531.

Messer, M., and Mossop, G. S. (1977). Milk carbohydrates of marsupials. I. Partial separation and characterization of neutral milk oligosaccharides of the eastern grey kangaroo. *Aust. J. Biol. Sci.* **30,** 379–388.

Michalowski, D. R. (1971). Hand-rearing a polar bear cub (*Thalarctos maritimus*) at Rochester Zoo. *Int. Zoo Ybk.* **11,** 107–109.

Michalowski, D. R. (1972). Hand-rearing a polar bear at Rochester Zoo. *Int. Zoo Ybk.* **12,** 175–176.

Miles, P. (1967). Notes on the rearing and development of a hand-reared spider monkey. *Int. Zoo Ybk.* **7,** 82–85.

Millar, J. S. (1977). Adaptive features of mammalian reproduction. *Evolution* **31,** 370–386.

Moen, A. N. (1973). "Wildlife Ecology: An Analytical Approach." Freeman, San Francisco.

Moll, W., Kunzel, W., and Ross, H. G. (1970). Gas exchange of the pregnant uterus of anaesthetized and unanesthetized guinea pigs. *Respir. Physiol.* **8,** 303–310.

Moore, G. T., and Cummins, L. B. (1979). Nursery rearing of infant baboons. *In* "Nursery Care of Non-human Primates" (G. C. Ruppenthal, ed.), pp. 145–151. Plenum, New York.

Mueller, C. C., and Sadleir, R. M. F. S. (1977). Changes in the nutrient composition of milk of black-tailed deer during lactation. *J. Mammal.* **58,** 421–423.

Muller-Schwarze, D., and Muller-Schwarze, C. (1973). Behavioural development of hand-reared pronghorn (*Antilocapra americana*). *Int. Zoo Ybk.* **13,** 217–220.

Murphy, D. A. (1960). Rearing and breeding white-tailed fawns in captivity. *J. Wildl. Manage.* **24,** 439–441.

Nakazato, R., Nakayama, T., and Nakagawa, S. (1971). Hand-rearing agile wallabies (*Protemnodon agilis*) at Ueno Zoo, Tokyo. *Int. Zoo Ybk.* **11,** 13–16.

Oftedal, O. T. (1981). Milk, protein and energy intakes of suckling mammalian young: A comparative study. Doctoral dissertation, Cornell University, Ithaca, New York.

Ogden, J. O. (1979). Hand rearing *Saguinus* and *Callithrix* genera of marmosets. *In* "Nursery Care of Non-human Primates" (G. C. Ruppenthal, ed.), pp. 313–319. Plenum, New York.

Orbell, E., and Orbell, J. (1976). Hand-rearing a Saiga antelope at the Highland Wildlife Park. *Int. Zoo Ybk.* **16,** 208–209.

Pattee, O. H., and Beasom, S. L. (1979). Supplemental feeding to increase wild turkey productivity. *J. Wildl. Manage.* **43,** 512–516.

Payne, P. R., and Wheeler, E. F. (1967). Comparative nutrition in pregnancy. *Nature* **215,** 1134–1136.

Payne, P. R., and Wheeler, E. F. (1968). Comparative nutrition in pregnancy and lactation. *Proc. Nutr. Soc.* **27,** 129–138.

Pinter, H. (1962). Artificial rearing of roe deer (*Capreolus capreolus*) and observations on their behaviour. *Int. Zoo Ybk.* **4,** 297–300.

Pond, C. M. (1977). The significance of lactation in the evolution of mammals. *Evolution* **31,** 177–199.

Prévost, J., and Vilter, V. (1963). Histologie de la sécrétion oesophagienne du manchot empereur. *Proc. Int. Ornithol. Congr.* **13,** 1085–1094.

Quick, R. (1969). Hand-rearing Kodiak bear (*Ursus arctos middendorffi*) at Houston Zoo. *Int. Zoo Ybk.* **9,** 160–163.

Rahn, H., Paganelli, C. V., and Ar, A. (1975). Relation of avian egg weight to body weight. *Auk* **92**, 750–765.

Randolph, P. A., Randolph, J. C., Mattingly, K., and Foster, M. M. (1977). Energy costs of reproduction in the cotton rat, *Sigmodon hispidus*. *Ecology* **58**, 31–45.

Raveling, D. G. (1979). The annual cycle of body composition of Canada geese with special reference to control of reproduction. *Auk* **96**, 234–252.

Reid, B. (1971). The weight of the kiwi and its egg. *Notornis* **18**, 245–249.

Reidman, M., and Ortiz, C. L. (1979). Changes in milk composition during lactation on the Northern elephant seal. *Physiol. Zool.* **52**, 240–249.

Reineck, M. (1962). The rearing of abandoned sucklings of *Phoca vitulina*. *Int. Zoo Ybk.* **4**, 293–294.

Reuther, R. T. (1969). Growth and diet of young elephants in captivity. *Int. Zoo Ybk.* **9**, 168–178.

Ricklefs, R. E. (1974). Energetics of reproduction in birds. *In* "Avian Energetics" (R. A. Paynter, Jr, ed.), pp. 152–297. Publ. Nuttall Ornith. Club No. 15, Cambridge, Mass.

Ricklefs, R. E. (1977). Composition of eggs of several bird species. *Auk* **94**, 350–356.

Ricklefs, R. E. (1980). Geographical variation in clutch size among passerine birds: Ashmole's hypothesis. *Auk* **97**, 38–49.

Robbins, C. T. (1973). The biological basis for the determination of carrying capacity. Doctoral dissertation, Cornell Univ., Ithaca, New York.

Robbins, C. T. (1981). Estimation of the relative protein cost of reproduction in birds. *Condor* **83**, 177–179.

Robbins, C. T., and Moen, A. N. (1975a). Uterine composition and growth in pregnant white-tailed deer. *J. Wildl. Manage.* **39**, 684–691.

Robbins, C. T., and Moen, A. N. (1975b). Milk consumption and weight gain of white-tailed deer. *J. Wildl. Manage.* **39**, 355–360.

Robbins, C. T., and Robbins, B. L. (1979). Fetal and neonatal growth patterns and maternal reproductive effort in ungulates and subungulates. *Am. Nat.* **114**, 101–116.

Robbins, C. T., and Stevens, V. (Unpublished). Mountain goat milk composition. Washington St. Univ., Pullman.

Robbins, C. T., Podbielancik-Norman, R. S., Wilson, D. L., and Mould, E. D. (1981). Growth and nutrient consumption of elk calves compared to other ungulate species. *J. Wildl. Manage.* **45**, 172–186.

Robinson, J. J., McDonald, I., Fraser, C., and Gordon, J. G. (1980). Studies on reproduction in prolific ewes. 6. The efficiency of energy utilization for conceptus growth. *J. Agric. Sci.* **94**, 331–338.

Rohwer, F. C. (1980). Blue-winged teal egg composition in relation to laying sequence, date of laying, and egg size. Master's thesis, Washington St. Univ., Pullman.

Romanoff, A. L., and Romanoff, A. J. (1949). "The Avian Egg." Wiley, New York.

Rosenthal, M. A. (1974). Hand-rearing Patagonian cavies or maras (*Dolichotis patagonum*) at Lincoln Park, Chicago. *Int. Zoo Ybk.* **14**, 214–215.

Rosenthal, M. A., and Meritt, D. A., Jr. (1971). Hand-rearing Grant's gazelle (*Gazella granti*) at Lincoln Park Zoo, Chicago. *Int. Zoo Ybk.* **11**, 130.

Rosenthal, M. A., and Meritt, D. A., Jr. (1973). Hand-rearing springhaas (*Pedetes capensis*) at Lincoln Park Zoo, Chicago. *Int. Zoo Ybk.* **13**, 135–137.

Roudybush, T. E., Grau, C. R., Petersen, M. R., Ainley, D. G., Kirsch, K. V., Gilman, A. P., and Patten, S. M. (1979). Yolk formation in some charadriiform birds. *Condor* **81**, 293–298.

Ruppenthal, G. C. (1979). Survey of protocols for nursery-rearing infant macaques. *In* "Nursery Care of Non-human Primates" (G. C. Ruppenthal, ed.), pp. 165–185. Plenum, New York.

Sadleir, R. M. F. S. (1980a). Milk yield of black-tailed deer. *J. Wildl. Manage.* **44**, 472–478.

Sadleir, R. M. F. S. (1980b). Energy and protein intake in relation to growth of suckling black-tailed deer fawns. *Can. J. Zool.* **58,** 1347–1354.

Sampsell, R. N. (1969). Hand rearing an aardvark (*Orycteropus afer*) at Crandon Park Zoo, Miami. *Int. Zoo Ybk.* **9,** 97–99.

Sasaki, T. (1962). Hand-rearing a baby gibbon (*Hylobates lar*). *Int. Zoo Ybk.* **4,** 289–290.

Sawicka-Kapusta, K. (1970). Changes in the gross body composition and the caloric value of the common vole during their postnatal development. *Acta Theriol.* **15,** 67–79.

Sawicka-Kapusta, K. (1974). Changes in the gross body composition and energy value of the bank vole during their postnatal development. *Acta Theriol.* **19,** 27–54.

Silver, H. (1961). Deer milk compared to substitute milk for fawns. *J. Wildl. Manage.* **25,** 66–70.

Stenseth, N., Framstad, E., Migula, P., Trojan, P., and Wojciechowska-Trojan, B. (1980). Energy models for the common vole *Microtus arvalis:* Energy as a limiting resource for reproductive output. *Oikos* **34,** 1–22.

Stephens, T. (1975). Nutrition of orphan marsupials. *Aust. Vet. J.* **51,** 453–458.

Stewart, R. E. A., and Lavigne, D. M. (1980). Neonatal growth of Northwest Atlantic harp seals, *Pagophilus groenlandicus. J. Mammal.* **61,** 670–680.

Stroman, H. R., and Slaughter, L. M. (1972). The care and breeding of the pygmy hippopotamus (*Choeropsis liberiensis*) in captivity. *Int. Zoo Ybk.* **12,** 126–131.

Swanson, G. A., Meyer, M. I., and Serie, J. R. (1974). Feeding ecology of breeding blue-winged teals. *J. Wildl. Manage.* **38,** 396–407.

Tener, J. S. (1956). Gross composition of musk-ox milk. *Can. J. Zool.* **34,** 569–571.

Thorne, E. T., Dean, R. E., and Hepworth, W. G. (1976). Nutrition during gestation in relation to successful reproduction in elk. *J. Wildl. Manage.* **40,** 330–335.

Tobias, J. (1981). Hand-raising a hippopotamus at the Denver Colorado Zoo. *Anim. Keepers' Forum* **8,** 87–88.

Toweill, D. E., and Toweill, D. B. (1978). Growth and development of captive ringtails (*Bassariscus astutus flavus*). *Carnivore* **1,** 46–53.

Treus, D. B., and Kravchenko, D. (1968). Methods of rearing and economic utilization of eland in Askanya-Nova Zoological Park. *Symp. Zool. Soc. London* **21,** 395–411.

Tryon, C. A. (1947). Behavior and post-natal development of a porcupine. *J. Wildl. Manage.* **11,** 282–283.

Vainisi, S. J., Edelhauser, H. F., Wolf, E. D., Cotlier, E., and Reeser, F. (1980). Nutritional cataracts in timber wolves. *In* "Proceedings of the First Annual Dr. Scholl Nutrition Conference" (E. R. Maschgan, M. E. Allen, and L. E. Fisher, eds.), pp. 3–21. Lincoln Park Zoo, Chicago.

Vandeputte-Poma, J. (1980). Feeding, growth and metabolism of the pigeon, *Columba livia domestica:* Duration and role of crop milk feeding. *J. Comp. Physiol.* **135,** 97–99.

Vanforeest, A. W., Greenwood, A. G., and Taylor, D. C. (1978). Hand-rearing an infant Nile hipopotamus (*Hippopotamus amphibius*). *Am. Assoc. Zoo Vet. Ann. Proc,* 170.

Van Zyl, J. H. M., and Wehmeyer, A. S. (1970). The composition of the milk of springbok (*Antidorcas marsupialis*), eland (*Taurotragus oryx*) and black wildebeest (*Connochaetus gnou*). *Zool. Afr.* **5,** 131–133.

Verme, L. J. (1967). Influence of experimental diets on white-tailed deer reproduction. *Trans. North Am. Wildl. Nat. Resour. Conf.* **32,** 405–420.

Vice, T. E., and Olin, F. H. (1969). A note on the milk analysis and hand-rearing of the greater kudu. *Int. Zoo Ybk.* **9,** 114.

Vleck, C. E. (1977). Energy cost of incubation in the zebra finch, *Poephila guttata. Am. Zool.* **17,** 930.

Vleck, C. M. (1981). Energetic cost of incubation in the zebra finch. *Condor* **83,** 229–237.

Vleck, C. M., Hoyt, D. F., and Vleck, D. (1979). Metabolism of avian embryos: patterns in altricial and precocial birds. *Physiol. Zool.* **52,** 363–377.

Vleck, C. M., Vleck, D., and Hoyt, D. F. (1980). Patterns of metabolism and growth in avian embryos. *Am. Zool.* **20,** 405–416.

Vogt, P., Schneidermann, Chr., and Schneidermann, B. (1980). Hand-rearing a red panda (*Ailurus fulgens*) at Krefeld Zoo. *Int. Zoo Ybk.* **20,** 280–281.

Walcott, P. J., and Messer, M. (1980). Intestinal lactase (*B*-galactosidase) and other glycosidase activities in suckling and adult tammar wallabies (*Macropus eugenii*). *Aust. J. Biol. Sci.* **33,** 521–530.

Wallach, J. D. (1969). Hand-rearing and observations of a white rhinoceros. (*Diceros s. simus*). *Int. Zoo Ybk.* **9,** 103–104.

Walsberg, G. E., and King, J. R. (1978). The energetic consequences of incubation for two passerine species. *Auk* **95,** 644–655.

Wayre, P. (1967). Artificial rearing of roe deer (*Capreolus capreolus*) and fallow deer (*Dama dama*) at Norfolk Wildlife Park. *Int. Zoo Ybk.* **7,** 168–171.

Weber, E. (1969). Notes on hand-rearing a Malayan sun bear (*Helarctos malayanus*) at Melbourne Zoo. *Int. Zoo Ybk.* **9,** 163.

Weber, E. (1970). Hand-rearing a hippopotamus (*Hippopotamus amphibius*) at Melbourne Zoo. *Int. Zoo Ybk.* **10,** 153.

Weiher, E. (1976). Hand-rearing fennec foxes (*Fennecus zerda*) at Melbourne Zoo. *Int. Zoo Ybk.* **16,** 200–202.

Wemmer, C. (1971). Birth, development and behaviour of a fanaloka (*Fossa fossa*) at the National Zoological Park, Washington, D.C. *Int. Zoo Ybk.* **11,** 113–115.

White, J. M. (1975). Milk yield in lines of mice selected for growth or maternal ability. *Can J. Genet. Cytol.* **17,** 263–268.

White, R. G., Thomson, B. R., Skogland, T., Person, S. J., Russell, D. E., Holleman, D. F., and Luick, J. R. (1975). Ecology of caribou at Prudhoe Bay, Alaska. *Univ. Alaska Biol. Pap. Spec. Rep.* **2.**

Widdowson, E. M. (1950). Chemical composition of newly born mammals. *Nature* **166,** 626–628.

Widdowson, E. M., and McCance, R. A. (1955). Physiological undernutrition in the newborn guinea-pig. *Brit. J. Nutr.* **9,** 316–321.

Wilson, P. (1971). Hand-rearing a Dama wallaby (*Protemnodon eugenii*) at Auckland Zoo. *Int. Zoo Ybk.* **11,** 20.

Wortman, J. D., and Larue, M. D. (1974). Hand-rearing polar bear cubs (*Thalarctos maritimus*) at Topeka Zoo. *Int. Zoo Ybk.* **14,** 215–218.

Wypkema, R. C. P., and Ankney, C. D. (1979). Nutrient reserve dynamics of lesser snow geese staging at James Bay, Ontario. *Can. J. Zool.* **57,** 213–219.

Yadav, M. (1971). The transmissions of antibodies across the gut of pouch-young marsupials. *Immunology* **21,** 839–851.

Yamamoto, S. (1962). Hand-rearing of Japanese squirrels (*Sciurus vulgaris lis*). *Int. Zoo Ybk.* **4,** 290–291.

Yamamoto, S. (1967). Notes on hand-rearing chimpanzee twins (*Pan troglodytes*) at Kobe Zoo. *Int. Zoo Ybk.* **7,** 97–98.

Yazan, Y., and Knorre, Y. (1964). Domesticating elk in a Russian National Park. *Oryx* **7,** 301–304.

Young, W. A. (1961). Rearing an American tapir (*Tapirus terrestris*). *Int. Zoo Ybk.* **3,** 94–95.

Youngson, R. W. (1970). Rearing red deer calves in captivity. *J. Wildl. Manage.* **34,** 467–470.

Zellmer, G. (1960). Hand-rearing of giraffe at Bristol Zoo. *Int. Zoo Ybk.* **2,** 90–93.

Zeveloff, S. I., and Boyce, M. S. (1980). Parental investment and mating systems in mammals. *Evolution* **34,** 973–982.

10

Productive Costs

During the last half century there has been a striking change in the study of wild animals and their relation to the environment. Descriptive natural history, however interesting, is increasingly being replaced by quantitative information on animal function. This change to a quantitative approach . . . is essential to a proper understanding . . . of animals in the wild. . . .

—Schmidt-Nielsen, 1977, p. 345

I. BODY GROWTH

Growth is the process whereby an animal incorporates into its molecular structure a portion of the external chemical environment. Growth in all animals, particularly resident wildlife occupying very seasonal environments, has both positive and negative phases. Nutrient requirements for growth are dependent on the rate and composition of the gain. Growth rates per se are not adequate to evaluate the requirement, since the composition of the gain or loss can vary dramatically. Consequently, while many field ecologists have monitored body weights, the nutritionist must also understand body composition.

A. Body Composition

The major components of the ingesta-free animal body are fat, water, protein, and ash or minerals. Their concentrations are usually determined by killing and mincing an animal to provide a homogeneous mass for sampling. Since it may not always be feasible or desirable to kill the animal, the dilution of injected water isotopes by the body water pool in the living animal can also be used to determine body composition. The water concentration determined by dilution procedures usually overestimates the gravimetrically determined water content by up to 13%, although errors of less than 5% are common (Donnelly and Freer, 1974; Holleman and Dieterich, 1975; Bakker and Main, 1980). The other constituents can be estimated from the water concentration.

Although carbohydrates are of immense importance in animal metabolism, their total concentration within the animal is less than 1%, even though there is significant variation with time of day, animal age, and latitude (Galster and Morrison, 1975; Thomas *et al.*, 1975; Okon and Ekanem, 1976; Bintz *et al.*, 1979; Freminet, 1981). The two major animal carbohydrates are glucose and glycogen. Their concentrations range from 0.08 to 0.18% in sparrows (Farner *et al.*, 1961; Dolnik, 1970) and from 0.06 to 0.56% in ground squirrels (Bintz *et al.*, 1979).

Body lipids function as an energy reserve, as structural elements in cell and organelle membranes, and as sterol hormones (Schemmel, 1976). Since lipids can be stored as relatively nonhydrated adipose tissue containing 2–15% free water (King, 1961; Odum *et al.*, 1965), eight times more calories per unit of weight can be stored as fat than as hydrated carbohydrates (Allen, 1976). Thus, whereas carbohydrates are a major energy reserve for plants, fat storage is far more useful to active animals.

Lipid energy reserves are primarily triglycerides. Triglycerides are synthesized from glycerol and three fatty acids (Fig. 10.1; Table 10.1). The energy content of fatty acids increases with molecular weight and degree of saturation. Saturation refers to the absence of internal double bonds and, correspondingly, increased hydrogenation. Thus, stearic acid (18 carbons:0 double bonds) has a higher energy content (9.53 kcal/g) than either oleic acid (18:1, 9.50 kcal/g) or palmitic acid (16:0, 9.35 kcal/g).

The physical and chemical characteristics of triglycerides are determined by the constituent fatty acids. The major fatty acids of body triglycerides are palmitic, stearic, oleic, linoleic, and linolenic acids. However, the fatty acid spectra of body triglycerides can be influenced by site of sample, diet, type of gastrointestinal digestion, environmental temperature, photoperiod, and animal species (Crawford, 1968; Gale *et al.*, 1969; Schultz and Ferguson, 1974; Blem, 1976; Zar, 1977; Cmelik and Ley, 1977; Ijas *et al.*, 1978; Yom-Tov and Tietz, 1978; West *et al.*, 1979). For example, the fatty acid content of appendage triglycerides in cold-adapted animals increases in saturation from distal to proximal segments (Meng *et al.*, 1969; West and Shaw, 1975; Zar, 1977; Turner, 1979). Since the melting point and therefore the softness of a fatty acid at a given temperature is dependent on its molecular weight and degree of saturation (melt-

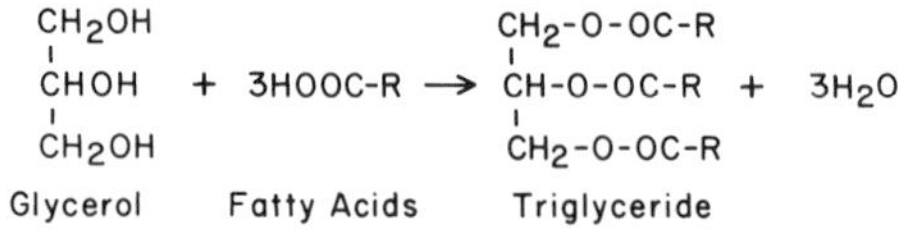

$$
\begin{array}{ccc}
\mathrm{CH_2OH} & & \mathrm{CH_2\text{-}O\text{-}OC\text{-}R} \\
| & & | \\
\mathrm{CHOH} \;\; + \;\; \mathrm{3HOOC\text{-}R} \;\rightarrow\; \mathrm{CH\text{-}O\text{-}OC\text{-}R} \;\; + \;\; \mathrm{3H_2O} \\
| & & | \\
\mathrm{CH_2OH} & & \mathrm{CH_2\text{-}O\text{-}OC\text{-}R} \\
\text{Glycerol} \quad \text{Fatty Acids} & & \text{Triglyceride}
\end{array}
$$

Fig. 10.1. Simplistic construction of a triglyceride. (*R* refers to the carbon chain of the fatty acid.)

TABLE 10.1
Major Naturally Occurring Fatty Acids

Molecular formula		Common name	Systematic name
Saturated fatty acids			
$C_2H_4O_2$		Acetic	
$C_3H_6O_2$		Propionic	
$C_4H_8O_2$		Butyric	Butanoic
$C_6H_{12}O_2$		Caproic	Hexanoic
$C_8H_{16}O_2$		Caprylic	Octanoic
$C_9H_{18}O_2$		Pelargonic	Nonanoic
$C_{10}H_{20}O_2$		Capric	Decanoic
$C_{12}H_{24}O_2$		Lauric	Dodecanoic
$C_{14}H_{28}O_2$		Myristic	Tetradecanoic
$C_{16}H_{32}O_2$		Palmitic	Hexadecanoic
$C_{18}H_{36}O_2$		Stearic	Octadecanoic
$C_{20}H_{40}O_2$		Arachidic	Eicosanoic
$C_{22}H_{44}O_2$		Behenic	Docosanoic
$C_{24}H_{48}O_2$		Lignoceric	Tetracosanoic
Unsaturated fatty acids			
$C_{16}H_{30}O_2$	C16:1	Palmitoleic	9-Hexadecenoic
$C_{18}H_{34}O_2$	C18:1	Oleic	Cis-9-Octadecenoic
$C_{18}H_{34}O_2$	C18:1	Elaidic	Trans-9-Octadecenoic
$C_{18}H_{34}O_2$	C18:1	Vaccenic	11-Octadecenoic
$C_{18}H_{32}O_2$	C18:2	Linoleic	Octadecadienoic
$C_{18}H_{30}O_2$	C18:3	Linolenic	Octadecatrienoic
$C_{20}H_{32}O_2$	C20:4	Arachidonic	Eicosatetraenoic

ing points ranging from 69.9°C for stearic acid [18:0] to −49.0°C for arachidonic acid [20:4]), increasing unsaturation of the fatty acids in heterothermic appendages is essential for the fats to remain soft and metabolizable.

Water and fat are inversely proportional to each other in their concentrations in the adult animal (Fig. 10.2). The inverse proportionality between fat and water is not true of very young animals. Young animals have a water content in the fat-free tissues that is disproportionately higher than that of adults, which subsequently decreases during growth to an asymptotic concentration termed chemical maturity (Moulton, 1923). The water content of the fat-free tissues of young animals is higher in altricial than in precocial species (Ricklefs and White, 1981). For example, the fat-free body of neonatal mice, voles, and starlings contains 87.08±1.82% water as compared to 76.97±3.06% for deer, scaup, geese, and quail (Sawicka-Kapusta, 1970, 1974; Sugden and Harris, 1972; Robbins et al., 1974; Robbins and Moen, 1975; Ricklefs, 1979; Campbell and Leatherland, 1980). These differences in water content can undoubtedly be related to the contrasting functional maturity at hatching or birth of the muscles in

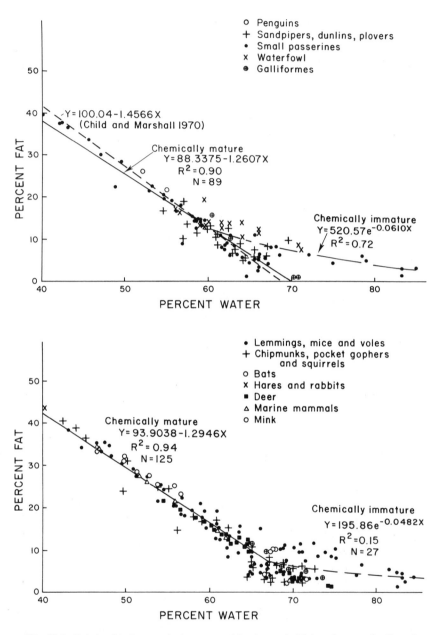

Fig. 10.2. Relationship between body water and fat content in birds and mammals. (Data from Odum, 1960; Johnston, 1964; Gifford and Odum, 1965; Hayward, 1965; Kale, 1965; Yarbrough and Johnston, 1965; Zimmerman, 1965; Helms *et al.,* 1967; Johnston and MacFarlane, 1967; Baker *et al.,* 1968; Brisbin, 1968; Myrcha, 1968; Sawicka-Kapusta, 1968, 1970, 1974; Helms and Smythe,

altricial and precocial neonates. The fat-free body of chemically mature mammals and birds averages 72.5 and 68.7–70.1% water, respectively (Fig. 10.2). The difference in the estimates for birds is largely due to the inclusion of the point at 49% water–22% fat for a series of migrating passerines collected along the Egyptian coast (Moreau and Dolp, 1970). If these birds were temporarily dehydrated, the regression of Child and Marshall (1970) would more appropriately represent the average bird. The inverse relationship between fat and water in the chemically mature animal is largely due to the accumulation of fat in adipose cells by hypertrophy rather than by hyperplasia and the corresponding weight dilution of the other components of the fat-free body (Odum et al., 1964; Young, 1976).

The remaining dry, fat-free body is largely protein and ash. The dry, fat-free, mature mammal contains 82.89±3.29% protein and 17.11% ash ($N = 9$), and the bird 87.98±3.22% protein and 12.02% ash ($N = 5$). Variation does occur as indicated by the standard deviations, with protein concentrations ranging from 75.10% in arctic ground squirrels to 87.29% in mink and from 85.79% in long-billed marsh wrens to 93.66% in rock-hopper penguins (Kale, 1965; Myrcha, 1968; Myrcha and Pinowski, 1970; Sawicka-Kapusta, 1970, 1974; Sugden and Harris, 1972; Robbins, 1973; Galster and Morrison, 1976; Husband, 1976; G. A. Kaufman and Kaufman, 1977; Williams et al., 1977; Harper et al., 1978).

Since the major ash-containing body tissue is the skeleton, similarities between skeletal mass and body ash content are expected. The ash content of chemically mature, terrestrial mammals (10 species, weighing from 6 g to 60 kg) and birds (3 species, weighing 10–500 g) are linear functions of body weight and average 4.37 and 3.33% respectively (Table 10.2) (Kale, 1965; Myrcha, 1968; Myrcha and Pinowski, 1970; Sawicka-Kapusta, 1970, 1974; Sugden and Harris, 1972; Robbins, 1973; Galster and Morrison, 1976; Husband, 1976; G. A. Kaufman and Kaufman, 1977; Harper et al., 1978). The lower concentration of ash in birds than in mammals and the linear relationship between body weight and ash content supports the contentions that (a) birds have a lighter skeletal mass than mammals because of pneumatization of avian bones (Reynolds, 1977) and (b) skeletal mass is a linear function of body weight (Prange et al., 1979). Interestingly, penguins have a lower body ash content (1.39–2.95%; Williams et al., 1977) than do terrestrial birds, which reflects the lessened need of skeletal mass for support and locomotion in the aquatic environment (Reynolds, 1977).

1969; Child and Marshall, 1970; Moreau and Dolp, 1970; Myrcha and Pinowski, 1970; Penney and Bailey, 1970; Yarbrough, 1970; Morton and Tung, 1971; Sugden and Harris, 1972; Robbins, 1973; Holleman and Dieterich, 1975; Schreiber and Johnson, 1975; Fehrenbacher and Fleharty, 1976; Galster and Morrison, 1976; Holmes, 1976; Husband, 1976; O'Conner, 1976; O'Farrell and Studier, 1976; Chilgren, 1977; Fedyk, 1977; G. A. Kaufman and Kaufman, 1977; Williams et al., 1977; Harper et al., 1978; Campbell and Leatherland, 1980.)

TABLE 10.2
Scaling of Skeletal Mass and Total Body Ash Content (Y in grams) Relative to Body Mass (W in grams) in Terrestrial Mammals and Birds

Group		Equation	Reference
Mammals	(skeletal mass)	$Y = 0.031W^{1.123}$	Reynolds (1977)
	(skeletal mass)	$Y = 0.061W^{1.090}$	Prange et al. (1979)
	(ash content)	$Y = 0.044W^{1.004}$	Current compilation
Birds	(skeletal mass)	$Y = 0.030W^{1.119}$	Reynolds (1977)
	(skeletal mass)	$Y = 0.065W^{1.071}$	Prange et al. (1979)
	(ash content)	$Y = 0.033W^{1.021}$	Current compilation

B. Growth Rates

Growth of an individual over time is often described by a sigmoid curve in which most of the growth occurs during a relatively linear intermediate phase. The maximum growth rates of young animals during the linear phase increase as power functions of adult body weight (Table 10.3; Fig. 10.3). Neonates of larger species grow at faster absolute rates but slower relative rates in comparison to adult body weight than do neonates of smaller species. Altricial birds grow at rates twice that of similarly sized precocial birds and placental mammals (Fig. 10.3). Although precocial and altricial mammals grow at approximately the same

TABLE 10.3
Relationship between Adult Body Weight (W in grams) and Growth Rate (Y in g/day) in Neonatal Birds and Mammals

Group	Regression equation	N	R^2	Reference
Altricial land birds	$Y = 0.21W^{0.72}$	13	0.97	Ricklefs (1968, 1973)
Precocial land birds	$Y = 0.02W^{0.91}$	4	0.91	Ricklefs (1968, 1973)
Marsupials	$Y = 0.0033W^{0.82}$	4	0.97	Case (1978)
Placental Mammals	$Y = 0.0326W^{0.75}$	160	0.94	Case (1978), Robbins et al. (1981)
Cetaceans	$Y = 0.0447W^{0.75}$	3	0.99	Case (1978)
Chiropterans	$Y = 0.0526W^{0.65}$	10	0.63	Case (1978)
Carnivores				
Fissipeds	$Y = 0.0543W^{0.70}$	23	0.86	Case (1978)
Pinnipeds	$Y = 0.0194W^{0.84}$	11	0.48	Case (1978)
Insectivores	$Y = 0.0522W^{0.67}$	8	0.76	Case (1978)
Lagomorphs	$Y = 0.1262W^{0.61}$	9	0.82	Case (1978)
Primates	$Y = 0.1941W^{0.37}$	16	0.82	Case (1978)
Rodents	$Y = 0.0408W^{0.71}$	60	0.90	Case (1978)
Ungulates and subungulates	$Y = 0.0766W^{0.71}$	22	0.94	Robbins et al. (1981)

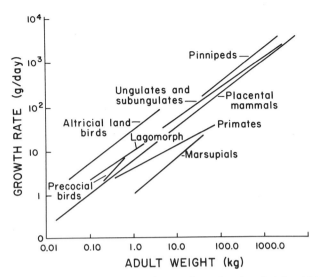

Fig. 10.3. Growth rates of neonatal birds and mammals as a function of adult weight. (Sources of data are from Table 10.3.)

rates, marsupials and primates have very slow growth rates and pinniped carnivores have very high rates relative to the adult body weight. The maximum growth rates of young animals are apparently established by genetically determined physiological limits to cellular metabolism (Ricklefs, 1968, 1973). Genetically determined growth rates within each species are evolved relative to the selective pressures of infant mortality and nutrient availability.

Changes in either food availability or the magnitude of other requirements will alter the rate at which growth can occur. Although the sigmoid curve is the idealized expression, neonatal growth rates will be below the maximum possible if food is limiting. Similarly, loss–gain cycles of more mature animals are often superimposed on the attainment of the asymptotic adult weight. Thus, while growth requirements can be predicted from body composition analyses and average growth rates, the actual daily requirement will vary as growth rates change throughout life.

C. Growth Requirements

Growth requirements are the mathematical description of the rate at which matter is accumulated in the developing organism. Energy requirements are a function of the chemical energy content of the different body constituents. Body water and ash have no available chemical energy. However, anhydrous body protein and fat average 5.43 ± 0.11 and 9.11 ± 0.23 kcal/g, respectively, in wild mammals (Baker *et al.*, 1968; Sawicka-Kapusta, 1968; Barrett, 1969; Ewing *et al.*, 1970, Pucek, 1973; Fedyk, 1974; Robbins *et al.*, 1974; D. W. Kaufman and Kaufman, 1975; Stirling and McEwan, 1975; Husband, 1976) and 5.41 ± 0.11 and 8.92 ± 0.29 kcal/g, respectively, in wild birds (Odum *et al.*, 1965; Johnston, 1970; Clay *et al.*, 1979). Consequently, the energy content of the gain or loss can range from 0 to 9 kcal/g depending on the composition of the weight change.

If the relative concentrations of body constituents change in an orderly progression during growth, first approximations of growth requirements can be generated from body composition analyses. For example, if 20 g of protein are accumulated for each 100 g of weight gained by a neonate (the other 80 g being a mixture of water, fat, and ash), then the protein requirement is 20% of the gain. If individuals of a species are analyzed throughout life and found to contain 20% protein, then the requirement must be constant. However, the accumulations of

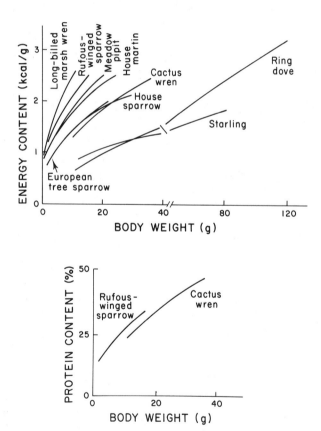

Fig. 10.4. Caloric and protein content of the gain in several altricial birds. (See Table 10.4 for references.)

most constituents are curvilinear functions, because their concentration in the gain and therefore in the entire body changes as the animal grows. The use of body composition analyses to estimate growth requirements assumes that composition is genetically regulated and therefore not determined by food composition. This is not strictly true since several exceptions will be discussed.

The caloric content of the gain in most wildlife increases curvilinearly as body weight increases (Figs. 10.4–10.8; Table 10.4). Since the body water content decreases and the protein and fat content generally increase during growth, the caloric content of the gain must correspondingly increase. Energy requirements for growth are generally bracketed between 1 and 3 kcal/g. The exception in mammals is the ringed seal, and probably other marine mammals, in which 75% of the biomass growth during nursing is subcutaneous fat deposition (Stewart and Lavigne, 1980). The very rapid growth of pinniped carnivores is largely due to

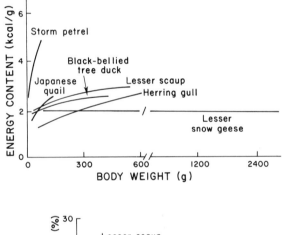

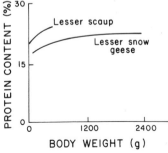

Fig. 10.5. Caloric and protein content of the gain in several precocial and semiprecocial birds. (See Table 10.4 for references.)

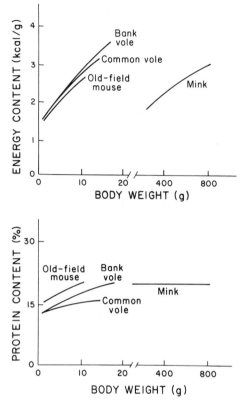

Fig. 10.6. Caloric and protein content of the gain in several small mammals. (See Table 10.4 for references.)

the direct deposition of milk triglycerides (Bailey *et al.*, 1980). The higher lipid reserves of storm petrel chicks, and probably other pelagic birds, provide an energy reserve for the long intervals between parental feeding and after fledging when foraging may be difficult (Ricklefs *et al.*, 1980).

The equations describing the energy content of the gain in birds are functions of mode of development, adult body weight, and stability or predictability of the food supply (Ricklefs, 1974). For example, the intercepts decrease in altricial species and increase in precocial Galliformes and Anseriformes as adult body weight increases. The net effect for the altricial birds is that the average requirement increases from 0.9 kcal/g at hatching to 2.5±0.5 kcal/g at maturity. However, three major groups of altricial birds are apparent (Fig. 10.8). The caloric content of the gain increases to 1.99±0.11 kcal/g in house sparrows, European tree sparrows, and starlings, to 2.52±0.05 kcal/g in the cactus wren, house martin, long-billed marsh wren, meadow pipit, and rufous-winged sparrow, and to 3.63 kcal/g in ring doves. The granivorous sparrows and omnivorous starling accumulate less fat than the insectivorous wrens, pipits, and martins that use the additional reserves while developing an efficient foraging capability and during cool inclement weather either at or before fledging when food supplies may be reduced. The food supply of the granivorous, desert-adapted rufous-winged spar-row may also be transitory, requiring the accumulation of moderate reserves. The very high energy content of the gain in ring doves (Fig. 10.8) is probably similar to that occurring in seals and is due to the direct deposition of "milk" triglycerides. The energy content of the gain in precocial birds covers a range similar to that occurring in altricial species, although the changes become less dramatic with increasing adult weight.

Protein requirements for wildlife range from 15 to 25% of the gain and are generally higher during the early growth of precocial than of altricial species (Figs. 10.4–10.7). The very high protein requirement in fledging passerines is due to protein deposition in both internal tissues and feathers. Theoretically, the protein requirement for internal tissue synthesis cannot exceed 30% since fat-free tissues contain 70% water. The major exception to the constant or increasing

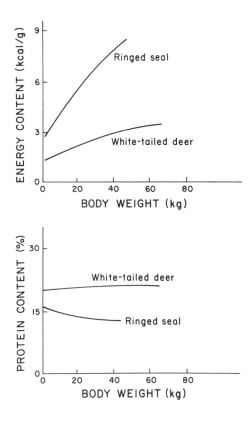

Fig. 10.7. Caloric and protein content of the gain in several large mammals. (See Table 10.4 for references.)

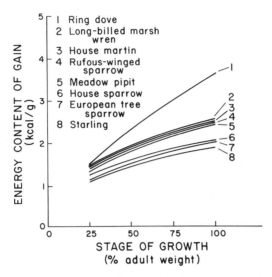

Fig. 10.8. Caloric content of the gain in altricial birds at comparable stages of growth.

protein deposition during growth is the ringed seal, since its relative protein requirement decreases because of the enormous amounts of fat being deposited.

The generalized requirement equations (Table 10.4) should be used with caution since they do not adequately represent daily or seasonal alterations in body constituents. For example, requirement estimates are confounded in precocial birds by the utilization of residual yolk while simultaneously depositing body fat. More importantly, weight loss–weight gain cycles are not included. However, one of the more important observations from the equations is that the caloric content of the gain generally does not approach the maximum of 9 kcal/g for fat deposition. Very few wild animals accumulate significant levels of fat during growth. Fat is accumulated seasonally in hibernators, in birds that migrate over long distances without feeding, and in some inactive captive wildlife. Although large fat reserves would seem advantageous for survival during prolonged periods of food scarcity, the selective pressures of the additional energy required to maintain these tissues, the increased cost of transport, and, perhaps, decreased capacities for predator avoidance reduce the accumulation of fat to a more minimal level (King, 1972). Selection for fat deposition in wildlife is extremely conservative (Pond, 1978). In fact, many birds store only enough reserve energy as body tissue or gastrointestinal contents to survive overnight and part or all of the next day (West and Meng, 1968; King, 1972; Blem, 1976; Ketterson and Nolan, 1978; Wyndham, 1980). Small, nonhibernating mammals are similar in that their energy stores last from only a few hours in shrews to a couple of days in some rodents (Gyug and Millar, 1980). Consequently, ice storms, crusted snow,

TABLE 10.4

Regression Equations Describing the Energy and Protein Requirements per Unit of Gain (Y in kcal of Energy or g of Protein/g of Weight Gain) as a Function of Body Weight (W) in Several Wild Birds and Mammals[a]

Species or group	Weight range	Energy	Protein	Reference
Birds, altricial				
Long-billed marsh wren	1–11g	$Y = 0.94W^{0.42}$	$Y = 0.11W^{0.40}$	Kale (1965)
Rufous-winged sparrow	2–16g	$Y = 0.77W^{0.43}$		Austin and Ricklefs (1977)
Meadow pipit	4–19g	$Y = 0.67W^{0.44}$		Skar et al. (1975)
European tree sparrow	2–22g	$Y = 0.55W^{0.42}$		Myrcha and Pinowski (1969)
House martin	2–25g	$Y = 0.77W^{0.37}$		Bryant and Gardiner (1979)
House sparrow	2–28g	$Y = 0.61W^{0.37}$		Blem (1975)
Cactus wren	12–36g	$Y = 0.40W^{0.51}$	$Y = 0.06W^{0.57}$	Ricklefs (1975)
Starling	12–79g	$Y = 0.34W^{0.39}$		Myrcha et al. (1973)
Ring dove	11–150g	$Y = 0.14W^{0.65}$		Brisbin (1969)
Birds, semiprecocial				
Storm petrel	11–67g	$Y = 1.07W^{0.36}$		Ricklefs et al. (1980)
Herring gulls	70–800g	$Y = 0.32W^{0.33}$		Dunn and Brisbin (1980)
Birds, precocial				
Japanese quail	11–133g	$Y = 0.71W^{0.26}$		Brisbin and Tally (1973)
Black-bellied tree duck	30–440g	$Y = 1.17W^{0.13}$		Cain (1976)
Lesser scaup	29–538g	$Y = 1.29W^{0.13}$	$Y = 0.14W^{0.09}$	Sugden and Harris (1972)
Lesser snow geese	85–2500g	$Y = 2.25W^{-0.02}$	$Y = 0.13W^{0.07}$	Campbell and Leatherland (1980)
Mammals				
Old-field mouse	2–11g	$Y = 1.00W^{0.44}$	$Y = 0.14W^{0.16}$	D. W. Kaufman and Kaufman (1975)
				G. A. Kaufman and Kaufman (1977)
Common vole	2–16g	$Y = 1.08W^{0.40}$	$Y = 0.12W^{0.11}$	Sawicka-Kapusta (1974)
Bank vole	2–18g	$Y = 1.05W^{0.43}$	$Y = 0.11W^{0.21}$	Sawicka-Kapusta (1974), Fedyk (1974)
Mink	274–750g	$Y = 0.21W^{0.40}$	$Y = 0.20W^{0.00}$	Harper et al. (1978)
Ringed seal	3–45kg	$Y = 1.28W^{0.49}$	$Y = 0.19W^{0.10}$	Stirling and McEwan (1975)
White-tailed deer	3–60kg	$Y = 0.76W^{0.37}$	$Y = 0.20W^{0.01}$	Robbins (1973)

[a]Since all the sources except Robbins (1973) included hair or feathers in their body composition analyses, the equations include the net energy and protein required for hair or feather molt.

219

or other environmental vicissitudes that remove the food supply of small birds and mammals can produce sudden and extensive mortality.

D. Weight Loss and Starvation

Weight loss by wildlife has been studied far less than has weight gain. This neglect is unfortunate, since virtually all wildlife must contend with daily or seasonal reductions in mass. The lack of studies is, in part, due to the assumption that, when energy intake is inadequate, fat is mobilized to yield 9 kcal/g of weight loss. The principal supporting argument for this assumption is that the RQ of fasted animals during basal metabolism studies (0.7) suggests fat metabolism. However, it is often forgotten that such measurements in birds can indicate either fat or protein metabolism and that RQs were classically corrected for nitrogen excretion (Brody, 1945). Furthermore, hibernating animals can indeed utilize fat almost exclusively (Nelson et al., 1975).

However, the prolonged utilization of only fat by the active animal is physiologically and biochemically difficult to justify since (a) the complete oxidation of fatty acids requires simultaneous carbohydrate metabolism and (b) the brain requires glucose (Krebs, 1965; Young, 1976). Carbohydrates can come directly from the diet, from reserve glycogen, from protein via gluconeogenesis, or from glycerol. In contrast to weight loss in the hibernating black bear, summer weight loss in bears produces extensive protein catabolism (Nelson et al., 1975). Similarly, weight loss in either resting or active, fasted birds is a mixture of water, protein, and fat containing as little as 2 kcal/g (Brisbin, 1969; Dolnik and Gavrilov, 1973; Williams et al., 1977; Jones, 1980). The caloric content of the nocturnal weight loss in small passerines decreases as ambient temperature increases (Dolnik and Gavrilov, 1973; Dargol'ts, 1973) (Fig. 10.9). The decrease is, in part, due to increasing water losses for temperature regulation (Kendeigh et al., 1977).

Weight loss characteristics in small mammals are also dependent on season and water intake. For example, starved black-tailed prairie dogs lost far more protein in summer than in winter (Pfeiffer et al., 1979). Water deprivation of starving ground squirrels increased by 33% the relative amount of protein lost per gram of fat metabolized (Bintz and Mackin, 1980). Thus, while fat is lost in all animals during starvation, the weight loss in nonhibernating mammals can contain a great deal of protein and water.

The recognition that not all fat can be mobilized is necessary in estimating survival times of starved animals. Nonrecoverable, structural lipids have ranged from 0.3 to 1.3% of the body weight in birds (Newton, 1969; P. Ward, 1969; Wyndham, 1980) and up to 3.1% of body weight in mammals (Rock and Williams, 1979).

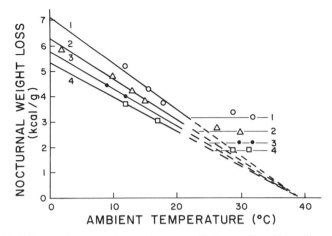

Fig. 10.9. Calories in the nocturnal weight loss of chaffinches as affected by ambient temperature during (1) fall migration, (2) winter, (3) summer, and (4) molt. (Adapted from Dolnik and Gavrilov, 1973, courtesy of the Israel Program for Scientific Translations and Keter Publishing House, Jerusalem.)

E. Compensatory Growth

Loss–gain cycles can have very subtle long-term effects on body composition and fitness of individuals. For example, increased rates of growth and protein accretion and decreased fat accumulation have occurred after weight loss and refeeding in domestic livestock (termed compensatory growth) when compared to continuously growing animals. Even though the fat levels in well-fed, captive wild ungulates are below those occurring in domestic livestock, compensatory growth may still occur. For example, a very large, mature, captive male white-tailed deer killed at his peak yearly weight contained only 8.9% fat—27% less than smaller, younger males, which averaged 12.1±1.1% fat (Robbins *et al.*, 1974). Since white-tailed deer voluntarily restrict feed intake during winter even when food is available ad libitum, the male deer described here had experienced several gain–loss cycles. Similarly, a captive female white-tailed deer that had been purposefully fed a restricted ration during the preceding winter had a 24% lower body fat content (10.2%) the following fall than females fed ad libitum year round (13.4±2.5%). Both animals at slaughter would have been classified as obese by the casual observer.

Compensatory growth could either increase or decrease an individual's fitness (Bandy *et al.*, 1970). For example, an increased efficiency in food utilization and protein accretion (Reid *et al.*, 1968; Fox *et al.*, 1972) would maximize seasonal growth rates. A trade-off may occur between the advantages of increased growth rates and size, such as increased dominance, breeding opportunities, or improved surface area to weight ratios, and the negative effects on survival of reduced fat and energy reserves. However, the assumption that reduced fat levels in older, larger animals would reduce survival capabilities is probably incorrect, since energy expenditure is a function of body weight to the 0.75 power rather than 1. For example, when the fat levels in the deer described earlier are expressed as functions of metabolic body weight, the fat concentrations are equal. Similarly, the reduced fat concentrations in captive deer may be accentuated if, as is widely assumed, free-ranging deer are able to accumulate directly the triglycerides of acorns and other deciduous tree nuts in their fall diet (McCullough and Ullrey, unpublished). Since firm evidence that compensatory growth occurs in any wild-life is lacking, further speculation on its significance is unwarranted.

F. Indices of Body Condition or Composition

Since body composition is difficult to determine in many field studies, numer-ous ecologists have attempted to develop indices of body fat or protein content to relate ultimately to survival and reproductive success. A general review of the most common methods and their problems has been published by Kirkpatrick (1980). The indices include measurements of back fat, leg fat, kidney fat, bone marrow fat, visceral fat, dry and fresh weights of many organs, blood or urinary chemical analyses, and mass and linear dimensions of the animal. Unfortunately, few ecologists have tested the relationship between an index and total body composition (Bamford, 1970; Smith, 1970; Anderson et al., 1972; R. O. Bailey, 1979; Bakker and Main, 1980; Finger et al., 1981).

An index that will be a valuable indicator of body composition will have either a linear or slightly curvilinear relationship, with minimal variation. An index that measures an on–off process, such as a body fat reserve that is the last to be mobilized and first to be replenished and maintains a constant value over a wide range of possible conditions, will be far less useful than one that provides a continuous, accurate, and easily determined indication of body composition. For example, one of the more common indices of body fat used on North American cervids is the kidney fat index, which is a ratio between the weight of the kidney and the adhering fat deposit. However, since the kidney fat index has a cur-vilinear, highly variable relationship to total body fat content (Bamford, 1970; Finger et al., 1981), kidney fat indices are virtually meaningless for predicting body composition (Fig. 10.10). While it is tempting to use the linear expression in Fig. 10.10 because of the identical correlation, the curvilinear relationship is the biologically correct expression. Thus, use of the various indices requires a recognition of their shortcomings, a distinction between statistical and biological significance (Flux, 1971), and a basic understanding of animal functioning. It is unfortunate that literally hundreds of field studies have used the various indices

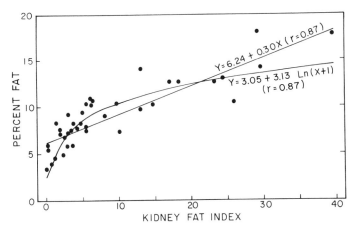

Fig. 10.10. Relationship between the kidney fat index and total body fat content in white-tailed deer. (From Finger *et al.*, 1981, courtesy of the *Journal of Wildlife Management.*)

of body condition or composition, while only a few have attempted to determine whether a given index has even the slightest biological validity.

II. PELAGE AND PLUMAGE

Hair and feathers are of interest to the nutritionist because of their role in thermoregulation and the need to quantify the additional nutrients necessary for their growth and replacement. Molts, or the periodic replacement of pelage or plumage, can be partial or complete, with from one to three molts per year occurring in various birds or mammals (Ling, 1970; King, 1973). Single annual molts are largely confined to aquatic forms in which wear or abrasion are minimal and seasonal variations in environmental temperature are limited. Most animals have two molts per year.

Both hair and feathers are primarily protein. For example, white-tailed deer hair is 89.0% protein (Robbins *et al.*, 1974) and feathers are 93–98.5% protein (McCasland and Richardson, 1966; Nitsan *et al.*, 1981). The principal protein of these structures is keratin. Keratins are resistant to digestion by vertebrate gastrointestinal enzymes and are insoluble in dilute acids and alkalies, water, and organic solvents. One of the unique characteristics of many keratins is their high concentration relative to other plant and animal proteins of the sulfur-containing amino acid cystine—6.7–8.2% cystine in feather keratin, 8.0–9.5% in porcupine quills, and 10.5–15.7% in horn keratins as compared to 0–2.9% in plant proteins and 0–6.3% in other animal proteins (Block and Bolling, 1951; Ward and Lundgren, 1954; Harrap and Woods, 1967; Newton, 1968; Frenkel and Gillespie, 1976; Nitsan *et al.,* 1981).

Feathers average 5.37 ± 0.23 kcal/g (Myrcha and Pinowski, 1970; Chilgren, 1975). Mammal hair often has a higher caloric content, such as 5.94 kcal/g for white-tailed deer hair (Robbins *et al.,* 1974), because of larger amounts of surface lipids, which contain 9.74 kcal/g (Paladines *et al.,* 1964). Since the growth rates of hair and feathers are relatively constant during the daily cycle and throughout the molt (Lillie and Juhn, 1932; Newton, 1968; Jacobsen, 1973; Chilgren, 1975; Owen and Ogilvie, 1979), the requirement can be estimated by dividing the total energy or protein accumulated by the length of the molt.

For a complete avian molt, feather weight can be predicted from the equation, $Y = 0.09W^{0.95}$, where Y is plumage weight in grams and W is body weight in grams (Turček, 1966). A 60-day, complete molt in a 20 g passerine represents a 2% increase in energy requirements relative to basal metabolism (Aschoff and Pohl, 1970, inactive phase), but a 17% increase in nitrogen requirements relative to maintenance expenditures (Dolnik and Gavrilov, 1979; Robbins, 1981). Because of the higher exponent of feather weight (0.95) than of either energy or nitrogen maintenance costs (0.61–0.75), a larger bird would have a proportionately higher daily cost than would a smaller bird if the molt were of similar completeness and duration. Partial molts, such as the postbreeding molt of many waterfowl, would be far less costly than a complete molt because the flight feathers average only 23% of the entire feather mass (Newton, 1966; Chilgren, 1975; Owen and Ogilvie, 1979).

The maximum daily productive cost of molting in mammals averages approximately 4.7% of the interspecific basal metabolic rate and approximately 85% of the endogenous urinary nitrogen losses (Robbins *et al.,* 1974; Holleman and Dieterich, 1978). For example, hair growth in the fall molt of 25- and 60-kg white-tailed deer averages 35.0–79.2 kcal/day (4.5 and 5.3% of BMR) and 0.86–2.08 g N/day (73–92% of the EUN loss, where EUN is 105 mg $N/W_{kg}^{0.75}/$ day). Similarly, a 1-week-old brown lemming weighing approximately 10 g retains 0.11 kcal/day in hair (4.62% of BMR) and 3.1 mg N/day (88% of EUN). These cost estimates based solely on hair or feather weights underestimate the total productive cost since they exclude the cost of integument reorganization preparatory to hair or feather synthesis (Stettenheim, 1972). While this cost is probably quite minimal, increased thermoregulatory costs due to altered insulation relative to the nonmolting animal can be far more important. For example, whole-body conductances were increased 55% in molting chaffinches relative to the nonmolting bird (Dolnik and Gavrilov, 1979). Malnourishment during follicle formation in the fetus or neonate can decrease the total number of hair follicles and thus possibly alter the thermoregulatory requirement throughout life (Schinckel and Short, 1961).

The duration and completeness of the molt can be profoundly affected by nutrient intake. Nutritionally restricted mammals molt later, prolong the molt, and have a rough, protruding coat in comparison to the long, sleek coat of well-fed animals (French *et al.*, 1956; Ling, 1970; Cowan and Raddi, 1972). As an indication of the nutritional importance of the molt requirement, hair and feather molts are often timed to occur when other requirements are minimal or food resources are abundant (Payne, 1972). While the increased energy requirements of molting are often compensated for by reductions in activity (Kendeigh *et al.*, 1977), the additional protein costs must be met ultimately by dietary intake (Ankney, 1979).

III. ANTLERS

Dead, oven-dried antlers contain 54.0 ± 2.3% ash, 45.0% protein, 1.0% fat, and 2.53 kcal/g (Rush, 1932; Chaddock, 1940; Ullrey *et al.*, 1975; Hyvarinen *et al.*, 1977). As observed relative to hair and feathers, the incremental protein requirement is potentially far more significant than the energy requirement (Fig. 10.11). However, with the exception of European roe deer, which grow antlers from October to May (Chapman, 1975), antler growth occurs during the north temperate summer, when forage resources are most abundant.

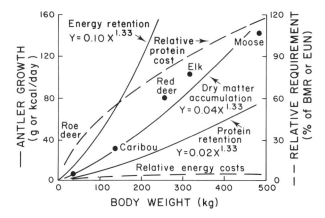

Fig. 10.11. Antler growth, energy and protein accumulation, and relative energy and nitrogen costs in several cervids. Antler and body weights were the larger sizes reported for each species. (Data from Huxley, 1931; Chapman, 1975.)

IV. DISEASE AND INJURY

The interaction of nutrition and disease has not been investigated thoroughly in any species (Chandra and Newberne, 1977). While scientists working with domestic animals can often medicate sick animals to minimize the disease process, the wildlife scientist frequently can do little to control or reduce diseases. Since virtually all wildlife carry disease or parasitic organisms throughout life, wildlife nutritionists must increase their attempts to understand the nutritional and ecological implications. Since the additional energy or matter requirements directly associated with the very small weight of most disease organisms is minimal, alterations in host metabolic processes will be far more important. In most cases, malnutrition and the course of an infectious disease act synergistically to reduce host resistance and increase the detrimental consequences.

Most disease or injury processes markedly increase energy and nitrogen metabolism. For example, fever is the overt expression of the altered energy metabolism (Richard *et al.,* 1978). Energy expenditure increases from 7 to 33% during the healing of bone fractures, and up to 125% following severe burns. Hartebeest infected with *Salmonella typhimurium* doubled urinary nitrogen excretion, while dietary nitrogen digestibility decreased by 2.2% relative to the healthy norm (Arman and Hopcraft, 1971). Although such increases in the requirements are significant, they are often exacerbated by reductions in food intake. Thus, understanding the nutritional impact of a disease process in wildlife is far more complex than simply identifying and quantifying infectious organisms, and will require the synergistic expertise of both nutritionists and those studying infectious diseases (Chandra and Newberne, 1977).

REFERENCES

Allen, W. V. (1976). Biochemical aspects of lipid storage and utilization in animals. *Am. Zool.* **16,** 631–647.
Anderson, A. E., Medin, D. E., and Bowden, D. C. (1972). Indices of carcass fat in a Colorado mule deer population. *J. Wildl. Manage.* **36,** 579–594.
Ankney, C. D. (1979). Does the wing molt cause nutritional stress in lesser snow geese. *Auk* **96,** 68–73.
Arman, P., and Hopcraft, D. (1971). The effect of disease on nitrogen excretion in the hartebeest. *Proc. Nutr. Soc.* **30,** 65A–66A.
Aschoff, J., and Pohl, H. (1970). Rhythmic variations in energy metabolism. *Fed. Proc.* **29,** 1541–1552.
Austin, G. T., and Ricklefs, R. E. (1977). Growth and development of the rufous-winged sparrow (*Aimophila carpalis*). *Condor* **79,** 37–50.
Bailey, B. A., Downer, R. G. H., and Lavigne, D. M. (1980). Neonatal changes in tissue levels of carbohydrate and lipid in the harp seal *Pagophilus groenlandicus. Comp. Biochem. Physiol.* **67B,** 179–182.

Bailey, R. O. (1979). Methods of estimating total lipid content in the redhead duck (*Aythya americana*) and an evaluation of condition indices. *Can. J. Zool.* **57**, 1830–1833.

Baker, W. W., Marshall, S. G., and Baker, V. B. (1968). Autumn fat deposition in the evening bat (*Nycticeius humeralis*). *J. Mammal.* **49**, 314–317.

Bakker, H. R., and Main, A. R. (1980). Condition, body composition and total body water estimation in the quokka, *Setonix brachyurus* (Macropodidae). *Aust. J. Zool.* **28**, 395–406.

Bamford, J. (1970). Estimating fat reserves in the brush-tailed possum, *Trichosurus vulpecula* Kerr (Marsupialia: Phalangeridae). *Aust. J. Zool.* **18**, 415–425.

Bandy, P. J., Cowan, I. McT., and Wood, A. J. (1970). Comparative growth in four races of black-tailed deer (*Odocoileus hemionus*). Part I. Growth in body weight. *Can. J. Zool.* **48**, 1401–1410.

Barrett, G. W. (1969). Bioenergetics of a captive least shrew, *Cryptotis parva*. *J. Mammal.* **50**, 629–630.

Bintz, G. L., and Mackin, W. W. (1980). The effect of water availability on tissue catabolism during starvation in Richardson's ground squirrels. *Comp. Biochem. Physiol.* **65A**, 181–186.

Bintz, G. L., Rosebery, H. W., and Bintz, L. B. (1979). Glycogen levels in field and laboratory-acclimated Richardson ground squirrels. *Comp. Biochem. Physiol.* **62A**, 339–342.

Blem, C. R. (1975). Energetics of nestling house sparrows *Passer domesticus*. *Comp. Biochem. Physiol.* **52A**, 305–312.

Blem, C. R. (1976). Patterns of lipid storage and utilization in birds. *Am. Zool.* **16**, 671–684.

Block, J. R., and Bolling, D. (1951). "The Amino Acid Composition of Proteins and Foods: Analytical Methods and Results." Thomas, Springfield, Ill.

Brisbin, I. L., Jr. (1968). A determination of the caloric density and major body components of large birds. *Ecology* **49**, 792–794.

Brisbin, I. L., Jr. (1969). Bioenergetics of the breeding cycle of the ring dove. *Auk* **86**, 54–74.

Brisbin, I. L., Jr., and Tally, L. J. (1973). Age-specific changes in the major body components and caloric value of growing Japanese quail. *Auk* **90**, 624–635.

Brody, S. (1945). "Bioenergetics and Growth." Hafner, New York.

Bryant, D. M., and Gardiner, A. (1979). Energetics of growth in house martins (*Delichon urbica*). *J. Zool.* **189**, 275–304.

Cain, B. W. (1976). Energetics of growth for black-bellied tree ducks. *Condor* **78**, 124–128.

Campbell, R. R., and Leatherland, J. F. (1980). Estimating body protein and fat from water content in lesser snow geese. *J. Wildl. Manage.* **44**, 438–446.

Case, T. J. (1978). On the evolution and adaptive significance of postnatal growth rates in the terrestrial vertebrates. *Q. Rev. Biol.* **53**, 243–282.

Chaddock, T. T. (1940). Chemical analysis of deer antlers. *Wisc. Cons. Bull.* **5**, 42.

Chandra, R. K., and Newberne, P. M. (1977). "Nutrition, Immunity, and Infection: Mechanisms of Interactions." Plenum, New York.

Chapman, D. I. (1975). Antlers—bones of contention. *Mammal. Rev.* **5**, 121–172.

Child, G. I., and Marshall, S. G. (1970). A method of estimating carcass fat and fat-free weight in migrant birds from water content of specimens. *Condor* **72**, 116–119.

Chilgren, J. D. (1975). Dynamics and bioenergetics of post-nuptial molt in captive white-crowned sparrows (*Zonotrichia leucophrys gambelii*). Doctoral dissertation, Washington State Univ., Pullman.

Chilgren, J. D. (1977). Body composition of captive white-crowned sparrows during postnuptial molt. *Auk* **94**, 677–688.

Clay, D. I., Brisbin, I. L. Jr., and Youngstrom, K. A. (1979). Age-specific changes in the major body components and caloric values of growing wood ducks. *Auk* **96**, 296–305.

Cmelik, S. H. W., and Ley, H. (1977). Composition of adrenal lipids from some domestic and wild ruminants. *Comp. Biochem. Physiol.* **56B**, 267–270.

Cowan, I. McT., and Raddi, A. G. (1972). Pelage and molt in the black-tailed deer. *Can. J. Zool.* **50,** 639–647.

Crawford, M. A. (1968). Fatty acid ratios in free-living and domestic animals. *Lancet* **7556,** 1329–1333.

Dargol'ts, V. G. (1973). The caloric value equivalent of body weight changes of homiothermal animals: Relationship between oxidized substances and heat loss through evaporation. *Zhurnal Obshchei Biologii* **34,** 887–899.

Dolnik, V. R. (1970). The water storation by the migratory fat deposition in *Passer domesticus bactrianus* Zar. et Kud.—the arid zone migrant. *In* "Productivity, Population Dynamics and Systematics of Granivorous Birds" (S. C. Kendeigh and J. Pinowski, eds.), pp. 103–109. Polish Scientific Publ., Warsaw.

Dolnik, V. R., and Gavrilov, V. M. (1973). Caloric equivalent of body weight variations in chaffinches (*Fringilla coelebs*). *In* "Bird Migrations: Ecological and Physiological Factors" (B. E. Bykhovskii, ed.), pp. 273–287. Wiley, New York.

Dolnik, V. R., and Gavrilov, V. M. (1979). Bioenergetics of molt in the chaffinch (*Fringilla coelebs*). *Auk* **96,** 253–264.

Donelly, J. R., and Freer, M. (1974). Prediction of body composition in live sheep. *Aust. J. Agric. Res.* **25,** 825–834.

Dunn, E. H., and Brisbin, I. L., Jr. (1980). Age-specific changes in the major body components and caloric values of herring gull chicks. *Condor* **82,** 398–401.

Ewing, W. G., Studier, E. H., and O'Farrell, M. J. (1970). Autumn fat deposition and gross body composition in three species of *Myotis. Comp. Biochem. Physiol.* **36,** 119–129.

Farner, D. S., Oksche, A., Kamemoto, F. I., King, J. R., and Cheyney, H. E. (1961). A comparison of the effect of long daily photoperiods on the pattern of energy storage in migratory and non-migratory finches. *Comp. Biochem. Physiol.* **2,** 125–142.

Fedyk, A. (1974). Gross body composition in postnatal development of the bank vole. I. Growth under laboratory conditions. *Acta Theriol.* **19,** 381–401.

Fedyk, A. (1977). Seasonal changes in the water content and level in the bank vole against the background of other gross body components. *Acta Theriol.* **22,** 355–363.

Fehrenbacher, L. H., and Fleharty, E. D. (1976). Body composition, energy content, and lipid cycles of two species of pocket gophers (*Geomys bursarius* and *Pappogeomys castonops*) in Kansas. *Southwest. Nat.* **21,** 185–198.

Finger, S. E., Brisbin, I. L., Jr., Smith, M. H., and Urbston, F. D. (1981). Kidney fat as a predictor of body condition in white-tailed deer. *J. Wildl. Manage.* **45,** 964–968.

Flux, J. E. C. (1971). Validity of the kidney fat index for estimating the condition of hares: A discussion. *New Zealand J. Sci.* **14,** 238–244.

Fox, D. G., Johnson, R. R., Preston, R. L., Dockerty, T. R., and Klosterman, E. W. (1972). Protein and energy utilization during compensatory growth in beef cattle. *J. Anim. Sci.* **34,** 310–318.

Freminet, A. (1981). Comparison of glycogen stores in fed and fasted rats and guinea-pigs in two-month and one-year-old animals. *Comp. Biochem. Physiol.* **69A,** 665–671.

French, C. E., McEwen, L. C., Magruder, N. D., Ingram, R. H., and Swift, R. W. (1956). Nutrient requirements for growth and antler development in the white-tailed deer. *J. Wildl. Manage.* **20,** 221–232.

Frenkel, M. J., and Gillespie, J. M. (1976). The proteins of the keratin component of the bird's beak. *Aust. J. Biol. Sci.* **29,** 467–479.

Gale, M. M., Crawford, M. A., and Woodford, M. (1969). The fatty acid composition of adipose and muscle tissue in domestic and free-living ruminants. *Biochem. J.* **113,** 6P.

Galster, W., and Morrison, P. (1975). Carbohydrate reserves of wild rodents from different latitudes. *Comp. Biochem. Physiol.* **50A,** 153–157.

Galster, W., and Morrison, P. (1976). Seasonal changes in body composition of the arctic ground squirrel, *Citellus undulatus*. *Can. J. Zool.* **54**, 74–78.

Gifford, C. E., and Odum, E. P. (1965). Bioenergetics of lipid deposition in the bobolink, a transequatorial migrant. *Condor* **67**, 383–403.

Gyug, L. W., and Millar, J. S. (1980). Fat levels in a subarctic population of *Peromyscus maniculatus*. *Can.J. Zool.* **58**, 1341–1346.

Harper, R. B., Travis, H. F., and Glinsky, M. S. (1978). Metabolizable energy requirement for maintenance and body composition of growing farm-raised male pastel mink (*Mustela vison*). *J. Nutr.* **108**, 1937–1943.

Harrap, B. S., and Woods, E. F. (1967). Species differences in the proteins of feathers. *Comp. Biochem. Physiol.* **20**, 449–460.

Hayward, J. S. (1965). The gross body composition of six geographic races of *Peromyscus*. *Can. J. Zool.* **43**, 297–308.

Helms, C. W., and Smythe, R. B. (1969). Variation in major body components of the tree sparrow (*Spizella arborea*) sampled within the winter range. *Wilson Bull.* **81**, 280–292.

Helms, C. W., Aussiker, W. H., Bower, E. B., and Fretwell, S. D. (1967). A biometric study of major body components of the slate-colored junco, *Junco hyemalis*. *Condor* **69**, 560–578.

Holleman, D. F., and Dieterich, R. A. (1975). An evaluation of the tritiated water method for estimating body water in small rodents. *Can. J. Zool.* **53**, 1376–1378.

Holleman, D. F., and Dieterich, R. A. (1978). Postnatal changes in body composition of laboratory maintained brown lemmings, *Lemmus sibiricus*. *Lab. Anim. Sci.* **28**, 529–535.

Holmes, R. T. (1976). Body composition, lipid reserves and caloric densities of summer birds in a northern deciduous forest. *Am. Midl. Nat.* **96**, 281–290.

Husband, T. P. (1976). Energy metabolism and body composition of the fox squirrel. *J. Wildl. Manage.* **40**, 255–263.

Huxley, J. (1931). The relative size of antlers in deer. *Proc. Zool. Soc. London* **1931**, 819–864.

Hyvarinen, H., Kay, R. N. B., and Hamilton, W. J. (1977). Variation in the weight, specific gravity and composition of the antlers of red deer (*Cervus elaphus* L.). *Brit. J. Nutr.* **38**, 301–311.

Ijas, L., Nuuja, I., Palokangas, R., Soimajarvi, J., and Valkeajarvi, P. (1978). Body lipid composition in the winter fed and control populations of the wild male black grouse (*Lyrurus tetrix* L.) in autumn and spring. *Comp. Biochem. Physiol.* **60A**, 313–317.

Jacobsen, N. K. (1973). Physiology, behavior, and thermal transactions of white-tailed deer. Doctoral dissertation, Cornell Univ., Ithaca, N.Y.

Johnston, D. W. (1964). Ecologic aspects of lipid secretion in some postbreeding arctic birds. *Ecology* **45**, 848–852.

Johnston, D. W. (1970). Caloric density of avian adipose tissue. *Comp. Biochem. Physiol.* **34A**, 827–832.

Johnston, D. W., and MacFarlane, R. W. (1967). Migration and bioenergetics of flight in the Pacific golden plover. *Condor* **69**, 156–168.

Jones, M. M. (1980). Nocturnal loss of muscle protein from house sparrows (*Passer domesticus*). *J. Zool.* **192**, 33–39.

Kale, H. W., II. (1965). Ecology and bioenergetics of the long-billed marsh wren in Georgia salt marshes. *Publ. Nuttall Ornith. Club No. 5*.

Kaufman, D. W., and Kaufman, G. A. (1975). Caloric density of the oldfield mouse during postnatal growth. *Acta Theriol.* **20**, 83–95.

Kaufman, G. A., and Kaufman, D. W. (1977). Body composition of the oldfield mouse (*Peromyscus polionotus*). *J. Mammal.* **58**, 429–434.

Kendeigh, S. C., Dolnik, V. R., and Gavrilov, V. M. (1977). Avian energetics. *In* "Granivorous Birds in Ecosystems" (J. Pinowski and S. C. Kendeigh, eds.), pp. 127–204. Cambridge Univ. Press, London.

Ketterson, E. D., and Nolan, V., Jr. (1978). Overnight weight loss in dark-eyed juncos (*Junco hyemalis*). *Auk* **95**, 755–758.

King, J. R. (1961). The bioenergetics of vernal premigratory fat deposition in the white-crowned sparrow. *Condor* **63**, 128–142.

King, J. R. (1972). Adaptive periodic fat storage of birds. *Proc. Int. Ornith. Congr.* **15**, 200–217.

King, J. R. (1973). Seasonal allocation of time and energy resources in birds. *In* "Avian Energetics" (R. A. Paynter, Jr., ed.), pp. 4–85. *Publ. Nuttal Ornith. Club* **15**, Cambridge, Mass.

Kirkpatrick, R. L. (1980). Physiological indices in wildlife management. *In* "Wildlife Management Techniques Manual" (S. D. Schemnitz, ed.), pp. 99–112. Wildl. Soc., Washington, D.C.

Krebs, H. A. (1965). Some aspects of gluconeogenesis (the relations between gluconeogenesis and ketogenesis). *In* "Energy Metabolism" (K. L. Blaxter, ed.), pp. 1–9. European Assoc. Anim. Prod. Publ. No. 11. Academic Press, London.

Lillie, F. R., and Juhn, M. (1932). The physiology of development of feathers. I. Growth-rate and pattern in the individual feather. *Physiol. Zool.* **5**, 124–184.

Ling, J. K. (1970). Pelage and molting in wild animals with special reference to aquatic forms. *Q. Rev. Biol.* **45**, 16–54.

McCasland, W. E., and Richardson, L. R. (1966). Methods for determining the nutritive value of feather meals. *Poult. Sci.* **45**, 1231–1236.

McCullough, D. R., and Ullrey, D. E. (1981). Tissue, organ and whole body composition and gross energy. Univ. Calif., Berkeley, Unpublished manuscript.

Meng, M. S., West, G. C., and Irving, L. (1969). Fatty acid composition of caribou bone marrow. *Comp. Biochem. Physiol.* **30**, 187–191.

Moreau, R. E., and Dolp, R. M. (1970). Fat, water, weights, and wing lengths of autumn migrants in transit on the northwest coast of Egypt. *Ibis* 112, 209–228.

Morton, M. L., and Tung, H. L. (1971). The relationship of total body lipid to fat depot weight and body weight in the Belding ground squirrel. *J. Mammal.* **52**, 839–842.

Moulton, C. R. (1923). Age and chemical development in mammals. *J. Biol. Chem.* **57**, 79–97.

Myrcha, A. (1968). Caloric value and chemical composition of the body of the European hare. *Acta Theriol.* **13**, 65–71.

Myrcha, A., and Pinowski, J. (1969). Variations in the body composition and caloric value of nestling tree sparrows (*Passer m. montanus* L.). *Bull. Pol. Acad. Sci. Cl. II Ser. Sci. Biol.* **17**, 475–480.

Myrcha, A., and Pinowski, J. (1970). Weights, body composition, and caloric value of postjuvenal molting European tree sparrows (*Passer montanus*). *Condor* 72, 175–181.

Myrcha, A., Pinowski, J., and Tomek, T. (1973). Variations in the water and ash contents and in the caloric value of nestling starlings (*Sturnus vulgaris* L.) during their development. *Bull. Pol. Acad. Sci. Cl. II Ser. Sci. Biol.* **21**, 649–655.

Nelson, R. A., Jones, J. D., Wahner, H. W., McGill, D. B., and Code, C. R. (1975). Nitrogen metabolism in bears: Urea metabolism in summer starvation and in winter sleep and role of urinary bladder in water and nitrogen conservation. *Mayo Clin. Proc.* **50**, 141–146.

Newton, I. (1966). The moult of the bullfinch *Pyrrhula pyrrhula*. *Ibis* **108**, 41–67.

Newton, I. (1968). The temperatures, weights, and body composition of molting bullfinches. *Condor* **70**, 323–332.

Newton, I. (1969). Winter fattening in the bullfinch. *Physiol. Zool.* **42**, 96–107.

Nitsan, Z., Dvorin, A., and Nir, I. (1981). Composition and amino acid content of carcass, skin and feathers of the growing gosling. *Brit. Poult. Sci.* **22**, 79–84.

O'Conner, R. J. (1976). Weight and body composition in nestling blue tits, *Parus caeruleus*. *Ibis* **118**, 108–112.

Odum, E. P. (1960). Lipid deposition in nocturnal migrant birds. *Proc. XII Int. Ornith. Congr.,* *Helsinki,* **2,** 563–576.

Odum, E. P., Rogers, D. T., and Hicks, D. L. (1964). Homeostasis of the nonfat components of migrating birds. *Science* **143,** 1037–1039.

Odum, E. P., Marshall, S. G., and Marples, T. G. (1965). The caloric content of migrating birds. *Ecology* **46,** 901–904.

O'Farrell, M. J., and Studier, E. H. (1976). Seasonal changes in wing loading, body composition, and organ weights in *Myotis thysanodes* and *M. lucifugus* (Chiroptera: Vespertilionidae). *Bull. South. Calif. Acad. Sci.* **75,** 258–266.

Okon, E. E., and Ekanem, R. T. (1976). Diurnal variations of the glycogen and fat stores in the liver and breast muscle of the insect bat, *Tadarida nigeriae. Physiol. Behav.* **23,** 659–661.

Owen, M., and Ogilvie, M. A. (1979). Wing molt and weights of barnacle geese in Spitsbergen. *Condor* **81,** 42–52.

Paladines, O. L., Reid, J. T., Bensadoun, A., and Van Niekerk, B. D. H. (1964). Heat of combustion values of the protein and fat in the body and wool of sheep. *J. Nutr.* **82,** 145–149.

Payne, R. B. (1972). Mechanisms and control of molt. *In* "Avian Biology," Vol. II (D. S. Farner and J. R. King, eds.), pp. 103–155. Academic Press, New York.

Penney, J. G., and Bailey, E. D. (1970). Comparison of the energy requirements of fledging black ducks and American coots. *J. Wildl. Manage.* **34,** 105–114.

Pfeiffer, W. E., Reinking, L. N., and Hamilton, J. D. (1979). Some effects of food and water deprivation on metabolism in black-tailed prairie dogs, *Cynomys ludovicianus. Comp. Biochem. Physiol.* **63A,** 19–22.

Pond, C. M. (1978). Morphological aspects and the ecological and mechanical consequences of fat deposition in wild vertebrates. *Ann. Rev. Ecol. Syst.* **9,** 519–570.

Prange, H. D., Anderson, J. F., and Rahn, H. (1979). Scaling of skeletal mass to body mass in birds and mammals. *Am. Nat.* **113,** 103–122.

Pucek, M. (1973). Variability of fat and water content in two rodent species. *Acta Theriol.* **18,** 57–80.

Reid, J. T., Bensadoun, A., Bull, L. S., Burton, J. H., Gleeson, P. A., Han, I. K., Joo, Y. D., Johnson, D. E., McManus, W. R., Paladines, O. L., Stroud, J. W., Tyrrell, H. F., Van Niekerk, B. D. H., and Wellington, G. W. (1968). Some peculiarities in the body composition of animals. *Nat. Acad. Sci. Pub.* **1598,** 19–44.

Reynolds, W. W. (1977). Skeleton weight allometry in aquatic and terrestrial vertebrates. *Hydrobiologia* **56,** 35–37.

Richards, J. R., Drury, J. K., Goll, C., Besent, R. G., and Al-Shamma, G. A. A. (1978). Energy exchanges and injury. *Proc. Nutr. Soc.* **37,** 39–43.

Ricklefs, R. E. (1968). Patterns of growth in birds. *Ibis* **110,** 419–451.

Ricklefs, R. E. (1973). Patterns of growth in birds. II. Growth rate and mode of development. *Ibis* **115,** 177–201.

Ricklefs, R. E. (1974). Energetics of reproduction in birds. *In* "Avian Energetics" (R. A. Paynter, Jr., ed.), pp. 152–297. Publ. Nuttall Ornith. Club. No. 15, Cambridge, Mass.

Ricklefs, R. E. (1975). Patterns of growth in birds. III. Growth and development of the cactus wren. *Condor* **77,** 34–45.

Ricklefs, R. E. (1979). Patterns of growth in birds. V. A comparative study of development in the starling, common tern, and Japanese quail. *Auk* **96,** 10–30.

Ricklefs, R. E., and White, S. C. (1981). Growth and energetics of chicks of the sooty tern (*Sterna fuscata*) and common tern (*S. hirundo*). *Auk* **98,** 361–378.

Ricklefs, R. E., White, S. C., and Cullen, J. (1980). Energetics of postnatal growth in Leach's storm-petrel. *Auk* **97,** 566–575.

Robbins, C. T. (1973). The biological basis for the determination of carrying capacity. Doctoral dissertation, Cornell Univ., Ithaca, N.Y.

Robbins, C. T. (1981). Estimation of the relative protein cost of reproduction in birds. *Condor* **83**, 177–179.

Robbins, C. T., and Moen, A. N. (1975). Uterine composition and growth in pregnant white-tailed deer. *J. Wildl. Manage.* **39**, 684–691.

Robbins, C. T., Moen, A. N., and Reid, J. T. (1974). Body composition of white-tailed deer. *J. Anim. Sci.* **38**, 871–876.

Robbins, C. T., Podbielancik-Norman, R. S., Wilson, D. L., and Mould, E. D. (1981). Growth and nutrient consumption of elk calves compared to other ungulate species. *J. Wildl. Manage.* **45**, 172–186.

Rock, P., and Williams, O. (1979). Changes in lipid content of the Montane vole. *Acta Theriol.* **24**, 237–247.

Rush, W. M. (1932). Northern Yellowstone elk study. *Montana Fish and Game Comm.*

Sawicka-Kapusta, K. (1968). Annual fat cycle of field mice *Apodemus flavicollus* (Melchoir, 1834). *Acta Theriol.* **13**, 329–339.

Sawicka-Kapusta, K. (1970). Changes in the gross body composition and the caloric value of the common voles during their postnatal development. *Acta Theriol.* **15**, 67–79.

Sawicka-Kapusta, K. (1974). Changes in the gross body composition and energy value of the bank voles during their postnatal development. *Acta Theriol.* **19**, 27–54.

Schemmel, R. (1976). Physiological considerations of lipid storage and utilization. *Am. Zool.* **16**, 661–670.

Schinckel, P. G., and Short, B. F. (1961). The influence of nutritional level during pre-natal and early post-natal life on adult fleece and body characters. *Aust. J. Agric. Res.* **12**, 176–202.

Schmidt-Nielsen, K. (1977). The physiology of wild animals. *Proc. Royal Soc. London* **199B**, 345–360.

Schultz, T. D., and Ferguson, J. H. (1974). Influence of dietary fatty acids on the composition of plasma fatty acids in tundra wolf (*Canis lupus tundrarum*). *Comp. Biochem. Physiol.* **49A**, 575–581.

Schreiber, R. K., and Johnson, D. R. (1975). Seasonal changes in body composition and caloric content of Great Basin rodents. *Acta Theriol.* **20**, 343–364.

Skar, H. J., Hagvar, S., Hagen, A., and Ostbye, E. (1975). Food habits and body composition of adult and juvenile meadow pipit [*Anthus pratensis* (L.)]. *In* "Fennoscandian Tundra Ecosystems. Part 2. Animals and Systems Analysis" (F. W. Wielgolaski, ed.), pp. 160–173. Springer-Verlag, New York.

Smith, N. S. (1970). Appraisal of condition estimation methods of East African ungulates. *E. Afr. Wildl. J.* **8**, 123–129.

Stettenheim, P. (1972). The integument of birds. *In* "Avian Biology," Vol. II (D. S. Farner and J. R. King, eds.), pp. 2–63. Academic Press, New York.

Stewart, R. E. A., and Lavigne, D. M. (1980). Neonatal growth of northwest Atlantic harp seals, *Pagophilus groenlandicus. J. Mammal.* **61**, 670–680.

Stirling, I., and McEwan, E. H. (1975). The caloric value of whole ringed seals (*Phoca hispida*) in relation to polar bear (*Ursus maritimus*) ecology and hunting behavior. *Can. J. Zool.* **53**, 1021–1027.

Sugden, L. G., and Harris, L. E. (1972). Energy requirements and growth of captive lesser scaup. *Poult. Sci.* **51**, 625–633.

Thomas, V. G., Lumsden, H. G., and Price, D. H. (1975). Aspects of winter metabolism of ruffed grouse (*Bonasa umbellus*) with special reference to energy reserves. *Can. J. Zool.* **53**, 434–440.

Turček, F. J. (1966). On plumage quantity in birds. *Ekol. Pol., Ser. A* **14**, 617–634.

Turner, J. C. (1979). Adaptive strategies of selective fatty acid deposition in the bone marrow of desert bighorn sheep. *Comp. Biochem. Physiol.* **62A**, 589–604.

Ullrey, D. E., Youatt, W. G., Johnson, H. E., Cowan, A. B., Fay, L. D., Covert, R. L., Magee, W.

T., and Keahey, K. K. (1975). Phosphorus requirements of weaned white-tailed deer fawns. *J. Wildl. Manage.* **39**, 590–595.

Ward, P. (1969). Seasonal and diurnal changes in the fat content of an equitorial bird. *Physiol. Zool.* **42**, 85–95.

Ward, W. H., and Lundgren, H. P. (1954). The formation, composition and properties of the keratins. *In* "Advances in Protein Chemistry" (M. T. Anson, K. Bailey, and J. T. Edwall, eds.), pp. 243–297. Academic Press, New York.

West, G. C., and Meng, M. S. (1968). Seasonal changes in body weight and fat and the relation of fatty acid composition to diet in the willow ptarmigan. *Wilson Bull.* **80**, 426–441.

West, G. C., and Shaw, D. L. (1975). Fatty acid composition of dall sheep bone marrow. *Comp. Biochem. Physiol.* **50B**, 599–601.

West, G. C., Burns, J. J., and Modafferi, M. (1979). Fatty acid composition of Pacific walrus skin and blubber fats. *Can. J. Zool.* **57**, 1249–1255.

Williams, A. J., Siegfried, W. R., Burger, A. E., and Berruti, A. (1977). Body composition and energy metabolism of molting eudyptid penguins. *Comp. Biochem. Physiol.* **56A**, 27–30.

Wyndham, E. (1980). Total body lipids of the budgerigar *Melopsittacus undulatus* (Psittaciformes: Platycercidae) in inland mid-eastern Australia. *Aust. J. Zool.* **28**, 239–247.

Yarbrough, C. G. (1970). Summer lipid levels of some subarctic birds. *Auk* **87**, 100–110.

Yarbrough, C. G., and Johnston, D. W. (1965). Lipid deposition in wintering and premigratory myrtle warblers. *Wilson Bull.* **77**, 175–191.

Yom-Tov, Y., and Tietz, A. (1978). The effect of diet, ambient temperature and day length on the fatty acid composition in the depot fat of the European starling (*Sturnus vulgaris*) and the rock partridge (*Alectoris chucar*). *Comp. Biochem. Physiol.* **60A**, 161–164.

Young, R. A. (1976). Fat, energy, and mammalian survival. *Am. Zool.* **16**, 699–710.

Zar, J. H. (1977). Fatty acid composition of emperor penguin (*Aptenodytes forsteri*) lipids. *Comp. Biochem. Physiol.* **56B**, 109–112.

Zimmerman, J. L. (1965). Carcass analysis of wild and thermal stressed dickcissels. *Wilson Bull.* **77**, 55–70.

11

Food Resources and Their Utilization

Biology is by far a more difficult subject than physics. It is much more complex, and it has more variables. It is more difficult to do a controlled experiment and to understand the basic laws of nature which apply in biology than it is in physics. The complexity of problems in biology means that the subject should be attacked with all the skills and tools available.

—Gates, 1962, p. 2

The animal's nutritional environment is a vast array of chemical compounds conveniently ordered by wildlife scientists into aggregations from species to communities. While countless food habit studies have indicated the relative importance of plant or animal species in the nutritional environment of many animals, our increasing knowledge of food selection and metabolism dictates that far more basic and analytical studies are needed if we are to understand the food–animal interaction. While the simple listing of species ingested by animals is a first step in understanding an animal's natural history, such studies provide only the slightest glimmer of a nutritional understanding.

Latin binomials (animals) simply do not eat other latin binomials (plant or animal) (Janzen, 1979). Food selection by free-ranging animals is not based on the anatomical characteristics used by plant or animal taxonomists, but rather on the animal's perception of cost–benefit constraints imposed when different foods are sought and ingested (Arnold and Hill, 1972). Therefore, our quantification of the nutritional environment must be based on the animal's perception of that environment. The animal is linked to its nutritional environment by (a) olfactory, visual, and taste cues enabling food selection and (b) digestion–metabolism processes enabling absorption, distribution, and utilization of the specific chemicals within the ingested food. The integration of these processes in which external cues are associated with internal metabolic events provides the framework within which the animal can constantly evaluate its nutrient environment.

I. FOOD COMPOSITION

A. Carbohydrates

Most plant carbohydrates are polymers of five- and six-carbon monosaccharides. Plant carbohydrates can be divided into nonstructural and structural categories on the basis of their composition, position, and function. Nonstructural carbohydrates range from soluble mono- and disaccharides, such as sucrose in plant sap, to large polysaccharides, such as starch in seeds and roots. Nonstructural carbohydrates are important reserves for plant growth and respiration. They are used to initiate new growth and maintain plant respiration when photosynthesis is inadequate. Daily and seasonal rhythms in nonstructural carbohydrates represent a balancing between photosynthesis and respiration. For example, nonstructural carbohydrates are lowest in early morning after night respiration and prior to active photosynthesis and increase during the day as photosynthetic production exceeds respiration. On a seasonal basis, nonstructural carbohydrates within the vegetative structures decrease during rapid spring growth and during seed production, but often increase during late fall preparatory to plant dormancy (McConnell and Garrison, 1966). The annual cycle of reserve carbohydrates is extremely important in determining plant vigor, health, and regeneration after foliage removal. Although nonstructural carbohydrates occur in all living plant parts, highest concentrations occur in seeds, stem bases and roots of grasses, stems of legumes, and roots of most trees and shrubs (Kramer and Kozlowski, 1960; McConnell and Garrison, 1966; Smith, 1973; White, 1973).

The two principal structural carbohydrates of plant cell walls are cellulose and hemicellulose (Table 11.1). Cellulose, the predominant carbohydrate of higher plants, is a glucose polymer linked by $\beta1,4$ bonds. Unlike starch, which is a

TABLE 11.1
Carbohydrate Content of Plant Tissues[a]

Component	Function	Content (% of dry matter)			
		Seeds	Legumes	Grasses	Browses
Soluble sugars	Nonstructural	Negligible	2–16	5–15	5–15
Starch	Nonstructural	80	1–7	1–5	—
Pectin	Structural	Negligible	5–10	1–2	6–12
Hemicellulose	Structural	7–15	3–10	15–40	8–12
Cellulose	Structural	2–5	7–35	20–40	12–30

[a]Data from Bailey and Ulyatt, 1970; Mould and Robbins, 1981; Van Soest, 1981.

highly digestible glucose polymer, cellulose cannot be digested by vertebrates without a symbiotic gastrointestinal microflora. Hemicelluloses are predominantly polymers of five-carbon sugars, although six-carbon sugars, including glucose, also occur (Bailey, 1973; Wilkie, 1979). Because of the diversity of monosaccharide components and linkages, hemicelluloses are poorly understood. Hemicelluloses are hydrolyzed by both acid and alkaline solutions and thus are partially digested by stomach acidity. Cellulose and hemicellulose are occasionally grouped as holocellulose.

Pectin is the remaining principal structural carbohydrate and occurs primarily in the middle lamella between adjacent cell walls. Pectins are more prevalent in dicotyledonous plants than in grasses (Table 11.1). It is primarily composed of galacturonic acid linked $\alpha1,4$, with smaller amounts of arabinose and galactose (Van Soest, 1981). Pectins are readily digestible via fermentation. Because of the difficulty in definitively separating hemicelluloses and pectins on the basis of either their solubility or carbohydrate fractions, they are occasionally grouped simply as noncellulosic polysaccharides.

B. Protein

Dietary crude protein is usually determined by the Kjeldahl procedure, in which a weighed sample is boiled in a concentrated sulfuric acid-catalyst solution until all organic matter is destroyed and the nitrogen converted to ammonium sulfate. The solution is subsequently cooled, sodium hydroxide is added, and the ammonia is volatilized. The ammonia is captured in a weak acid solution and back titrated with a known normality acid to determine the amount of nitrogen in the original sample. As discussed earlier, nitrogen times 6.25 provides a crude estimate of dietary protein content.

True protein frequently accounts for 75–85% of the nitrogen in herbage and animal products, with the remainder being inorganic nitrogen, free amino acids, low molecular weight peptides, nucleic acids, and other nonprotein nitrogen compounds (Lyttleton, 1973; Luick et al., 1974). Most plant protein is found within the cytoplasm of the cell, with only small amounts associated with the cell wall (Lyttleton, 1973; Preston, 1974; Albersheim, 1975). Plant protein increases during early growth or regeneration when active anabolic processes are occurring, but then it decreases with increasing vegetative maturity as greater emphasis is placed on the production of nitrogen-free cell wall components for structural support (Greenwood and Barnes, 1978). For example, early growth in grasses, forbs, and browses can often be as high as 20–30% protein on a dry-weight basis, but the protein subsequently falls to as little as 3–4% at maturity (Dietz et al., 1962; Cowan et al., 1970; Kubota et al., 1970; Lyttleton, 1973; Cogswell and Kamstra, 1976).

The one exception to the generally decreasing protein content in plants with increasing maturity are relatively high levels in many seeds. For example, seeds consumed by ruffed grouse in New York contained 9.5–33.8% crude protein (Bump *et al.*, 1947). However, many commercial grains, such as wheat or corn, have relatively low levels of crude protein (generally 8–15%) since the high levels of endosperm carbohydrates dilute the higher protein concentrations in the germ. Many seeds also contain digestion inhibitors or toxins that reduce the amount of available protein or minimize their total intake (Bullard and Elias, 1979; McKey, 1979; Seigler, 1979).

Animal tissues are generally higher in total protein content on a dry-weight basis than are plants, and animal protein is generally either more available or more useful because of its amino acid spectra than is plant protein. For example, the dry matter of aquatic invertebrates contains 31–80% crude protein, with all species supplying essential amino acids in excess of the estimated requirements of growing waterfowl (Driver *et al.*, 1974). Similarly, while the vertebrate body usually contains 25% or less protein on a fresh-weight basis (Fig. 11.1), carnivores selectively feeding to avoid large fat deposits could consume diets as high as 70–75% protein per unit of dry weight.

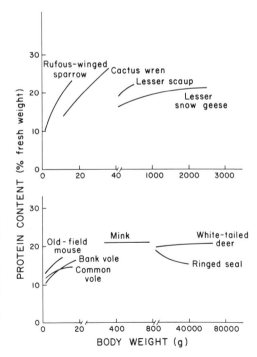

Fig. 11.1. Total body protein content of several wildlife species. (Data from Sugden and Harris, 1972; Robbins *et al.*, 1974; Fedyk, 1974; Sawicka-Kapusta, 1974; D. W. Kaufman and Kaufman, 1975; Ricklefs, 1975; Stirling and McEwan, 1975; G. A. Kaufman and Kaufman, 1977; Austin and Ricklefs, 1977; Harper *et al.*, 1978; Campbell and Leatherland, 1980.)

C. Fats

Total dietary lipids have traditionally been estimated by extracting a dried sample with an organic solvent, primarily anhydrous ethyl ether. Plant ether extract is a very heterogeneous mixture of generally high-energy compounds, such as glycerides, phospholipids, sterols, pigments, waxes, volatile oils, and resins. Very high concentrations of triglycerides in plants are confined to oil-bearing seeds and fruits, such as 75% of the dry weight in pecans, 62% in walnuts, and 64% in pinyon pine seeds (Kramer and Kozlowski, 1960). Conversely, the predominant glycerides of plant leaves and stems are largely mono- and digalactosyldiglycerides having a lower caloric content than a triglyceride (Lough and Garton, 1968; Hawke, 1973; Karis and Hudson, 1974). Rather than the three fatty acids of a triglyceride, plant glycerides often have either one or two galactose units replacing the fatty acids of a triglyceride. Since galactolipids are most prevalent in actively metabolizing leaves, their concentration decreases with increasing leaf maturity and increasing stem to leaf ratios (Van Soest, 1981).

Lipid analyses are confounded by their propensity to oxidize, volatilize, or form insoluble soaps. Unsaturated lipids are particularly prone to oxidation during storage to form insoluble, indigestible resins and gels. The aromatic fragrances of evergreens and sagebrush are volatile oils that can easily be lost during drying. Lipids in feces are often underestimated, since they commonly form calcium or magnesium soaps (Van Soest, 1981). Thus, nonspecific lipid extractions often have little nutritional meaning.

II. PROTECTIVE AND DEFENSIVE AGENTS

Since plant carbohydrates, proteins, and lipids are easily degraded, the development of protective and defensive agents has been an important step in plant evolution. Major evolutionary emphasis has been at the chemical level. Plants contain substances that physically impede digestive enzymes or microorganism attack as well as a far more diverse and active chemical defense arsenal of secondary plant compounds. Lignin, cutin, suberin, and biogenic silica are the principal structural agents that physically prevent degradation. Secondary plant constituents are a heterogeneous mixture of over 12,000 compounds of relatively small molecular weight that can interfere with growth, neurological and tissue functioning, reproduction, and digestion of organisms that ingest them (Freeland and Janzen, 1974; Scott, 1974).

A. Lignin

Lignin is a high molecular weight, aromatic, nonsaccharide polymer that adds rigidity to the plant cell wall (Harkin, 1973). Structural rigidity is probably lignin's most important evolutionary function since it is confined to the higher plants and does not occur in algae, fungi, and mosses. Lignin concentrations increase with advancing cellular maturity and can reach as much as 15–20% of the cell wall.

Lignin is resistant to normal enzymatic and acid hydrolysis and is therefore indigestible. Since lignin degradation is dependent on oxidation, it is also resistant to gastrointestinal fermentation. Although lignin is the main factor limiting the digestibility of cell wall polysaccharides, its mode of action is poorly understood. The most plausible explanation is that lignin–carbohydrate bonds throughout the cell wall impede enzymatic hydrolysis (Rhoades, 1979) because delignification of cell walls does increase their digestibility (Darcy and Belyea, 1980).

B. Cutin–Suberin

Cutin, an aliphatic polymer, and suberin, a mixed aromatic–aliphatic polymer, are closely related chemical compounds (Kolattukudy, 1977). Cutin is the structural component of the plant cuticle and is thus extracellular at the plant surface. Suberin occurs between the cell wall and cytoplasm and is a major constituent of bark (Martin and Juniper, 1970). Cutin and suberin are important to the plant in wound healing, in reducing water loss, and in physically preventing entry by microorganisms.

Cutin and suberin, as a percentage of plant dry matter, are quantitatively insignificant in grasses. However, browse cell walls contain up to 15% cutin–suberin (Robbins and Moen, 1975). Although cutin–suberin are slightly digestible (Brown and Kolattukudy, 1978), they may physically block the digestibility of more readily available cell wall polysaccharides.

C. Biogenic Silica

Silica can be incorporated in plant tissue as distinct, localized concretions. Grasses, sedges, and many of the lower plants, such as *Equisetum,* accumulate silica at many times the rate of dicotyledonous plants (Jones and Handreck, 1967). Silica has only recently been recognized as an important nutritional component of monocot cell walls because it reduces their digestibility (Van Soest and Jones, 1968).

D. Secondary Plant Compounds

Secondary plant compounds are an extremely diverse assemblage of antiherbivory chemicals. Their effects can range from feeding deterrency to antibiosis. They have been labeled "secondary" because few have primary metabolic functions within the plant and at one time they were viewed as secondary end products of other metabolic systems. They include alkaloids, pyrethrins, rotenoids, cyanogenic glycosides, long-chain unsaturated isobutylamides, cardenolides and saponins, sesquiterpene lactones, nonprotein amino acids, glucosinolates and isothiocyanates, oxalates, protoanemonin, hypericin, fluoro fatty acids, selenoamino acids, gossypol, condensed tannins, phenolic resin and phenol oxidase, proteinase inhibitors, and phytohemagglutinins (Harborne, 1972; Richardson, 1977; Rhoades, 1979). While most of these compounds have a broad spectrum of activity in providing defense against all enemies, selective toxicities do occur. For example, the pyrethrins are potent insecticides, with minimal mammalian toxicity (Mabry and Gill, 1979).

Since plants and their varied consumers are "involved in an incessant evolutionary struggle," evolution will favor the continued development of molecular uniqueness in secondary plant constituents (Levin, 1976b, p. 146). Plants must constantly evolve new defenses as their consumers evolve more elaborate detoxification pathways (Brattsten, 1979). Because many of these compounds are also toxic to plants, they must be either isolated in plant organelles or stored as inactive forms that are released or activated when consumed. For example, inactive cyanogenic glycosides occur in over 1000 plant species and are activated upon consumption to hydrogen cyanide. Both the glycoside and the activating glycosidase occur within the plant but are brought into contact only when the plant tissue is mascerated (Conn, 1979).

The concentrations of these compounds vary in both space and time within and between plants (Feeny, 1976; Rhoades and Cates, 1976; Mattson, 1980). For example, plants often increase the concentration of defensive compounds following stress or damage by herbivores (Rhoades, 1979). However, the increases after damage are often localized to adjacent plant parts. Similarly, although many unripened fruits contain very high levels of secondary plant compounds, plants that depend on vertebrate ingestion for seed scattering detoxify these compounds in the ripening fruit while continuing to concentrate them in the seed itself. Thus, the fruit is ingested only after the seed has matured and will pass undigested through the gastrointestinal tract. The three most prevalent groups of plant secondary compounds are soluble phenolics, alkaloids, and terpenoids.

1. Soluble Phenolics

Soluble plant phenolics include flavonoids, isoflavonoids, and hydrolyzable and condensed tannins. The defensive properties of these compounds are primarily due to their abilities to bind with proteins and other macromolecules, thereby precipitating cellular proteins, inactivating digestive enzymes, and possibly forming indigestible macromolecules with cell wall carbohydrates (Rhoades and Cates, 1976; Lohan et al., 1981). The flavonoids are particularly well known for the brilliant flower, fruit, and leaf colors that they confer. Flavonoids usually occur as soluble nontoxic glycosides in the plant. Although flavonoids and isoflavonoids are relatively nontoxic to vertebrate consumers, they are capable of reducing protein digestion and altering reproductive patterns in both birds and mammals. Several of the isoflavonoids, such as coumestrol and genistein, are similar in molecular configuration to estrogen (Harborne, 1979).

The term *tannin* derives from the ability of these soluble polyphenols to tan animal hides into leather. Tannins are the most ancient, most widespread, and most successful generalized defensive plant compound (Swain, 1979). They occur in about 17% of the nonwoody annuals, 14% of herbaceous perennials, 79% of deciduous woody perennials, and 87% of evergreen woody perennials (Rhoades and Cates, 1976). Because tannins are extremely nonspecific in their ability to precipitate proteins, they are equally effective in preventing fungal, viral, and bacterial attacks as well as herbivory by both vertebrates and invertebrates. Some of the smaller, hydrolyzable tannins can be absorbed by vertebrates that ingest them and thereby produce extensive physiological damage (McLeod, 1974; Mould and Robbins, 1982). The deterrence of soluble phenolics to herbivory depends on either innate or acquired association of their astringent taste with the incurred metabolic aberrations when ingested (Rhoades, 1979).

2. Alkaloids

Alkaloids differ from the other secondary plant compounds by having one or more nitrogen atoms in a heterocyclic ring (Culvenor, 1973; Levin, 1976b; Robinson, 1979). The term *alkaloid* was derived at a time when plant compounds were considered as either neutral or acidic from the observation that these nitrogen-containing compounds had alkaline properties (Wasacz, 1981). Approximately 4000 of the 12,000 known secondary plant compounds are alkaloids that are found in at least 7% of all flowering plants (Culvenor, 1973). The alkaloids include many of the plant-derived pharmacological and hallucinogenic drugs used by humans. Examples of well-known alkaloids are nicotine and various morphine drugs, delphinine of larkspur, conine of poison hemlock, tomatine of tomato plants, atropine of deadly nightshade, and lupinine of lupine. Plant alkaloid concentrations vary during the day. For example, opium sap has been collected for centuries during the morning, because the sap has its highest morphine content between 9 and 10 A.M. (Wasacz, 1981).

The prevalence of secondary plant compounds, and particularly alkaloids, also increases from arctic to tropical latitudes. Just as plant diversity and abundance increase as the tropics are approached, so do the pressures from plant pests that dictate the need for more extensive plant defenses. For example, the percentage of alkaloid-bearing plants increases from approximately 15% at 40° north or south latitude to 40% at the equator (Levin, 1976a; Moody, 1978; Robinson, 1979). Alkaloids also are more common in plants growing in nitrogen-rich environments (Mattson, 1980).

3. Terpenoids

Terpenoids are one of the largest classes of plant secondary constituents. They are all low molecular weight, generally cyclic compounds that are soluble in organic solvents. They are all structurally related through their repeating isopentenoid units, $[CH_2 = C(—CH_3) — CH = CH_2]$. Examples of commonly encountered terpenes include the volatile, aromatic oils of sagebrush and evergreens, carotenes and vitamin A, the insecticide pyrethrin from *Chrysanthemum*, eucalyptol of euclayptus, peel oils of citrus fruits, and gossypol of cottonseed.

Terpenoids are best known for their insecticidal and antimicrobial activities. Terpenoid toxicity to gastrointestinal bacteria in which small amounts can be tolerated is dose dependent. However, cellulolytic rumen bacteria are more susceptible to terpenoid toxicity than are starch-digesting bacteria (Oh *et al.*, 1967; Schwartz *et al.*, 1980a). Because many of the terpenoids are either volatile or bitter in taste and thus can be readily associated with the metabolic aberrations, they are effective feeding deterrents (Morrow *et al.*, 1976; Mabry and Gill, 1979; Schwartz *et al.*, 1980b).

III. ENERGY CONTENT

The caloric content, or gross energy, of a food is dependent on its relative composition and caloric content of the different constituents. Prey items consumed by carnivores can vary extensively in their relative energy content. For example, the larger individual of a vertebrate species generally has a higher energy content because of the increasing fat and decreasing water deposition during growth (Figs. 11.2 and 11.3). Thus, the live weight of the necessary prey consumed to meet a predator's requirement could vary by at least twofold depending on the relative growth stage of the prey item. Similarly, selective feeding on only portions of prey can alter the energy content of the diet. For example, the selective feeding by polar bears on the fat deposits of ringed seals dramatically increases the average energy content of their diet (Stirling and McEwan, 1975). Whole fish range from 0.83 to 1.50 kcal/g fresh weight and average 5.0 kcal/g dry weight (MacKinnon, 1972; Jensen, 1979). The average caloric content of fish on a dry-weight basis indicates primarily protein, although species with significant lipid deposition cycles, such as preparatory to spawning or migration, can contain significantly more energy. The highest caloric content of invertebrates, ranging from 3.7 to 7.1 kcal/g dry ash-free weight, with most between 5 and 6, is in reproductively active, egg-laden females, advancing prepupal stages, and seasonally dormant species because of extensive fat accumulation (Slobodkin and Richman, 1961; Wiegert, 1965; Wissing and Hasler, 1971; Driver *et al.*, 1974; Norberg, 1978).

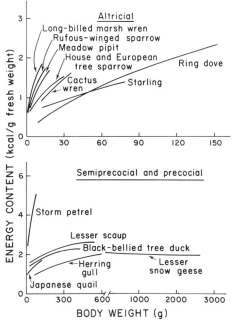

Fig. 11.2. The average body energy content of several wild birds. (Data from Kale, 1965; Brisbin, 1969; Myrcha and Pinowski, 1969; Sugden and Harris, 1972; Myrcha *et al.,* 1973; Brisbin and Tally, 1973; Blem, 1975; Ricklefs, 1975; Skar *et al.,* 1975; Cain, 1976; Austin and Ricklefs, 1977; Bryant and Gardiner, 1979; Campbell and Leatherland, 1980; Dunn and Brisbin, 1980; Ricklefs *et al.,* 1980.)

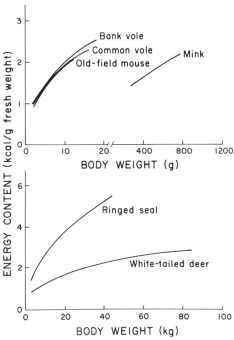

Fig. 11.3. The average body energy content of several wild mammals. (Data from Fedyk, 1974; Robbins *et al.,* 1974; Sawicka-Kapusta, 1974; Stirling and McEwan, 1975; D. W. Kaufman and Kaufman, 1975; G. A. Kaufman and Kaufman, 1977; Harper *et al.,* 1978.)

The energy content of plant seeds, ranging from 4.3 to 6.8 kcal/g dry weight, increases from averages of 4.5 kcal/g for grass seeds, 5.0 kcal/g for deciduous tree and shrub seeds, to 6.1 kcal/g for coniferous tree seeds (Kendeigh and West, 1965; Johnson and Robel, 1968; Grodzinski and Sawicka-Kapusta, 1970; Robel *et al.*, 1979a, 1979b). The increasing energy content of the seeds is directly related to their fat content. Plant leaves and stems are generally less variable than are seeds in energy content and range from 3.9 to 5.1 kcal/g dry weight (Golley, 1961; Bliss, 1962).

IV. ANALYTICAL METHODS FOR PLANT FIBER

Nutritionists have long attempted to develop analytical systems to predict the nutritive value of plant matter. Since plant cytoplasmic proteins, carbohydrates, and fats are usually highly digestible, much of the emphasis has been on understanding plant cell walls. The nutritive value of the plant cell wall is a function of its digestibility. Plant cell wall digestibility depends on its chemical composition, the internal molecular configuration of each component, the three-dimensional relationships of all components, and the digestive system of the ingesting animal. The statistical comparison of cell wall chemical analyses and cell wall digestibility in order to predict nutritive value rests on the development of biologically meaningful chemical analyses.

A. Sample Preparation

Care of samples during collection, storage, and analysis is of the utmost importance in the development of analyses that reflect the nutritive value of the animal's diet. Sample care and subsequent analysis must ensure both maximum accuracy in component extraction and identification, as well as practicality. Food samples are routinely dried and ground prior to analysis. Such procedures are justifiable only as long as the chemical composition is unaltered.

The effects of drying are dependent on temperature and the forage chemical composition. Drying of forages above 50°C often increases the lignin content, because proteins and carbohydrates complex in the nonenzymatic browning or Maillard reaction to form insoluble lignin-like complexes. Conversely, air drying of forages can increase cell wall content, because cell solubles are metabolized during the time the fresh plant sample continues respiration. Thus, the preferred drying temperature for most forages is 40°C, which is rapid and hot enough to minimize cell soluble catabolism yet cool enough to minimize artifact lignin formation (Van Soest, 1965b; Goering *et al.*, 1973; Culberson *et al.*, 1977; Acosta and Kothmann, 1978; Mould and Robbins, 1981).

Forages that contain significant levels of soluble phenolics, volatile terpenoids, or unsaturated lipids often need special care before or during drying. Since the analyses of these compounds in the dried sample will underestimate their content in the original sample because of volatilization, formation of insoluble complexes, or oxidation, immediate extraction of the fresh sample may be necessary. Volatile oils are commonly obtained through steam distillation of fresh or frozen samples, and soluble phenolics by aqueous-organic solvent extraction of fresh samples (Lofti *et al.*, 1976; Feeny and Bostock, 1968; Swain, 1979; Schwartz *et al.*, 1980a; Mould and Robbins, 1981).

When forages are dried, ground, and stored, all hydrophilic compounds when cooled will equilibrate with atmospheric water. Dried forage samples that have cooled to room temperature commonly contain 6–10% water. Consequently, a small subsample of the previously dried material is redried at 100°C in order to express all subsequent analyses on a 100% dry-matter basis.

B. Fiber Analyses

Fiber analyses have attempted either to quantify chemically discrete components, such as cellulose, hemicellulose, or lignin, or to separate digestible from nondigestible fractions. The two aims of fiber analyses are not synonymous, since cellulose and hemicellulose are partially digestible. Attempts to develop simple chemical procedures to quantify the nondigestible fraction have been futile.

First attempts at fiber analyses for nutritional purposes began in 1803, when plant samples were merely washed with water. Subsequently, alcohol, ether, and weak acid and alkaline solutions were used in attempts to remove digestible fractions more thoroughly while retaining fiber (Tyler, 1975). Further development and standardization of methods to define nutritional entities led in the 1850s and 1860s to what is now known as the proximate system of analysis (Fonnesbeck, 1977; Van Soest, 1981). It consists of the following steps:

1. Determine dry matter at 100°C.
2. Extract with ether the dried sample to estimate lipid content.
3. Reflux the ether-extracted residue for 30 min in 1.25% sulfuric acid, followed by a similar boiling in 1.25% sodium hydroxide. The insoluble organic residues are reported as crude fiber.
4. Determine nitrogen by the Kjeldahl method and multiplied by 6.25 to estimate crude protein content.
5. Determine ash content at 500°C.
6. Subtract the percentage of ether extract, crude fiber, crude protein, and ash from 100 to estimate the nitrogen-free extract (NFE).

Such analyses are definitive, since the method of isolation arbitrarily defines the residue (Van Soest, 1969). The components at each extraction are not purified and identified on the basis of their chemical properties. Such analyses are useful only if the method of isolation depends on a unique molecular property that allows a quantitatively discrete chemical determination relatively free of interfering compounds or provides the ability to predict the outcome of a specific nutritional interaction, such as digestion.

Since each step in the proximate analysis determines the content of a heterogeneous mixture, the value of proximate analysis must rest on its ability to predict the sample's nutritive value. Crude fiber has been considered the least digestible fibrous matter, and NFE has been considered highly digestible carbohydrates. However, the digestibilities of crude fiber range from 0 to 90% in ruminants (Maynard and Loosli, 1969). Because lignin is partially soluble in alkaline solutions and hemicellulose in both acid and alkali (Van Soest, 1969), NFE often has a lower digestibility than does crude fiber. Thus, the two major components of proximate analysis have neither a uniform composition nor a predictable digestibility. Although crude fiber–NFE analyses continue to be used because of their simplicity and sizable data base generated during the last 100 years, further progress in understanding the role of fiber in animal nutrition is dependent on the development of better analytical procedures.

Detergent analyses were proposed during the early 1960s as a replacement for crude fiber–NFE determinations (Van Soest, 1963a, 1963b, 1965a, 1967; Goering and Van Soest, 1970). The plant sample is theoretically divided into cell wall–cell soluble fractions on the basis of their solubility in detergent solutions of differing acidity. Detergents are used to emulsify and bind cell solubles (cytoplasm), particularly protein that must be removed to minimize artifact lignin formation. The residual cell wall can be further partitioned into its components: hemicellulose, cellulose, lignin, and cutin–suberin. The basic steps in detergent analyses are:

1. Reflux a plant sample in neutral detergent to obtain the cell wall or neutral detergent fiber residue (NDF). Cell solubles or neutral detergent solubles (NDS) are equal to the weight loss between the original sample and the NDF. The NDF contains hemicellulose, cellulose, lignin, and cutin– suberin. Separate amylase treatments are necessary when high levels of starch occur, as in seeds or vegetables (McQueen and Nicholson, 1979).
2. Reflux the NDF in acid detergent to remove hemicellulose and determine acid detergent fiber (ADF) by difference.
3. Treat ADF with saturated potassium permanganate to oxidize and remove lignin and determine lignin by difference.
4. Treat the delignified ADF with 72% sulfuric acid to solubilize cellulose and determine cellulose by difference.
5. Ash the sulfuric acid residue to gravimetrically determine cutin–suberin. When the samples contain minimal cutin–suberin, Step 3 can be omitted,

since the ADF cellulose is removed with 72% H_2SO_4. The residue is subsequently ashed to estimate lignin.

6. Biogenic silica is determined by treating an ashed separate ADF residue with hydrobromic acid.

Detergent analysis does not attempt in itself to determine the nutritive value or digestibility of the various plant components. However, it does provide a useful definitive separation of the plant sample into its basic components, which ultimately may be related to nutritive value through statistical comparisons. While detergent analyses are appealing after decades of stagnation in forage analyses, intriguing difficulties still remain because of the chemical complexity of plants relative to the use of definitive methods.

Detergent analyses have changed significantly since their inception. For example, sodium sulfite, which was initially added to the neutral detergent solution to improve protein removal, is no longer used since it also removed lignin (Van Soest and Robertson, 1980, 1981; Mould and Robbins, 1981). Neutral detergent fiber and ADF were originally isolated from separate samples of the intact plant material rather than sequentially as outlined here. However, it was not uncommon to obtain higher ADF than NDF, resulting in negative hemicellulose estimates by subtraction (Milchunas et al., 1978). The use of sequential analyses eliminates the potential for negative hemicellulose values, but it does not necessarily provide an accurate hemicellulose estimate (Mould and Robbins, 1981, 1982). Similarly, pectins are removed by neutral detergent (Bailey and Ulyatt, 1970; Bailey et al., 1978; Mould and Robbins, 1981). Since pectins are a major structural carbohydrate of the plant cell wall, neutral detergent fiber and cell wall are not synonyms.

Secondary plant constituents form intriguing paradoxes in detergent analyses because they are usually extracted with the neutral detergent solubles. Consequently, current detergent analyses should be viewed with a healthy degree of skepticism. Although the system offers far greater potential for understanding the nutritional value of plant matter than did crude fiber–NFE analyses, further development is necessary. Ultimately, it may be unrealistic to expect that one analytical scheme will be suitable for all forages. More extensive studies of cell wall carbohydrates and secondary plant compounds are needed if meaningful development of forage analyses is to continue.

REFERENCES

Acosta, R. A., and Kothmann, M. M. (1978). Chemical composition of esophageal-fistula forage samples as influenced by drying method and salivary leaching. J. Anim. Sci. 47, 691–698.

Albersheim, P. (1975). The walls of growing plant cells. Sci. Am. 232, 80–95.

Arnold, G. W., and Hill, J. L. (1972). Chemical factors affecting selection of food plants by ruminants. *In* "Phytochemical Ecology" (J. B. Harborne, ed.), pp. 72–101. Academic Press, New York.

Austin, G. T., and Ricklefs, R. E. (1977). Growth and development of the rufous-winged sparrow (*Aimophila carpalis*). *Condor* **79**, 37–50.

Bailey, R. W. (1973). Structural carbohydrates. *In* "Chemistry and Biochemistry of Herbage," Vol. I (G. W. Butler and R. W. Bailey, eds.), pp. 157–211. Academic Press, New York.

Bailey, R. W., and Ulyatt, M. J. (1970). Pasture quality and ruminant nutrition. II. Carbohydrate and lignin composition of detergent-extracted residues from pasture grasses and legumes. *N. Z. J. Agric. Res.* **13**, 591–604.

Bailey, R. W., Chesson, A., and Munroe, J. (1978). Plant cell wall fractionation and structural analysis. *Am. J. Clin. Nutr.* **31**, S77–S81.

Blem, C. R. (1975). Energetics of nestling house sparrows *Passer domesticus*. *Comp. Biochem. Physiol.* **52A**, 305–312.

Bliss, L. C. (1962). Caloric and lipid content in alpine tundra plants. *Ecology* **43**, 753–754.

Brattsten, L. B. (1979). Biochemical defense mechanisms in herbivores against plant allelochemicals. *In* "Herbivores: Their Interaction with Secondary Plant Metabolites" (G. A. Rosenthal and D. H. Janzen, eds.), pp. 199–270. Academic Press, New York.

Brisbin, I. L., Jr. (1969). Bioenergetics of the breeding cycle of the ring dove. *Auk* **86**, 54–74.

Brisbin, I. L., Jr., and Tally, L. J. (1973). Age-specific changes in the major body components and caloric value of growing Japanese quail. *Auk* **90**, 624–635.

Brown, A. J., and Kolattukudy, P. E. (1978). Evidence that pancreatic lipase is responsible for the hydrolysis of cutin, a biopolyester present in mammalian diet, and the role of bile salt and colipase in this hydrolysis. *Arch. Biochem. Biophys.* **190**, 17–26.

Bryant, D. M., and Gardiner, A. (1979). Energetics of growth in house martins (*Delichon urbica*). *J. Zool.* **189**, 275–304.

Bullard, R. W., and Elias, D. J. (1979). Sorghum polyphenols and bird resistance. *Proc. Inst. Food Tech.* **36**, 43–49.

Bump, G., Darrow, R. W., Edminster, F. C., and Crissey, W. F. (1947). "The Ruffed Grouse: Life History, Propagation, Management." New York Cons. Dept., Albany.

Cain, B. W. (1976). Energetics of growth for black-bellied tree ducks. *Condor* **78**, 124–128.

Campbell, R. R., and Leatherland, J. F. (1980). Estimating body protein and fat from water content in lesser snow geese. *Q. Rev. Biol.* **53**, 243–282.

Cogswell, C., and Kamstra, L. D. (1976). The stage of maturity and its effect upon the chemical composition of four native range species. *J. Range Manage.* **29**, 460–463.

Conn, E. E. (1979). Cyanide and cyanogenic glycosides. *In* "Herbivores: Their Interaction with Secondary Plant Metabolites" (G. A. Rosenthal and D. H. Janzen, eds.), pp. 387–412. Academic Press, New York.

Cowan, R. L., Jordan, J. S., Grimes, J. L., and Gill, J. D. (1970). Comparative nutritive values of forage species. *Range and Wildlife Habitat Evaluation, U.S.D.A. Misc. Publ. No.* **1147**, 48–56.

Culberson, C. R., Culberson, W. L., and Johnson, A. (1977). Thermally induced chemical artifacts in lichens. *Phytochemistry* **16**, 127–130.

Culvenor, C. C. (1973). Alkaloids. *In* "Chemistry and Biochemistry of Herbage," Vol. I (G. W. Butler and R. W. Bailey, eds.), pp. 375–446. Academic Press, New York.

Darcy, B. K., and Belyea, R. L. (1980). Effect of delignification upon *in vitro* digestion of forage cellulose. *J. Anim. Sci.* **51**, 798–803.

Dietz, D. R., Udall, R. H., and Yeager, L. E. (1962). Chemical composition and digestibility by mule deer of selected forage species, Cache La Poudre Range, Colorado. *Colo. Game Fish Tech. Publ. No.* **14**.

Driver, E. A., Sugden, L. G., and Kovach, R. J. (1974). Calorific, chemical and physical values of potential duck foods. *Freshwater Biol.* **4**, 281–292.

Dunn, E. H., and Brisbin, I. L., Jr. (1980). Age-specific changes in the major body components and caloric values of herring gull chicks. *Condor* **82,** 398–401.

Fedyk, A. (1974). Gross body composition in postnatal development of the bank vole. I. Growth under laboratory conditions. *Acta Theriol.* **19,** 381–401.

Feeny, P. (1976). Plant apparency and chemical defenses. *Recent Adv. Phytochem.* **10,** 1–42.

Feeny, P. P., and Bostock, H. (1968). Seasonal changes in the tannin content of oak leaves. *Phytochemistry* **7,** 871–880.

Fonnesbeck, P. V. (1977). Estimating nutritive value from chemical analyses. *Proc. Int. Symp. Feed Composition, Animal Nutr. Reg., and Computerization of Diets* **1,** 219–227.

Freeland, W. J., and Janzen, D. H. (1974). Strategies in herbivory by mammals: The role of secondary plant compounds. *Am. Nat.* **108,** 269–289.

Gates, D. M. (1962). "Energy Exchange in the Biosphere." Harper, New York.

Goering, H. K., and Van Soest, P. J. (1970). Forage fiber analyses (apparatus, reagents, procedures, and some applications). *U.S.D.A. Agric. Handbook* **379.**

Goering, H. K., Van Soest, P. J., and Hemken, R. W. (1973). Relative susceptibility of forages to heat damage as affected by moisture, temperature, and pH. *J. Dairy Sci.* **56,** 137–143.

Golley, F. B. (1961). Energy values of ecological materials. *Ecology* **42,** 581–584.

Greenwood, D. J., and Barnes, A. (1978). A theoretical model for the decline in the protein content of plants during growth. *J. Agric. Sci.* **91,** 461–466.

Grodzinski, W., and Sawicka-Kapusta, K. (1970). Energy values of tree-seeds eaten by small mammals. *Oikos* **21,** 52–58.

Harborne, J. B. (Ed). (1972). "Phytochemical Ecology." Academic Press, London.

Harborne, J. B. (1979). Flavonoid pigments. *In* "Herbivores: Their Interaction with Secondary Plant Metabolites" (G. A. Rosenthal and D. H. Janzen, eds.), pp. 619–655. Academic Press, New York.

Harkin, J. M. (1973). Lignin. *In* "Chemistry and Biochemistry of Herbage," Vol. I (G. W. Butler and R. W. Bailey, eds.), pp. 323–373. Academic Press, New York.

Harper, R. B., Travis, H. F., and Glinsky, M. S. (1978). Metabolizable energy requirement for maintenance and body composition of growing farm-raised male pastel mink (*Mustela vison*). *J. Nutr.* **108,** 1937–1943.

Hawke, J. C. (1973). Lipids. *In* "Chemistry and Biochemistry of Herbage," Vol. I (G. W. Butler and R. W. Bailey, eds.), pp. 213–263. Academic Press, New York.

Janzen, D. H. (1979). New horizons in the biology of plant defenses. *In* "Herbivores: Their Interaction with Secondary Plant Metabolites" (G. A. Rosenthal and D. H. Janzen, eds.), pp. 331–350. Academic Press, New York.

Jensen, A. J. (1979). Energy content analysis from weight and liver index measurements of immature pollock (*Pollachius virens*). *J. Fish. Res. Bd. Can.* **36,** 1207–1213.

Johnson, S. R., and Robel, R. J. (1968). Caloric values of seeds from four range sites in northeastern Kansas. *Ecology* **49,** 956–961.

Jones, L. H. P., and Handreck, K. A. (1967). Silica in soils, plants, and animals. *Adv. Agron.* **19,** 107–149.

Kale, H. W., II. (1965). Ecology and bioenergetics of the long-billed marsh wren in Georgia salt marshes. *Publ. Nuttall Ornith. Club No.* **5.**

Karis, I. G., and Hudson, B. J. F. (1974). Effect of crop maturity on leaf lipids. *J. Sci. Food Agric.* **25,** 885–886.

Kaufman, D. W., and Kaufman, G. A. (1975). Caloric density of the oldfield mouse during postnatal growth. *Acta Theriol.* **20,** 83–95.

Kaufman, G. A., and Kaufman, D. W. (1977). Body composition of the oldfield mouse (*Peromyscus polionotus*). *J. Mammal.* **58,** 429–434.

Kendeigh, S. C., and West, G. C. (1965). Caloric values of plant seeds eaten by birds. *Ecology* **46,** 553–555.

Kolattukudy, P. E. (1977). Lipid polymers and associated phenols, their chemistry, biosynthesis, and role in pathogenesis. In "The Structure, Biosynthesis, and Degradation of Wood. Recent Advances in Phytochemistry," Vol. II (F. A. Loewus and V. C. Runeckles, eds.), pp. 185–246. Plenum, New York.

Kramer, P. J., and Kozlowski, T. T. (1960). "Physiology of Trees." McGraw-Hill, New York.

Kubota, J., Rieger, S., and Lazer, V. A. (1970). Mineral composition of herbage browsed by moose in Alaska. J. Wildl. Manage. 34, 565–569.

Levin, D. A. (1976a). Alkaloid-bearing plants: An ecogeographic perspective. Am. Nat. 110, 261–284.

Levin, D. A. (1976b). The chemical defenses of plants to pathogens and herbivores. Ann. Rev. Ecol. Syst. 7, 121–159.

Lofti, M., MacDonald, I. A., and Stock, M. J. (1976). Energy losses associated with oven-drying and the preparation of rat carcasses for analysis. Brit. J. Nutr. 36, 305–309.

Lohan, O. P., Lall, D., Makkar, H. P. S., and Negi, S. S. (1981). Inhibition of rumen urease activity by tannins in oak leaves. Indian J. Anim. Sci. 51, 279–281.

Lough, A. K., and Garton, G. A. (1968). Digestion and metabolism of feed lipids in ruminants and non-ruminants. Symp. Zool. Soc. London 21, 163–173.

Luick, J. R., White, R. G., Gau, A. M., and Jenness, R. (1974). Compositional changes in the milk secreted by grazing reindeer. I. Gross composition and ash. J. Dairy Sci. 57, 1325–1333.

Lyttleton, J. W. (1973). Proteins and nucleic acids. In "Chemistry and Biochemistry of Herbage," Vol. I (G. W. Butler and R. W. Bailey, eds.), pp. 63–103. Academic Press, New York.

Mabry, T. J., and Gill, J. E. (1979). Sesquiterpene lactones and other terpenoids. In "Herbivores: Their Interaction with Secondary Plant Metabolites" (G. A. Rosenthal and D. H. Janzen, eds.), pp. 501–537. Academic Press, New York.

McConnell, B. R., and Garrison, G. A. (1966). Seasonal variations of available carbohydrates in bitterbrush. J. Wildl. Manage. 30, 168–172.

McKey, D. (1979). The distribution of secondary compounds within plants. In "Herbivores: Their Interaction with Secondary Plant Metabolites" (G. A. Rosenthal and D. H. Janzen, eds.), pp. 55–133. Academic Press, New York.

MacKinnon, J. C. (1972). Summer storage of energy and its use for winter metabolism and gonad maturation in American plaice (Hippoglossoides platessoides). J. Fish. Res. Bd. Can. 29, 1749–1759.

McLeod, M. N. (1974). Plant tannins—their role in forage quality. Nutr. Abstr. Rev. 44, 803–815.

McQueen, R. E., and Nicholson, J. W. G. (1979). Modification of the neutral-detergent fiber procedure for cereals and vegetables by using α-amylase. J. Assoc. Off. Anal. Chem. 62, 676–680.

Martin, J. T., and Juniper, B. E. (1970). "The Cuticles of Plants." St. Martin's, New York.

Mattson, W. J., Jr. (1980). Herbivory in relation to plant nitrogen content. Ann. Rev. Ecol. Syst. 11, 119–161.

Maynard, L. A., and Loosli, J. K. (1969). "Animal Nutrition." McGraw-Hill, New York.

Milchunas, D. G., Dyer, M. I., Wallmo, O. C., and Johnson, D. E. (1978). In vivo/in vitro relationships of Colorado mule deer forages. Spec. Rep. 43, Colo. Div. Wildl., Ft. Collins.

Moody, S. (1978). Latitude, continental drift, and the percentage of alkaloid-bearing plants in floras. Am. Nat. 112, 965–968.

Morrow, P. A., Bellas, T. E., and Eisner, T. (1976). Eucalyptus oils in the defensive oral discharge of Australian sawfly larvae (Hymenoptera: Pergidae). Oecologia 24, 193–206.

Mould, E. D., and Robbins, C. T. (1981). Evaluation of detergent analysis in estimating nutritional value of browse. J. Wildl. Manage. 45, 937–947.

Mould, E. D., and Robbins, C. T. (1982). Digestive capabilities in elk compared to white-tailed deer. J. Wildl. Manage. 46, 22–29.

Myrcha, A., and Pinowski, J. (1969). Variations in the body composition and caloric value of nestling tree sparrows (*Passer m. montanus* L.). *Bull. Pol. Acad. Sci. Cl. II Ser. Sci. Biol.* **17**, 475–480.

Myrcha, A., Pinowski, J., and Tomek, T. (1973). Variations in the water and ash contents and in the caloric value of nestling starlings (*Sturnus vulgaris* L.) during their development. *Bull. Pol. Acad. Sci. Cl. II Ser. Sci. Biol.* **21**, 649–655.

Norberg, R. A. (1978). Energy content of some spiders and insects on branches of spruce (*Picea abies*) in winter; prey of certain passerine birds. *Oikos* **31**, 222–229.

Oh, H. K., Sakai, T., Jones, M. B., and Longhurst, W. M. (1967). Effects of various essential oils isolated from Douglas fir needles upon sheep and deer rumen microbial activity. *Appl. Microbiol.* **15**, 777–784.

Preston, R. D. (1974). "The Physical Biology of Plant Cell Walls." Chapman & Hall, London.

Rhoades, D. F. (1979). Evolution of plant chemical defense against herbivores. *In* "Herbivores: Their Interaction with Secondary Plant Metabolites" (G. A. Rosenthal and D. H. Janzen, eds.), pp. 3–54. Academic Press, New York.

Rhoades, D. F., and Cates, R. G. (1976). Toward a general theory of plant antiherbivore chemistry. *In* "Recent Advances in Phytochemistry, Vol. 10. Biochemical Interaction Between Plants and Insects" (J. W. Wallace and R. L. Mansell, eds.), pp. 168–213. Plenum, New York.

Richardson, M. (1977). The proteinase inhibitors of plants and microorganisms. *Phytochemistry* **16**, 159–169.

Ricklefs, R. E. (1975). Patterns of growth in birds. III. Growth and development of the cactus wren. *Condor* **77**, 34–45.

Ricklefs, R. E., White, S. C., and Cullen, J. (1980). Energetics of postnatal growth in Leach's storm petrel. *Auk* **97**, 566–575.

Robbins, C. T., and Moen, A. N. (1975). Composition and digestibility of several deciduous browses in the Northeast. *J. Wildl. Manage.* **39**, 337–341.

Robbins, C. T., Moen, A. N., and Reid, J. J. (1974). Body composition of white-tailed deer. *J. Anim. Sci.* **38**, 871–876.

Robel, R. J., Bisset, A. R., Clement, T. M., Jr., and Dayton, A. D. (1979a). Metabolizable energy of important foods in bobwhites in Kansas. *J. Wildl. Manage.* **43**, 982–987.

Robel, R. J., Bisset, A. R., Dayton, A. D., and Kemp, K. E. (1979b). Comparative energetics of bobwhites on six different foods. *J. Wildl. Manage.* **43**, 987–992.

Robinson, T. (1979). The evolutionary ecology of alkaloids. *In* "Herbivores: Their Interaction with Secondary Plant Metabolites" (G. A. Rosenthal and D. H. Janzen, eds.), pp. 413–448. Academic Press, New York.

Sawicka-Kapusta, K. (1974). Changes in the gross body composition and energy value of the bank voles during their postnatal development. *Acta Theriol.* **19**, 27–54.

Schwartz, C. C., Nagy, J. G., and Regelin, W. L. (1980a). Juniper oil yield, terpenoid cncentration, and antimicrobial effects on deer. *J. Wildl. Manage.* **44**, 107–113.

Schwartz, C. C., Regelin, W. L., and Nagy, J. G. (1980b). Deer preference for juniper forage and volatile oil treated foods. *J. Wildl. Manage.* **44**, 114–120.

Scott, A. I. (1974). Biosynthesis of natural products. *Science* **184**, 760–764.

Seigler, D. S. (1979). Toxic seed lipids. *In* "Herbivores: Their Interaction with Secondary Plant Metabolites" (G. A. Rosenthal and D. H. Janzen, eds.), pp. 449–470. Academic Press, New York.

Skar, H. J., Hagvar, S., Hagen, A., and Ostbye, E. (1975). Food habits and body composition of adult juvenile meadow pipit [*Anthus pratensis* (L.)]. *In* "Fennoscandian Tundra Ecosystems. Part 2. Animals and Systems Analysis" (F. W. Wielgolaski, ed.), pp. 160–173. Springer-Verlag, New York.

Slobodkin, L. B., and Richman, S. (1961). Calories/gm. in species of animals. *Nature* **191**, 299.

Smith, D. (1973). The nonstructural carbohydrates. In "Chemistry and Biochemistry of Herbage," Vol. I (G. W. Butler and R. W. Bailey, eds.), pp. 105–155. Academic Press, New York.

Stirling, I., and McEwan, E. H. (1975). The caloric value of whole-ringed seals (Phoca hispida) in relation to polar bear (Ursus maritimus) ecology and hunting behavior. Can. J. Zool. 53, 1021–1027.

Sugden, L. G., and Harris, L. E. (1972). Energy requirements and growth of captive lesser scaup. Poult. Sci. 51, 625–633.

Swain, T. (1979). Tannins and lipids. In "Herbivores: Their Interaction with Secondary Plant Metabolites" (G. A. Rosenthal and D. H. Janzen, eds.), pp. 657–682. Academic Press, New York.

Tyler, C. (1975). Albrecht Thaer's hay equivalents: Fact or fiction. Nutr. Abstr. Rev. 45, 1–11.

Van Soest, P. J. (1963a). Use of detergents in the analysis of fibrous feeds. I. Preparation of fiber residues of low nitrogen content. J. Assoc. Off. Agric. Chem. 46, 825–829.

Van Soest, P. J. (1963b). Use of detergents in the analysis of fibrous feeds. II. A rapid method for the determination of fiber and lignin. J. Assoc. Off. Agric. Chem. 46, 829–835.

Van Soest, P. J. (1965a). Non-nutritive residues: A system of analysis for the replacement of crude fiber. J. Assoc. Off. Agric. Chem. 49, 546–551.

Van Soest, P. J. (1965b). Use of detergents in the analysis of fibrous feeds. III. Study of effects of heating and drying on yield of fiber and lignin in forages. J. Assoc. Off. Agric. Chem. 48, 785–790.

Van Soest, P. J. (1967). Development of a comprehensive system of feed analysis and its application to forages. J. Anim. Sci. 26, 119–128.

Van Soest, P. J. (1969). The chemical basis for the nutritive evaluation of forages. Proc. Nat. Conf. Forage Qual. Eval. Util.

Van Soest, P. J. (1981). "Nutritional Ecology of the Ruminant." O and B Books, Inc., Corvallis, Oreg.

Van Soest, P. J., and Jones, L. H. P. (1968). Effect of silica in forages upon digestibility. J. Dairy Sci. 51, 1644–1648.

Van Soest, P. J., and Robertson, J. B. (1980). Systems of analysis for evaluating fibrous feeds. In "Standardization of Analytical Methodology for Feeds" (W. J. Pigden, C. C. Balch, and M. Graham, eds.), pp. 49–60. Int. Develop Center and Int. Union Nutr. Sci., Ottawa, Canada.

Van Soest, P. J., and Robertson, J. B. (1981). The detergent system of analysis and its application to human foods. In "Analysis of Dietary Fiber in Food" (P. James and O. Theander, eds.). M Dekker, New York.

Wasacz, J. (1981). Natural and synthetic narcotic drugs. Am. Sci. 69, 318–324.

White, L. M. (1973). Carbohydrate reserves of grasses: A review. J. Range Manage. 26, 13–18.

Wiegert, R. G. (1965). Intraspecific variation in calories/g of meadow spittlebugs (Philaenus spumarius L.). BioScience 15, 543–545.

Wilkie, K. C. B. (1979). The hemicellulose of grasses and cereals. Adv. Carbohydrate Chem. Biochem. 36, 215–264.

Wissing, T. E., and Hasler, A. D. (1971). Intraseasonal change in caloric content of some freshwater invertebrates. Ecology 52, 371–373.

12

Gastrointestinal Anatomy and Function

The field of nutrition has advanced significantly in recent years because of
experimental work conducted on a broad spectrum of animal species.
—Chandra and Newberne, 1977, p. 127

Knowledge of the general morphology and function of the gastrointestinal
tract is essential for the understanding of nutrient utilization. Digestive systems
provide many diverse examples of adaptive radiation in form and function that
correspond to observed diets and nutritional strategies. All birds and mammals
are dependent on the hydrolysis of ingested organic molecules by gastrointestinal
digestive enzymes and the subsequent absorption of smaller fragments. The
following discussion is largely based on the extensive reviews of Code (1968),
Moir and Hungate (1968), Ziswiler and Farner (1972), McDonald and Warner
(1975), and Hungate (1975).

I. BUCCAL CAVITY

The buccal cavity is the first structure of the generalized digestive tract. Major
differences between species occur in the teeth and lips of mammals, the bills of
birds, and tongues, taste buds, and saliva of both birds and mammals. For
example, the tongues of birds can be long and sticky, as in woodpeckers, or
tubular or semitubular, as in hummingbirds, or they can contain posterior-di-
rected horny hooks for holding food tightly, as in penguins, or filtering processes, as in ducks and geese. Mammalian teeth are particularly important in
reducing particle size for swallowing, as in large carnivores, and in providing
maximum particle surface area for digestion, as in ruminants.

Saliva's major roles include lubricating food for swallowing, providing nutrients and buffers for gastric fermentation in many herbivores, solubilizing
water-soluble components for tasting, and initiating digestion through salivary
enzymes. Saliva is produced primarily by the parotid, mandibular, and sub-
lingual glands. Species that ingest slippery aquatic food or those that do little

chewing because they ingest relatively moist food in large chunks, such as many carnivores, secrete little saliva, whereas those, such as ruminant and nonruminant herbivores, that ingest drier fibrous food needing both mastication and lubrication secrete large volumes. Large ruminants can secrete from tens to hundreds of liters of saliva daily (Chauncey and Quintarelli, 1961). The relative size of the salivary glands in wild ruminants increases from grazers to concentrate feeders because the buffers are increasingly needed to control the rapid fermentation of neutral detergent solubles (Kay *et al.*, 1980).

The identification of salivary enzymes is often difficult because of the contamination of saliva samples by food remnants, bacteria and protozoa, and oral tissue debris (Ellison, 1967). However, several digestive enzymes do occur in the saliva and include amylase and lipase. Salivary amylase does not occur in birds (Ziswiler and Farner, 1972), and it is restricted in mammals to rodents, primates, and the prototherian echidna (Chauncey and Quintarelli, 1961; Karn, 1978; Hjorth *et al.*, 1979). Salivary amylase is probably inconsequential in digestion, because the food is masticated and exposed to the enzyme for only a brief period. Although amylase is inactivated by the acidity of the mammalian glandular stomach, the enzyme could remain active in the more neutral pH of nonglandular stomach fermentation compartments of herbivorous rodents and primates (Carlton, 1973; Karn and Malacinski, 1978). Taste, and consequently food selection, may be affected by salivary amylase, because it hydrolyzes starch and glycogen to glucose, maltose, and small oligosaccharides. Pharyngeally secreted lipase for lipid digestion occurs in the neonate, but it decreases with age as pancreatic lipase activity increases (Moore and Noble, 1975).

II. AVIAN CROP AND ESOPHAGUS

The esophagus conducts food from the mouth to the stomach. Passage can be delayed in birds because food is stored in an expandable, nondifferentiated portion of the esophagus (e.g., in grebes, penguins, gulls, owls, and woodpeckers) or in an expandable, differentiated portion called a crop (Fig. 12.1). Although the crop usually protrudes ventrally to the vertebral column, the crop of several hummingbirds extends dorsally (Hainsworth and Wolf, 1972).

The functions of the crop include storing food, softening and swelling of hard food particles, providing nourishment for nestlings, and possibly initiating digestion. The storage capacity of the crop enables birds to consume more food than the stomach could handle efficiently at one time and thereby minimizes the frequency of feeding bouts. Because of the intense metabolism of hummingbirds and the very dilute nectar and sap consumed, the storage capacities of crops are a major determinant of many life strategy parameters (Hainsworth and Wolf, 1972; Hainsworth, 1978; Southwick and Southwick, 1980). Stored food is regurgitated

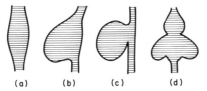

Fig. 12.1. Various forms of avian crops: (a) spindle-shaped (*Phalacrocorax:* cormorants), (b) unilateral out-pocketing (*Gyps:* vultures), (c) unilateral sac (*Pavo:* peafowl), and (d) double sac (*Columba:* pigeons). (From Ziswiler and Farner, 1972, courtesy of Academic Press, Inc.)

from the esophagus or crop to nourish the young in such diverse groups as hawks, storks, penguins, pelicans, doves and pigeons, and parrots. Digestion in the crop is probably minimal, since the crop does not secrete digestive enzymes. However, crop pH, ranging from 4.5 to 7.5 with an approximate mean of 6.0, is suitable for many plant and microbial enzymes, so some hydrolysis may occur.

III. STOMACH

A. Forestomach Fermentation Sites

Because vertebrates do not secrete enzymes to digest plant cell wall carbohydrates, symbiotic relationships with fiber-digesting bacteria enable herbivores to utilize otherwise nondigestible plant fiber. Bacteria occur in both aerobic and anaerobic forms. Symbiotic relationships with aerobic bacteria would yield only bacterial cells and oxidized end products, primarily carbon dioxide and water. However, anaerobic fermentation in which oxidation is limited by the molecular oxygen content of ingested organic matter yields bacterial cells, reduced bacterial end products, such as volatile fatty acids that can be further oxidized by the host, as well as oxygen-depleted, easily digestible food components, such as long-chain saturated fatty acids of dietary fat that cannot be metabolized by anaerobic bacteria. The complete oxidation of 1 mole of glucose from cellulose yields 38 moles of high-energy ATP. Because anaerobic bacteria fermenting glucose can capture only 2–6 moles of ATP (Van Soest, 1981), the residual 32–36 moles are available to the host via oxidative metabolism.

Animals with forestomach fermentation include the cervids; giraffes; bovids; tragulids or mouse deer; camels; peccaries and hippopotamuses; macropod marsupials, such as kangaroos and wallabies; leaf-eating monkeys, such as langurs; numerous herbivorous rodents; rock hyraxes; and leaf-eating tree sloths (Moir *et al.*, 1956; Golley, 1960; Williams, 1963; Moir, 1965; P. A. Vallenas, 1965; Bauchop and Martucci, 1968; Arman and Field, 1973; Lintern-Moore, 1973; Langer, 1974, 1978, 1979; Van Hoven, 1978; Leon, 1980; Perrin and Curtis,

1980). The anatomical complexity of the forestomach fermentation pouches ranges from a single diverticulum with minimal specialization communicating directly with the true stomach, as in many rodents, to the four-chambered, highly specialized ruminant stomach (Fig. 12.2). Very extensive forestomach pouches also occur in many cetaceans (whales, porpoises, and dolphins) in which fermentation would be completely maladapted because of the very digestible nature of the diet. The cetacean forestomach(s) is apparently partially analogous to the avian crop in storing large quantities of food taken opportunistically, although acidic digestion may begin (Gaskin, 1978; Chivers and Hladik, 1980).

The four chambers of the ruminant stomach are the rumen, reticulum, omasum, and abomasum (Fig. 12.3). The rumen is a large, thin-walled, saclike structure lined with finger-like projections termed papillae. The papillae increase the absorptive surface area from 16- to 38-fold in wild ruminants relative to the theoretically nonpapillated rumen wall (Hoppe *et al.*, 1977). Finer food particles and microorganisms pass into the reticulum, which has a honey-combed, reticulated epithelium. The omasum is a finely partitioned, weirlike structure that (*a*) separates the highly acidic abomasal contents from the fermenting contents of the rumen–reticulum; (*b*) provides for the passage of smaller food particles into the abomasum while retaining less digested, larger particles in the rumen–reticulum;

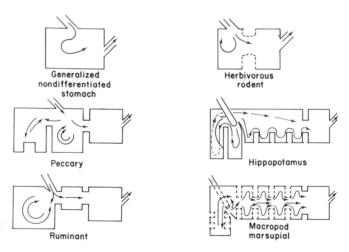

Fig. 12.2. Schematic diagram of several stomach types, indicating flow, general compartments, and separations. The folds of the macropod stomach are changeable and therefore drawn with broken lines. The division in the herbivorous rodent stomach is indicated with broken lines, since stomach specialization can range from a unilobular stomach with minimal differentiation to complex, multichambered stomachs having cornified epithelium and papillae in areas removed from the glandular stomach by constrictions or folds. The compartment on the right of each herbivore stomach diagram is the true stomach, or abomasum. The rumen and reticulum of the ruminant stomach are illustrated as a single fermentation chamber. (Adapted from Langer, 1979, courtesy of Verlag Paul Parey.)

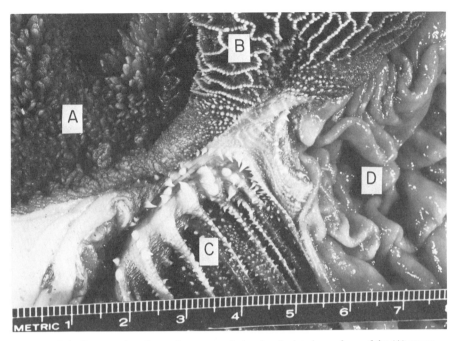

Fig. 12.3. Cross-section of a ruminant stomach showing the interior surfaces of the (A) rumen, (B) reticulum, (C) omasum, and (D) abomasum. (From Ullrey, 1980, courtesy of O and B Books, Inc., Corvallis, Oregon.)

and (c) absorbs water and soluble food and microbial products (Prins et al., 1972). Microbial cells, small food particles, and previously nonabsorbed metabolites pass into the abomasum, or true stomach, for enzymatic and acid hydrolysis.

The extensive evolution of forestomach fermentation is associated with the development of widespread grass and forb plant communities (Moir and Hungate, 1968; Langer, 1974, 1979; Janis, 1976; Kinnear et al., 1979; Van Soest, 1981). Fermentation enabled the animal to meet its energy requirements from previously nondigestible plant fiber. Perhaps equally important, since the Perissodactyla (horses, rhinoceroses, and other odd-toed ungulates) and many other animals had already developed an extensive lower-tract fermentation capable of digesting plant fiber, was the capability to conserve and synthesize microbial protein from nonprotein nitrogen, synthesize vitamins, and detoxify many secondary plant compounds *anterior* to the normal site of host enzymatic digestion and absorption. However, the benefits of forestomach fermentation can be balanced by a reduced rate of food passage and microbial losses of easily digestible, plant cellular contents.

The volumes of forestomach fermentation chambers as a percentage of total stomach volume in large adult herbivores average approximately 90%, ranging from 98% in the llama and guanaco, 95% in the hippopotamus, 92% in macropod marsupials, 90% in wild ruminants, to 86% in peccaries (Short, 1964; A. Vallenas et al., 1971; Arman and Field, 1973; Hofmann, 1973; Robbins and Moen, 1975; Langer, 1974, 1978, 1979). However, the volume and functionality of the compartments change as the animal grows. For example, the relative forestomach fermentation volume in white-tailed deer increases from approximately 10% of the stomach at birth to 90% in the adult (Short, 1964).

The neonatal stomach of forestomach fermenters functions as if the animal were a simple-stomached animal. Because milk is totally digested by enzymes of the true stomach and small intestine, fermentation of milk would be both unnecessary and detrimental. Consequently, a necessity of advanced forestomach fermenters has been the concomitant evolution of a mechanism to pass milk directly from the esophagus into the acidic stomach (Moir et al., 1956; Black and Sharkey, 1970; Arman and Field, 1973). This mechanism is called the reticular groove. In ruminants, the reticular groove is a bilipped channel passing along the reticulum wall from the esophageal orifice to the reticula–omasal orifice (Fig. 12.4). As milk passes down the esophagus, the lips of the groove close reflexively to form a rumen–reticulum bypass, which continues to function as long as milk is ingested (Titchen and Newhook, 1975). The groove will close even when milk is consumed from a bucket (Matthews and Kilgour, 1980). Such a bypass does not occur in rodents, some small macropod marsupials, and the rock hyrax because of the close juxtaposition of the esophageal opening to the acidic stomach (Kinnear et al., 1979; Leon, 1980).

Rumen development is dependent on the ingestion and fermentation of plant matter. Bacterial inoculation of the rumen is normally not a problem, with suitable bacteria potentially being transmitted in maternal saliva during licking of the neonate, on feed mouthed but not consumed by the mother, and on maternal feces touched or consumed by the neonate. Young deer and elk removed from the mother 6–18 hr after birth and raised in isolation develop normally functioning rumens at approximately the same time as maternal-raised neonates. Thus, either inoculation is occurring very shortly after birth or suitable bacteria are being encountered throughout the environment. Papillary growth is stimulated by the metabolic products of fermentation, particularly volatile fatty acids. Butyrate, because of its metabolism by the rumen epithelium, provides the greatest stimulant to papillary growth, followed by propionate and acetate. Reduced milk intake increases the rate of rumen development by stimulating dry-feed intake and therefore fermentation (Blaxter et al., 1952; Warner et al., 1955; Flatt et al., 1958; Lengemann and Allen, 1959; Tamate et al., 1962; Poe et al., 1971).

Only ruminants, camels, and macropod marsupials regurgitate the contents of the fermenting stomach in order to reduce particle size through further mastica-

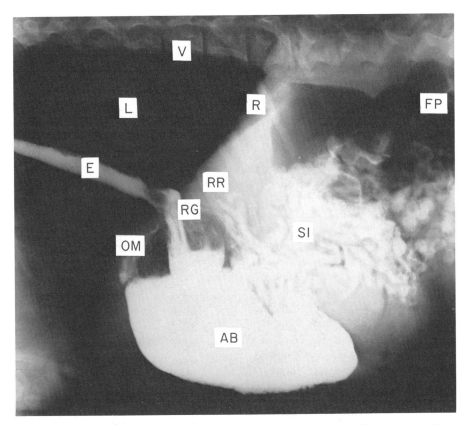

Fig. 12.4. Radiograph of the passage of a milk-contrast medium consumed by a bottle-fed, 3-week-old elk calf. The mixture, consumed approximately 20–30 min prior to the radiograph exposure, can be seen as it moves into the small intestine. Structures indicated are the esophagus (E), reticular groove (RG), omasum (OM), abomasum (AB), small intestine (SI), rumen–reticulum (RR), fecal pellets in the large intestine (FP), lungs (L), ribs (R), and vertebral column (V).

tion. This process is called rumination in the ruminant and merycism in nonruminants (Langer, 1979).

Many anatomical and functional differences occur among the stomachs of wild ruminants. Hofmann (1973) has done more than any other author to clarify these differences. Stomach structure, volume, mouth anatomy, and feeding habits are interrelated. On the basis of stomach characteristics and diet, Hofmann divides the ruminants into three major groups: concentrate feeders (fruit and dicotyledonous foliage eaters), bulk and roughage feeders (grass eaters), and intermediate feeders (mixed eaters). Bulk and roughage feeders and intermediate feeders have rumen–reticulum volumes that are 53% and 22% greater, respectively, than those of concentrate feeders. The concentrate feeders select food that is readily fermentable and low in refractory cell walls. Fermentation rates of plant cell solubles are even higher in concentrate feeders than in bulk and roughage feeders (Hungate *et al.*, 1959; Short, 1963; Prins and Geelen, 1971). Conversely, the larger forestomach of the grazer is more adapted to slower cellulolytic fermentation. Bacterial cellulase is released into an adjacent, extracellular zone from which the saccharide units of cell wall carbohydrates are hydrolyzed (Fig. 12.5). Many examples of captive wildlife doing very poorly or dying when fed diets that are inappropriate relative to stomach anatomy and function testify to the physiological significance of the anatomical differences (Hofmann, 1973; Schoonveld *et al.*, 1974; Nagy and Regelin, 1975; Chivers and Hladik, 1980).

Rumen fluid in mule deer and elk contained from 11×10^9 to 67×10^9 bacteria per milliliter of rumen fluid (McBee *et al.*, 1969; Pearson, 1969). The major volatile fatty acids (VFAs) produced by bacteria during fermentation are acetic, propionic, and butyric acids. Acetic acid commonly occurs in highest concentrations and can range from 50 to 80% of the total volatile fatty acids. However, the exact proportions of the VFAs vary with the diet. For example, increased percentages of acetic and butyric acids are characteristic of extensive plant cell wall fermentation, whereas high propionic acid concentrations represent soluble carbohydrate fermentation. The volatile fatty acids absorbed by the forestomach account for 21–75% of the digestible energy intake in ruminants and macropod marsupials (Stewart *et al.*, 1958; Gray *et al.*, 1967; Faichney, 1968; Leng *et al.*, 1968; Hume, 1977; Van Hoven and Boomker, 1981).

The pH of the forestomach contents is dependent on the balance between fermentation rate and the production of volatile fatty acids, their rate of absorption or outflow, salivary buffering capacity, and the completeness of separation between the forestomach and the true stomach. For example, rumen pH in red deer decreased from 6.5 prior to feeding to 5.5 90 min after feeding when the rate of acid production exceeded buffering and absorption capacities (Maloiy *et al.*, 1968). Forestomach pH has generally varied from 5.4 to 6.9 in wild ruminants, macropod marsupials, and hippopotamuses (Prins and Geelen, 1971; Dean *et al.*, 1975; Hoppe *et al.*, 1977; Van Hoven, 1978; Kinnear *et al.*, 1979; Kreulen and Hoppe, 1979; Leon, 1980; Van Hoven and Boomker, 1981).

Plant nitrogen in the rumen or forestomach following a meal decreases as bacterial, protozoal, and soluble nitrogen increase. Bacteria largely degrade feed amino acids into their respective organic acid and ammonia elements, but subsequently resynthesize amino acids relative to their needs from the same precursors. From 64 to 85% of the plant nitrogen is incorporated into microbial nitro-

Fig. 12.5. Electron micrograph of rumen bacteria degrading plant cell walls. Sharp zones of degradation are noted around the bacteria. (From Akin and Amos, 1975, courtesy of the American Society of Microbiology.)

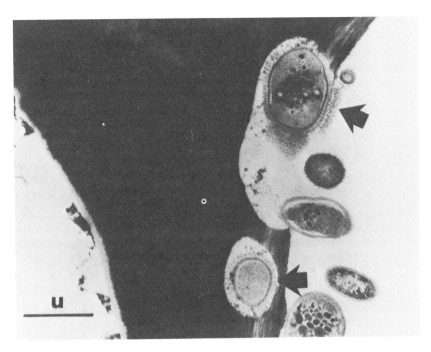

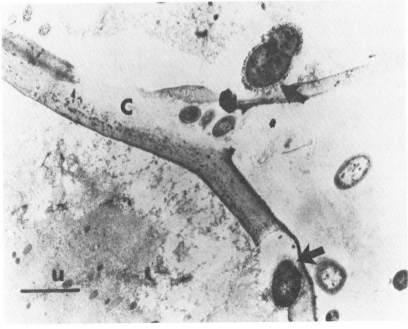

gen in the ruminant and marsupial forestomach (Weller *et al.*, 1962; Pilgrim *et al.*, 1970; Lintern-Moore, 1973). Fortunately, the synthesis of amino acids balances or even exceeds the degradation rate if nonprotein nitrogen, such as urea, is consumed (Silver, 1968; Klein and Schonheyder, 1970). Thus, despite differences in true protein intake and its amino acid composition, rumen bacteria tend to have a leveling effect on both dietary protein quality and quantity (Gray *et al.*, 1958; Purser, 1970; Leibholz, 1972).

B. Mammalian Abomasum and Avian Proventriculus–Gizzard

Acidic digestion occurs in the abomasum of mammals and the proventriculus and gizzard of birds. The avian proventriculus primarily secretes hydrochloric acid and digestive enzymes, while the muscular stomach mechanically reduces particle size concurrently with enzymatic and acid hydrolysis. The gizzard can have a very strong musculature for efficient grinding, such as in granivores, or it can be very weak, in which case digestion is entirely chemical, such as in hawks, vultures, pelicans, and loons (Ziswiler and Farner, 1972; Houston and Cooper, 1975). The acidic stomach of both mammals and birds can also be a storage organ even though digestion is occurring. Many carnivores are well known for their rapid ingestion and stomach storage of large quantities of food. The stomach can also be a filter in which bones, hairs, and feathers are retained and often regurgitated, as in hawks and owls, while the more digestible components move into the small intestine.

The pH of pure gastric acid in both birds and mammals ranges from 0.2 to 1.2 (Ziswiler and Farner, 1972). However, the observed pH of the stomach contents generally ranges from 1.0 to 3.0 in birds and from 2.4 to 4.0 in mammals (Moir *et al.*, 1956; Ziswiler and Farner, 1972; Schoonveld *et al.*, 1974; Duke *et al.*, 1975; Houston and Cooper, 1975; Rhoades and Duke, 1975; Van Hoven, 1978; Kinnear *et al.*, 1979; Clemens and Phillips, 1980; Leon, 1980; Clemens and Maloiy, 1981). The higher pH of stomach contents than of pure gastric acid is due to the dilution and neutralization by food or ingested fluids, the internal regulation or optimization of gastric pH, and, in some species, the retrograde flow of more alkaline small-intestinal contents into the stomach (LePrince *et al.*, 1979). Acid secretion is often minimal in neonatal mammals, but it increases during the first few hours or days of life until stomach pH approaches the adult level. Although many of the differences in stomach pH of adult animals are simply due to the method of measurement, significant differences between species do occur. For example, the pH of the stomach contents in the Falconiformes (hawks, falcons, and eagles; average of 1.6) is much lower than in the Strigiformes (owls; average of 2.35) (Duke *et al.*, 1975; Rhoades and Duke, 1975). Since pH is a logarithmic function of hydrogen ion concentration, the Falconiformes have a hydrogen ion concentration six times that of Strigiformes.

These differences are reflected in higher dry-matter digestibilities in Falconiformes versus Strigiformes due to the more extensive corrosion of bones in the Falconiformes stomach (Duke et al., 1975). However, since the differences in dry-matter digestibility are largely due to the absorption of bone ash, energy digestibilities are virtually identical in hawks and owls (Kirkwood, 1979).

Digestion in the acidic stomach is primarly proteolysis. Pepsin is the primary gastric proteolytic enzyme occurring in both mammals and birds. Pepsin occurs in the abomasum, the proventriculus, and, possibly, the gizzard mucosa in an inactive form called pepsinogen. When pepsinogen is secreted into the gastric lumen in the presence of hydrochloric acid at a pH below 5, a low molecular weight polypeptide is cleaved to form active pepsin. Below pH 4, the conversion of pepsinogen to pepsin is catalyzed by pepsin; thus, the reaction is autocatalytic. Pepsin activity is generally strongest between pH 1 and 2, although a second peak may occur between pH 3.5 and 4.0. The acidic environment of the stomach generally denatures ingested protein, which exposes the bonds to enzymatic hydrolysis. Pepsins preferentially split peptide bonds involving phenylalanine, tyrosine, leucine, and glutamic acid (Taylor, 1968; Ziswiler and Farner, 1972).

An additional enzyme, rennin, occurs in the gastric juice of nursing mammals and has an optimum activity at pH 4.0. Both pepsin and rennin readily coagulate milk casein in the nursing animal. The digestion of the milk clot provides a continuous, prolonged flow of food material into the small intestine of the young animal and avoids flow surges or the excessive passage of fermentable substrates into the intestinal tract.

Chitinase, an enzyme capable of hydrolyzing the chitinous exoskeleton of invertebrates, occurs in the stomach of some birds. Its distribution is probably limited to those birds in which insects are a significant part of the diet (LePrince et al., 1979).

Many other enzymes have been identified as of gastric origin, particularly in birds. Since it is now recognized that digestion in birds involves the movement of digesta back and forth between the stomach and small intestine (Le Prince et al., 1979; Sklan, 1980), one cannot use the mere presence of an enzyme in stomach contents as an indicator of its origin. The reversible flow of digesta between the stomach and small intestine in birds maximizes digestive efficiency while minimizing intestinal weight and length.

IV. SMALL INTESTINE

The small intestine is the primary site of enzymatic digestion and absorption. The mammalian small intestine is morphologically divided into a proximal duodenum looping around the pancreas, intermediate jejunum, and distal ileum. The avian small intestine is divided only into the duodenum and ileum, with no major histological differences between the two segments. The length of all intestinal segments in both mammals and birds relative to body weight are longest in herbivores, intermediate in granivores and frugivores, and shortest in carnivores and insectivores. Relative intestinal length within each species varies with sex, age, seasonal food habits, and level of intake (Leopold, 1953; Myrcha, 1964; Fenna and Boag, 1974; Burton *et al.*, 1979; Staaland *et al.*, 1979; Smith *et al.*, 1980; Chivers and Hladik, 1980).

Digestive secretions enter the small intestine from the liver (bile salts: glycocholates and taurocholates), pancreas (buffering bicarbonate ions; proteases, including trypsinogen, chymotrypsinogen, and procarboxypeptidase; amylase; lipase; lecithinase; and nuclease), intestinal mucosa (primarily peptidases and saccharidases), and duodenal glands (alkaline fluid for pH regulation). The secretion of the various enzymes and buffering solutions is stimulated by the passage of stomach contents, called chyme, into the duodenum. Liver and pancreas ducts carrying digestive secretions open near the duodenal–abomasal orifice in mammals and more distally in the duodenum in birds (Sklan, 1980). The initial response must be to neutralize the chyme, since the intestinal enzymes function optimally between pH 6 and 8. The pH of intestinal tract contents posterior to the stomach in both birds and mammals range from 5.2 to 7.5, with a mean of approximately 6.5 (Farner, 1942; Moir *et al.*, 1956; Johnson and McBee, 1967; Hoover and Clarke, 1972; Houston and Cooper, 1975; Clemens and Phillips, 1980; Clemens and Maloiy, 1981).

The pancreatic proteases are secreted as inactive proenzymes. The proenzymes are converted to active enzymes by a series of reactions beginning with the activation of trypsinogen by enterokinase, an enzyme produced by the intestinal mucosa. Trypsin in turn activates chymotrypsinogen and procarboxypeptidase. Each pancreatic protease preferentially hydrolyzes specific peptide bonds. For example, trypsin hydrolyzes bonds involving lysine and arginine, and chymotrypsin hydrolyzes bonds involving aromatic amino acids. These enzymes are termed endopeptidases, since they hydrolyze the interior peptide bonds. Carboxypeptidase is an exopeptidase that hydrolyzes peptide bonds in sequence from the end of the chain. The additional intestinal mucosal proteases complete protein digestion and produce free amino acids for absorption.

The absorption of colostral immunoglobulins is contrary to the general mosaic of protein digestion because the molecule must be absorbed intact by pinocytosis. Important mechanisms enabling the passage and absorption of intact immunoglobulins include (*a*) the time lag in acid and enzyme production by the neonatal stomach and small intestine and (*b*) the incorporation of enzyme inhibitors in maternal colostrum. Because colostral immunoglobulins are relatively resistant to chymotrypsin but are destroyed by trypsin, trypsin inhibitors occur in the colostrum of many species. The trypsin inhibitor protects the immunoglobulin without completely inhibiting the digestion of other milk proteins by pancreatic and intestinal proteases (Morris, 1968; Canwell, 1977; Corring *et al.,* 1978; Esparza and Brock, 1978; Brown and Perry, 1981).

Numerous saccharidases are produced by the pancreas and intestinal mucosa. These include amylase, which hydrolyzes starch and glycogen; maltase, which hydrolyzes the glucose disaccharide maltose; sucrase, which hydrolyzes the glucose–fructose dissaccharide of plants; isomaltase, which hydrolyzes dextrans linked $\alpha 1,6$; and lactase, which hydrolyzes the galactose–glucose disaccharide of milk. The occurrence of these enzymes is related to the age of the animal and the composition of the food entering the small intestine. For example, those species in which the milk does not contain lactose normally do not produce lactase. Even though marsupial milk in its natural form contains only traces of lactose, the pouch young do produce lactose equal to that occurring in eutherian neonates, apparently because the cleavage of the galactose-rich oligo- and polysaccharides produces a lactose unit that must be further digested (Walcott and Messer, 1980). In lactose-producing species, lactase is the first disaccharidase to develop and remains high while milk is consumed, but it subsequently falls after weaning as amylase, maltase, sucrase, and other saccharidases needed to digest the adult diet increase. Since the carbohydrate composition of the food and intestinal contents are the same in animals with simple stomachs, the intestinal saccharidases reflect the carbohydrate composition of the normally ingested food. However, the intestinal dissaccharidases of forestomach fermenters often do not reflect food composition, but rather the carbohydrate spectra of the fermentation products. For example, although ruminants and macropod marsupials consume plant sucrose, sucrase does not occur in these animals. The ingested sucrose is readily fermented to volatile fatty acids, obviating any need for intestinal sucrase (Semenza, 1968; Kerry, 1969; Walcott and Messer, 1980).

Fat digestion is due to pancreatic lipase and emulsifying bile salts. Pancreatic lipase in simple-stomached animals preferentially hydrolyzes the 1 and 3 position fatty acids of a triglyceride to produce two free fatty acids and a 2-monoglyceride. Pancreatic lipase in forestomach fermenters can be less important than in simple-stomached animals as the microbes have often already hydrolyzed the fatty acids of ingested glycerides and fermented the glycerol component. The bile salts and hydrolyzed fats form micelles whose diameter relative to the fat droplet is reduced 100-fold and the surface area increased 10,000-fold. The micelle is the functional unit of fat absorption at the intestinal epithelium. The intestinal reabsorption of bile salts and their transport to the liver, termed the enterohepatic circulation, is important because bile salts are often secreted in excess of hepatic production (Johnston, 1968; Weiner and Lack, 1968; Patton and Carey, 1979; Sklan, 1980).

V. CECUM

Ceca are blind, intestinal diverticula usually occurring at the junction of the small and large intestine. They are usually paired in birds (ceca) and unilateral in mammals. The most extensive cecal development in birds occurs in the Tetraonidae, or grouse and ptarmigan, although ducks and geese also have very functional ceca. Rudimentary, nondigestive ceca occur in many carnivorous and insectivorous birds (Ziswiler and Farner, 1972; Duke et al., 1981). Functional digestive ceca in mammals are largely confined to ruminant and nonruminant herbivores and frugivores.

The functions of the ceca include the fermentation of plant fiber and soluble plant or endogenous matter that has not previously been absorbed; the absorption of water and therefore small water-soluble nutrients, such as electrolytes, ammonia, and possibly amino acids; and vitamin synthesis (Johnson and McBee, 1967; Hintz et al., 1971; Mattocks, 1971; Slade et al., 1971; Hoover and Clark, 1972; Moss and Parkinson, 1975; Allo et al., 1973; Inman, 1973; Miller, 1974; Ulyatt et al., 1975; Gasaway, 1976b; Gasaway et al., 1976a, 1976b; Kempton et al., 1976; Parker, 1976; Hume, 1977; Murray et al., 1977; McKenzie, 1978; Sudo and Duke, 1980; Duke et al., 1981). Significant amounts of energy in the form of volatile fatty acids are produced and absorbed from the ceca (Table 12.1). The cecal orifice preferentially admits the liquid–small particle digesta.

TABLE 12.1

Contents of the Ceca as a Percentage of Body Weight and the Estimated Energy Derived from Fermentation and the Production of Volatile Fatty Acids[a]

Species	Contents (% body weight)	Energy (% BMR)	Reference
Birds			
Rock ptarmigan	2.1	18	Gasaway (1976a)
Sharp-tailed grouse	1.4	14	Gasaway (1976b)
Willow ptarmigan	1.2	11	Gasaway (1976b)
Mammals			
Beaver	2.1	19	Hoover and Clark (1972)
Black-tailed deer		1	Allo et al. (1973)
Black wildebeest	0.89	5–7	Van Hoven and Boomker (1981)
Domestic pig	0.4–0.7	6	Farrell and Johnson (1972)
Domestic rabbit	1.8–2.5	10–12	Hoover and Heitmann (1972)
Domestic rabbit		30	Parker (1976)
Porcupine	4.4	16	Johnson and McBee (1967)
Rat	1.8–2.4	9	Yang et al. (1969, 1970)
Wallabies		3–6	Hume (1977)

[a]Expanded from Gasaway (1976b).

For example, only 18% of the dry matter passing the ileocecal–colic junction of rock ptarmigan entered the ceca as compared to 96% of liquid digesta (Gasaway *et al.*, 1975, 1976c). Similarly, more than 70% of the liquid–small particle phase digesta entered the cecum of white-tailed deer (Mautz, 1969). Cecectomized rabbits gained 23% slower and excreted more than five times as much sodium and potassium as intact rabbits (Herndon and Hove, 1955).

VI. LARGE INTESTINE

The large intestine extends from the ileo–cecal juncture to the rectum of mammals and the cloaca of birds. The cloaca in birds is a common route for the passage of feces, urine, and reproductive products. The relative length of the large intestine is dependent on the dietary regimen of the species. For example, the length of the large intestine relative to the small intestine averages 6% in small carnivorous mammals, 33% in omnivores, and 78% in herbivores as fiber digestion, bulk, and a reduced rate of passage become more important (Barry, 1977).

The functions of the large intestine are similar to those of the ceca and include fermentation, vitamin synthesis, and the absorption of water and soluble nutrients (Ulyatt *et al.*, 1975). Herbivores without forestomach fermentation often depend on cecal and large intestine fermentation to digest fiber (Hoover and Clarke, 1972; Murray *et al.*, 1977; Milton, 1981). Volatile fatty acids, minerals, lipids, ammonia, and possibly amino acids are absorbed from the large intestine. The energy derived from VFA absorption in the large intestine of the howler monkey contributes from 26 to 36% of its daily energy requirement (Milton, 1981). VFA production in the capacious elephant cecum and large intestine, whose contents account for as much as 11% of the body weight, can account for 100% of the energy needed for basal metabolism (Van Hoven *et al.*, 1981). The large intestine could be an even more important site of water, nitrogen, and electrolyte conservation in birds than in mammals, since urine can move from the cloaca into the large intestine by antiperistalsis (Skadhauge, 1976).

REFERENCES

Akin, D. E., and Amos, H. E. (1975). Rumen bacterial degradation of forage cell walls investigated by electron microscopy. *Appl. Microbiol.* **29**, 692–701.

Allo, A. A., Oh, J. H., Longhurst, W. M., and Connolly, G. E. (1973). VFA production in the digestive systems of deer and sheep. *J. Wildl. Manage.* **37**, 202–211.

Arman, P., and Field, C. R. (1973). Digestion in the hippopotamus. *E. Afr. Wildl. J.* **11**, 9–17.

Barry, R. E., Jr. (1977). Length and absorptive surface area apportionment of segments of the hindgut for eight species of small mammals. *J. Mammal.* **58**, 419–420.

Bauchop, T., and Martucci, R. W. (1968). Ruminant-like digestion of the langur monkey. *Science* **161,** 698–700.

Black, J. L., and Sharkey, M. J. (1970). Reticular groove (*Sulcus reticuli*): an obligatory adaptation in ruminant-like herbivores? *Mammalia* **34,** 294–302.

Blaxter, K. L., Hutcheson, M. K., Robertson, J. M., and Wilson, A. L. (1952). The influence of diet on the development of the alimentary tract of the calf. *Brit. J. Nutr.* **6,** i–ii.

Brown, J. J., and Perry, T. W. (1981). Trypsin and chymotrypsin development in the neonatal lamb. *J. Anim. Sci.* **52,** 359–362.

Burton, B. A., Hudson, R. J., and Bragg, D. D. (1979). Efficiency of utilization of bulrush rhizomes by lesser snow geese. *J. Wildl. Manage.* **43,** 728–735.

Canwell, P. D. (1977). Acid and pepsin secretion in young pigs reared solely by the sow or supplemented with solid food and weaned at 21 d. *Proc. Nutr. Soc.* **36,** 142A.

Carleton, M. D. (1973). A survey of gross stomach morphology of New World Cricetinae (Rodentia, Muroidea), with comments on functional interpretations. *Mus. Zool. Univ. Michigan* **146,** 1–42.

Chandra, R. K., and Newberne, P. M. (1977). "Nutrition, Immunity, and Infection: Mechanisms of Interactions. Plenum, New York.

Chauncey, H. H., and Quintarelli, G. (1961). Localization of acid phosphatase, nonspecific esterases and *B*-D-galactosidase in parotid and submaxillary glands of domestic and laboratory animals. *Am. J. Anat.* **108,** 263–294.

Chivers, D. J., and Hladik, C. M. (1980). Morphology of the gastrointestinal tract in primates: Comparisons with other mammals in relation to diet. *J. Morphol.* **166,** 337–386.

Clemens, E., and Phillips, B. (1980). Organic acid production and digesta movement in the gastrointestinal tract of the baboon and Sykes monkey. *Comp. Biochem. Physiol.* **66A,** 529–532.

Clemens, E. T., and Maloiy, G. M. O. (1981). Organic acid concentrations and digesta movement in the gastrointestinal tract of the bushbaby (*Galago crassicaudatus*) and Vervet monkey (*Cercopithecidae pygerythrus*). *J. Zool.* **193,** 487–497.

Code, C. F. (Ed.). (1968). "Alimentary Canal, Section 6. Handbook of Physiology." Am. Physiol. Soc., Washington, D.C.

Corring, T., Aumaitre, A., and Durand, D. (1978). Development of digestive enzymes in the piglet from birth to 8 weeks. *Nutr. Metab.* **22,** 231–243.

Dean, R. E., Strickland, M. D., Newman, J. L., Thorne, E. T., and Hepworth, W. G. (1975). Reticulo-rumen characteristics of malnourished mule deer. *J. Wildl. Manage.* **39,** 601–604.

Duke, G. E., Jegers, A. A., Loff, G., and Evanson, O. A. (1975). Gastric digestion in some raptors. *Comp. Biochem. Physiol.* **50A,** 649–656.

Duke, G. E., Bird, J. E., Daniels, K. A., and Bertoy, R. W. (1981). Food metabolizability and water balance in intact and cecectomized great-horned owls. *Comp. Biochem. Physiol.* **68A,** 237–240.

Ellison, S. A. (1967). Proteins and glycoproteins of saliva. *In* "Alimentary Canal, Vol. 2. Handbook of Physiology," pp. 531–559. Am. Physiol. Soc., Washington, D.C.

Esparza, I., and Brock, J. H. (1978). Inhibition of rat and bovine trypsins and chymotrypsins by soybean, bovine basic pancreatic, and bovine colostrum trypsin inhibitors. *Comp. Biochem. Physiol.* **61B,** 347–350.

Faichney, G. J. (1968). The production and absorption of volatile fatty acids from the rumen of the sheep. *Aust. J. Agric. Res.* **19,** 791–802.

Farner, D. S. (1942). The hydrogen ion concentration in avian digestive tracts. *Poult. Sci.* **21,** 445–450.

Farrell, D. J., and Johnson, K. A. (1972). Utilization of cellulose by pigs and its effects on caecal fermentation. *Anim. Prod.* **14,** 209–217.

Fenna, L., and Boag, D. A. (1974). Adaptive significance of the caeca in Japanese quail and spruce grouse (Galliformes). *Can. J. Zool.* **52,** 1577–1584.

Flatt, W. P., Warner, R. G., and Loosli, J. K. (1958). Influence of purified materials on the development of the ruminant stomach. *J. Dairy Sci.* **41,** 1593–1600.

Gasaway, W. C. (1976a). Seasonal variation in diet, volatile fatty acid production and size of the cecum of rock ptarmigan. *Comp. Biochem. Physiol.* **53A,** 109–114.

Gasaway, W. C. (1976b). Volatile fatty acids and metabolizable energy derived from cecal fermentation in the willow ptarmigan. *Comp. Biochem. Physiol.* **53A,** 115–121.

Gasaway, W. C. (1976c). Cellulose digestion and metabolism by captive rock ptarmigan. *Comp. Biochem. Physiol.* **54A,** 179–182.

Gasaway, W. C., Holleman, D. F., and White, R. G. (1975). Flow of digesta in the intestine and cecum of the rock ptarmigan. *Condor* **77,** 467–474.

Gasaway, W. C., White, R. G., and Holleman, D. F. (1976). Digestion of dry matter and absorption of water in the intestine and cecum of rock ptarmigan. *Condor* **78,** 77–84.

Gaskin, D. E. (1978). Form and function in the digestive tract and associated organs in cetacea, with a consideration of metabolic rates and specific energy budgets. *Oceanogr. Mar. Biol. Ann. Rev.* **16,** 313–345.

Golley, F. B. (1960). Anatomy of the digestive tract of *Microtus*. *J. Mammal.* **41,** 89–99.

Gray, F. V., Pilgrim, A. F., and Weller, R. A. (1958). The digestion of foodstuffs in the stomach of the sheep and the passage of digesta through its compartments. *Brit. J. Nutr.* **12,** 413–420.

Gray, F. V., Weller, R. A., Pilgrim, A. F., and Jones, G. B. (1967). Rates of production of volatile fatty acids in the rumen. V. Evaluation of fodders in terms of volatile fatty acid produced in the rumen of the sheep. *Aust. J. Agric. Res.* **18,** 625–634.

Hainsworth, F. R. (1978). Feeding: Models of costs and benefits in energy regulation. *Am. Zool.* **18,** 701–714.

Hainsworth, F. R., and Wolf, L. L. (1972). Crop volume, nectar concentration and hummingbird energetics. *Comp. Biochem. Physiol.* **42A,** 359–366.

Herndon, J. F., and Hove, E. L. (1955). Surgical removal of the cecum and its effect on digestion and growth in rabbits. *J. Nutr.* **57,** 261–270.

Hintz, H. F., Argenzio, R. A., and Schryver, H. F. (1971). Digestion coefficients, blood glucose levels and molar percentages of volatile acids in intestinal fluid of ponies fed varying forage-grain ratios. *J. Anim. Sci.* **33,** 992–995.

Hjorth, J. P., Meisler, M., and Nielsen, J. T. (1979). Genetic variation in amount of salivary amylase in the bank vole, *Clethrionomys glareola*. *Genetics* **92,** 915–930.

Hofmann, R. R. (1973). ''The Ruminant Stomach.'' E. Afr. Monogr. in Biol. Vol. II. E. Afr. Lit. Bureau, Nairobi.

Hoover, W. H., and Clarke, S. D. (1972). Fiber digestion in the beaver. *J. Nutr.* **102,** 9–16.

Hoover, W. H., and Heitmann, R. N. (1972). Effects of dietary fiber levels on weight gain, cecal volume and volatile fatty acid production in rabbits. *J. Nutr.* **102,** 375–380.

Hoppe, P. P., Qvortrup, S. A., and Woodford, M. H. (1977). Rumen fermentation and food selection in East African Zebu cattle, wildebeest, Coke's hartebeest and topi. *J. Zool.* **181,** 1–9.

Houston, D. C., and Cooper, J. E. (1975). The digestive tract of the whiteback griffon vulture and its role in disease transmission among wild ungulates. *J. Wildl. Dis.* **11,** 306–313.

Hume, I. D. (1977). Production of volatile fatty acids in two species of wallaby and in sheep. *Comp. Biochem. Physiol.* **56A,** 299–304.

Hungate, R. E. (1966). ''The Rumen and Its Microbes.'' Academic Press, New York.

Hungate, R. E. (1975). The rumen microbial ecosystem. *Ann. Rev. Ecol. Syst.* **6,** 39–66.

Hungate, R. E., Phillips, G. D., McGregor, A., Hungate, D. P., and Buechner, H. K. (1959). Microbial fermentation in certain mammals. *Science* **130,** 1192–1194.

Inman, D. L. (1973). Cellulose digestion in ruffed grouse, chukar partridge, and bobwhite quail. *J. Wildl. Manage.* **37,** 114–121.

Janis, C. (1976). The evolutionary strategy of the Equidae and the origins of rumen and cecal digestion. *Evolution* **30**, 757–774.

Johnson, J. L., and McBee, R. H. (1967). The porcupine cecal fermentation. *J. Nutr.* **91**, 540–546.

Johnston, J. M. (1968). Mechanism of fat absorption. In "Alimentary Canal, Section 6. Handbook of Physiology" (C. F. Code, ed.), pp. 1353–1375. Am. Physiol. Soc., Washington, D.C.

Karn, R. C., and Malacinski, G. M. (1978). The comparative biochemistry, physiology, and genetics of animal a-amylases. *Adv. Comp. Physiol. Biochem.* **7**, 1–103.

Kay, R. N. B., Engelhardt, W. v., and White, R. G. (1980). The digestive physiology of wild ruminants. In "Digestive Physiology and Metabolism in Ruminants" (Y. Ruckebusch and P. Thivend, eds.), pp. 743–761. MTP Press, Lancaster, England.

Kempton, T. J., Murray, R. M., and Leng, R. A. (1976). Methane production and digestibility measurements in the grey kangaroo and sheep. *Aust. J. Biol. Sci.* **29**, 209–214.

Kerry, K. R. (1969). Intestinal disaccharidase activity in a monotreme and eight species of marsupials (with an added note on the disaccharidases of five species of sea birds). *Comp. Biochem. Physiol.* **29**, 1015–1022.

Kinnear, J. E., Cockson, A., Christensen, P., and Main, A. R. (1979). The nutritional biology of the ruminants and ruminant-like mammals—A new approach. *Comp. Biochem. Physiol.* **64A**, 357–365.

Kirkwood, J. K. (1979). The partition of food energy for existence in the kestrel (*Falco tinnunculus*) and the barn owl (*Tyto alba*). *Comp. Biochem. Physiol.* **63A**, 495–498.

Klein, D. R., and Schonheyder, F. (1970). Variation in ruminal nitrogen levels among some cervidae. *Can. J. Zool.* **48**, 1437–1442.

Kreulen, D. A., and Hoppe, P. P. (1979). Diurnal trends and relationship to forage quality of ruminal volatile fatty acid concentration, pH and osmolarity in wildebeest on dry range in Tanzania. *Afr. J. Ecol.* **17**, 53–63.

Langer, P. (1974). Stomach evolution in the Artiodactyla. *Mammalia* **38**, 295–314.

Langer, P. (1978). Anatomy of the stomach of the collared peccary, *Dicotyles tajacu* (L. 1758) (Artiodactyla: Mammalia). *Z. Saugetierkunde* **43**, 42–59.

Langer, P. (1979). Phylogenetic adaptation of the stomach of the Marcropodidae Owen, 1839, to food. *Z. Saugetierkunde* **44**, 321–333.

Leibholz, J. (1972). Nitrogen metabolism in sheep. II. The flow of amino acids into the duodenum from dietary and microbial sources. *Aust. J. Agric. Res.* **23**, 1073–1083.

Leng, R. A., Corbett, J. L., and Brett, D. J. (1968). Rates of production of volatile fatty acids in the rumen of grazing sheep and their relation to ruminal concentrations. *Brit. J. Nutr.* **22**, 57–68.

Lengemann, F. W., and Allen, N. N. (1959). Development of rumen function in the dairy calf. II. Effect of diet upon characteristics of the rumen flora and fauna of young calves. *J. Dairy Sci.* **42**, 1171–1181.

Leon, B. (1980). Fermentation and the production of volatile fatty acids in the alimentary tract of the rock hyrax, *Procavia capensis*. *Comp. Biochem. Physiol.* **65A**, 411–420.

Leopold, S. A. (1953). Intestinal morphology of gallinaceous birds in relation to food habits. *J. Wildl. Manage.* **17**, 197–203.

LePrince, P., Dandrifosse, G., and Schoffeniels, E. (1979). The digestive enzymes and acidity of the pellets regurgitated by raptors. *Biochem. Syst. Ecol.* **7**, 223–227.

Lintern-Moore, S. (1973). Incorporation of dietary nitrogen into microbial nitrogen in the forestomach of the Kangaroo Island wallaby *Protemnodon eugenii* (Desmarest). *Comp. Biochem. Physiol.* **44A**, 75–82.

McBee, R. H., Johnson, J. L., and Bryant, M. P. (1969). Ruminal microorganisms from elk. *J. Wildl. Manage.* **33**, 181–186.

McDonald, I. W., and Warner, A. C. I. (Eds.). (1975). "Digestion and Metabolism in the Ruminant." Proc. IV Int. Symp. Ruminant Physiol., Sydney, Australia.

McKenzie, R. A. (1978). The caecum of the koala, *Phascolarctos cinereus:* Light, scanning and transmission electron microscopic observations on its epithelium and flora. *Aust. J. Zool.* **26,** 249–256.

Maloiy, G. M. O., Kay, R. N. B., and Goodall, E. D. (1968). Studies on the physiology of digestion and metabolism of the red deer (*Cervus elaphus*). *Symp. Zool. Soc. London* **21,** 101–108.

Matthews, L. R., and Kilgour, R. (1980). Learning and associated factors in ruminant feeding behavior. *In* "Digestive Physiology and Metabolism in Ruminants" (Y. Ruckebusch and P. Thivend, eds.), pp. 123–144. MTP Press, Lancaster, England.

Mattocks, J. G. (1971). Goose feeding and cellulose digestion. *Wildfowl* **22,** 107–113.

Mautz, W. W. (1969). Investigation of some digestive parameters of the white-tailed deer using the radioisotope ^{51}Chromium. Doctoral dissertation, Michigan State Univ., East Lansing.

Miller, M. R. (1974). Digestive capabilities, gut morphology, and cecal fermentation in wild waterfowl (genus *Anas*) fed various diets. Master's thesis, Univ. of Calif., Davis.

Milton, K. (1981). Food choice and digestive strategies of two sympatric primate species. *Am. Nat.* **117,** 496–505.

Moir, R. J. (1965). The comparative physiology of ruminant-like animals. *In* "Physiology of Digestion in the Ruminant" (R. W. Dougherty, ed.), pp. 1–14. Butterworths, Washington, D.C.

Moir, R. J., and Hungate, R. W. (1968). Ruminant digestion and evolution. *In* "Handbook of Physiology," Vol. V (C. F. Code, ed.), pp. 2673–2694. Am. Physiol. Soc., Washington, D.C.

Moir, R. J., Somers, M., and Waring, H. (1956). Studies on marsupial nutrition. I. Ruminant-like digestion in a herbivorous marsupial (*Setonix brachyurus* Quoy and Gaimard). *Aust. J. Biol. Sci.* **9,** 293–304.

Moore, J. H., and Noble, R. C. (1975). Foetal and neonatal lipid metabolism. *In* "Digestion and Metabolism in the Ruminant" (I. W. McDonald and A. C. I. Warner, eds.), pp. 465–480. Proc. IV Int. Symp. Ruminant Physiol., Sydney, Australia.

Morris, I. G. (1968). Gamma globulin absorption in the newborn. *In* "Alimentary Canal, Section 6. Handbook of Physiology" (C. F. Code, ed.), pp. 1491–1512. Am. Physiol. Soc., Washington, D.C.

Moss, R., and Parkinson, J. A. (1975). The digestion of bulbils (*Polygonum viviparum* L.) and berries (*Vaccinium myrtillus* L. and *Empetrum* sp.) by captive ptarmigan (*Lagopus mutus*). *Brit. J. Nutr.* **33,** 197–206.

Murray, R. M., Marsh, H., Heinsohn, G. E., and Spain, A. V. (1977). The role of the midgut caecum and large intestine in the digestion of sea grasses by the dugong (Mammalia: Sirenia). *Comp. Biochem. Physiol.* **56A,** 7–10.

Myrcha, A. (1964). Variations in the length and weight of the alimentary tract of *Clethrionomys glareolus* (Schreber, 1780). *Acta Theriol.* **9,** 139–148.

Nagy, J. G., and Regelin, W. L. (1975). Comparison of digestive organ size of three deer species. *J. Wildl. Manage.* **39,** 621–624.

Parker, D. S. (1976). The measurement of production rates of volatile fatty acids in the caecum of the conscious rabbit. *Brit. J. Nutr.* **36,** 61–70.

Patton, J. S., and Carey, M. C. (1979). Watching fat digestion. *Science* **204,** 145–148.

Pearson, H. A. (1969). Rumen microbial ecology in mule deer. *Appl. Microbiol.* **17,** 819–824.

Perrin, M. R., and Curtis, B. A. (1980). Comparative morphology of the digestive system of 19 species of southern African myomorph rodents in relation to diet and evolution. *S. Afr. J. Zool.* **15,** 22–33.

Pilgrim, A. F., Gray, F. V., Weller, R. A., and Belling, C. B. (1970). Synthesis of microbial protein from ammonia in the sheep's rumen and the proportion of dietary nitrogen converted into microbial nitrogen. *Brit. J. Nutr.* **24,** 589–598.

Poe, S. E., Ely, D. G., Mitchell, G. E., Jr., Glimp, H. A., and Deweese, W. P. (1971). Rumen development in lambs. II. Rumen metabolite changes. *J. Anim. Sci.* **32, 989–993.**

Prins, R. A., and Geelen, M. J. H. (1971). Rumen characteristics of red deer, fallow deer, and roe deer. *J. Wildl. Manage.* **35,** 673–680.

Prins, R. A., Hungate, R. E., and Prast, E. R. (1972). Function of the omasum in several ruminant species. *Comp. Biochem. Physiol.* **43A,** 155–163.

Purser, D. B. (1970). Nitrogen metabolism in the rumen: Micro-organisms as a source of protein for the ruminant animal. *J. Anim. Sci.* **30,** 988–1001.

Rhoades, D. D., and Duke, G. E. (1975). Gastric function in a captive American bittern. *Auk* **92,** 786–792.

Robbins, C. T., and Moen, A. N. (1975). Milk consumption and weight gain of white-tailed deer. *J. Wildl. Manage.* **39,** 355–360.

Schoonveld, G. G., Nagy, J. G., and Bailey, J. A. (1974). Capability of mule deer to utilize fibrous alfalfa diets. *J. Wildl. Manage.* **38,** 823–829.

Semenza, G. (1968). Intestinal oligosaccharidases and disaccharidases. *In* "Alimentary Canal, Section 6. Handbook of Physiology" (C. F. Code, ed.), pp. 2543–2566. Am. Physiol. Soc., Washington, D.C.

Short, H. L. (1963). Rumen fermentations and energy relationships in white-tailed deer. *J. Wildl. Manage.* **27,** 184–195.

Short, H. L. (1964). Postnatal stomach development of white-tailed deer, *J. Wildl. Manage.* **28,** 445–458.

Silver, H. (1968). Deer nutrition studies. *In* "The White-tailed Deer of New Hampshire" (H. R. Siegler, ed.), pp. 182–196. New Hampshire Fish Game Department, Concord.

Skadhauge, E. (1976). Cloacal absorption of urine in birds. *Comp. Biochem. Physiol.* **55A,** 93–98.

Sklan, D. (1980). Site of digestion and absorption of lipids and bile acids in the rat and turkey. *Comp. Biochem. Physiol.* **65A,** 91–95.

Slade, L. M., Bishop, R., Morris, J. G., and Robinson, D. W. (1971). Digestion and absorption of ^{15}N-labelled microbial protein in the large intestine of the horse. *Brit. Vet. J.* **127,** xi–xiii.

Smith, R. L., Hubartt, D. J., and Shoemaker, R. L. (1980). Seasonal changes in weight, cecal length, and pancreatic function of snowshoe hares. *J. Wildl. Manage.* **44,** 719–724.

Southwick, E. E., and Southwick, A. K. (1980). Energetics of feeding on tree sap by ruby-throated hummingbirds in Michigan. *Am. Midl. Nat.* **104,** 328–334.

Staaland, H., Jacobsen, E., and White, R. G. (1979). Comparison of the digestive tract in Svalbard and Norwegian reindeer. *Arctic and Alpine Res.* **11,** 457–466.

Stewart, W. E., Stewart, D. G., and Schultz, L. H. (1958). Rates of volatile fatty acid production in the bovine rumen. *J. Anim. Sci.* **17,** 723–736.

Sudo, S. Z., and Duke, G. E. (1980). Kinetics of absorption of volatile fatty acids from the ceca of domestic turkeys. *Comp. Biochem. Physiol.* **67A,** 231–237.

Tamate, H., McGilliard, A. D., Jacobson, N. L., and Getty, R. (1962). Effect of various dietaries on the anatomical development of the stomach in the calf. *J. Dairy Sci.* **45,** 408–420.

Taylor, W. H. (1968). Biochemistry of pepsins. *In* "Alimentary Canal, Section 6. Handbook of Physiology" (C. F. Code, ed.), pp. 2567–2587. Am. Physiol. Soc., Washington, D.C.

Titchen, D. A., and Newhook, J. C. (1975). Physiological aspects of sucking and the passage of milk through the ruminant stomach. *In* "Digestion and Metabolism in the Ruminant" (I. W. McDonald and A. C. I. Warner, eds.), pp. 15–29. Proc. IV Int. Symp. Ruminant Physiol., Sydney, Australia.

Ullrey, D. E. (1980). The nutrition of captive wild ruminants. *In* "Digestive Physiology and Nutrition of Ruminants. Vol. 3. Practical Nutrition" (D. C. Church, ed.), pp. 306–320. O. and B. Books, Corvallis, Oreg.

Ulyatt, M. J., Dellow, D. W., Reid, C. S. W., and Bauchop, T. (1975). Structure and function of the large intestine of ruminants. *In* "Digestion and Metabolism in the Ruminant" (I. W.

McDonald and A. C. I. Warner, eds.), pp. 119–133. Proc. IV Int. Symp. Ruminant Physiol., Sidney, Australia.

Vallenas, A., Cummings, J. F., and Munnell, J. F. (1971). A gross study of the compartmentalized stomach of two new-world camelids, the llama and guanaco. *J. Morph.* **134,** 399–424.

Vallenas, P. A. (1965). Some physiological aspects of digestion in the alpaca (*Lama pacos*). *In* "Physiology of Digestion in the Ruminant" (R. W. Dougherty, ed.), pp. 147–158. Butterworths, Washington, D.C.

Van Hoven, W. (1978). Digestion physiology in the stomach complex and hindgut of the hippopotamus (*Hippopotamus amphibius*). *S. Afr. J. Wildl. Res.* **8,** 59–64.

Van Hoven, W., and Boomker, E. A. (1981). Feed utilization and digestion in the black wildebeest (*Connochaetes gnou,* Zimmerman, 1780) in the Golden Gate Highlands National Park. *S. Afr. J. Wildl. Res.* **11,** 35–40.

Van Hoven, W., Prins, R. A., and Lankhorst, A. (1981). Fermentative digestion in the African elephant. *S. Afr. J. Wildl. Res.* **11,** 78–86.

Van Soest, P. J. (1981). "Nutritional Ecology of the Ruminant." O. and B. Books, Corvallis, Oreg.

Walcott, P. J., and Messer, M. (1980). Intestinal lactase (*B*-galactosidase) and other glycosidase activities in suckling and adult tammar wallabies. *Aust. J. Biol. Sci.* **33,** 521–530.

Warner, R. G., Grippin, C. H., Flatt, W. P., and Loosli, J. K. (1955). Further studies on the influence of diet on the development of the ruminant stomach. *J. Dairy Sci.* **38,** 605.

Weiner, I. M., and Lack, L. (1968). Bile salt absorption: Enterohepatic circulation. *In* "Alimentary Canal, Section 6. Handbook of Physiology" (C. F. Code, ed.), pp. 1439–1455. Am. Physiol. Soc., Washington, D.C.

Weller, R. A., Pilgrim, A. F., and Gray, F. V. (1962). Digestion of foodstuffs in the rumen of the sheep and the passage of digesta through its compartments. *Brit. J. Nutr.* **16,** 83–90.

Williams, V. J. (1963). Rumen function in the camel. *Nature* **197,** 1221.

Yang, M. G., Manoharan, K., and Young, A. K. (1969). Influence and degradation of dietary cellulose in cecum of rats. *J. Nutr.* **97,** 260–264.

Yang, M. G., Manoharan, K., and Mickelsen, O. (1970). Nutritional contribution of volatile fatty acids from the cecum of rats. *J. Nutr.* **100,** 545–550.

Ziswiler, V., and Farner, D. S. (1972). Digestion and the digestive system. *In* "Avian Biology," Vol. II (D. S. Farner and J. R. King, eds.), pp. 343–430. Academic Press, New York.

13

Digestion and Nutrient Metabolism

Ingested food must be digested, absorbed, and metabolized if the animal is to meet its requirements. The amount of food that an animal must ingest to meet a fixed requirement is directly proportional to the losses in digestion and metabolism. For example, if an animal requires 250 nutritional units (such as kilocalories of energy, grams of protein, or milligrams of a mineral or vitamin) and the ingested food contains five units per gram, 50 g of food must be consumed if all five units are available. However, if only 40% of the five units are available because of losses in digestion and metabolism, then 125 g must be consumed $[250 \div (5 \times 0.4) = 125]$.

Efficiencies of nutrient utilization are dependent on the interaction of food composition and digestive and metabolic capabilities of the ingesting animal. For example, animals with fermenting gastrointestinal bacteria can utilize plant cell wall carbohydrates more efficiently than can animals without such bacteria. Losses of most nutrients include fecal and urinary excretion. Two additional losses, methane and excess heat production, are relevant in determining energy efficiencies. Consequently, the quantification and understanding of the efficiencies of nutrient utilization are necessary to relate requirements ecologically to available food resources.

I. DIGESTION

A. Methods and Terminology

Digestibility coefficients are broadly defined as the relative amount of ingested matter or energy that does not appear in the feces. Digestibilities determined by feeding the live animal and collecting the feces are termed *in vivo*. *In vivo* digestion trials typically use captive animals because of the ease with which feed and feces can be quantified. An underlying assumption in using captive animals is that the basic physiological processes of digestion are comparable between free-ranging and captive states. The assumption has never been adequately tested because of the difficulty in conducting digestion trials on unrestrained animals.

Dry-matter digestibilities in white-tailed deer were initially reduced when the animals were confined to digestion crates, but they subsequently rose to preconfinement levels after 12–16 days of confinement (Mautz, 1971b). Thus, training or habituation of wildlife to accept the confinement imposed by *in vivo* digestion trials may be necessary.

Most *in vivo* digestion trials use the total balance method in which all feed ingested and feces produced are weighed and analyzed. Usually, the animal is confined to a crate in which the only food available is that provided by the experimenter. Urine and feces separation is necessary, since digestibilities refer only to the efficiencies of processes occuring within the digestive tract and do not include the internal losses represented by urine. Urinary contamination of feces will often reduce digestibility estimates, particularly of nitrogen and minerals. While stanchioning or fitting fecal bags to male animals can ensure total separation of feces and urine, many wild animals simply will not accept such restraint or manipulation. Consequently, most wildlife digestion crates are large enough that the animal can turn around as it walks on a porous metal screen through which both urine and feces fall (Fig. 13.1). Feces in such crates are collected on

Fig. 13.1. Internal view of a deer digestion crate. Notice the expanded metal floor on which the deer walks and through which feces and urine fall to lower collection screens or funnels. Feeder and urine collection funnel are also visible. (From Cowan *et al.*, 1969, courtesy of the *Journal of Wildlife Management.*)

a second screen just below the floor, and the urine runs down plastic or metal sheeting to collection jars. Digestion trials can not be conducted on birds without surgical manipulation, since urine and feces mix in the cloaca and large intestine and therefore are physically inseparable.

Total collection trials typically have a 3- to 10-day pretrial period followed by the 5- to 7-day trial. The pretrial adaptation period provides time for (a) the microbial and enzymatic digestion processes to equilibrate with the diet; (b) the residues from previous diets to be eliminated; and (c) the feces to equilibrate quantitatively with the feed. Shorter pretrial and trial periods are characteristic of small simple-stomached animals that pass food very quickly through the gastrointestinal tract, as compared to longer periods for the large ruminant and nonruminant herbivores (Mothershead et al., 1972; Holter et al., 1974; Moors, 1977).

Total balance trials with small mammals are relatively inexpensive but become inceasingly expensive and laborious with larger animals. Collection of the few grams of feed consumed by a small rodent or carnivore may be a trivial problem when compared to collecting the hundreds of kilograms of small twigs or leaves consumed by several elk or moose. Because of the time, labor, and cost necessary for total balance trials using captive animals and the desire by many field ecologists to estimate directly the digestibility of mixed diets consumed by free-ranging animals, several attempts have been made to estimate digestibility by using nondigestible feed markers, such as plant lignin, silica, or added metals (Mautz, 1971a; Arman and Field, 1973; Murray et al., 1977; Ruggiero and Whelan, 1977). For example, the dry-matter digestibility of a diet could be determined by only two analyses of representative feed and fecal samples using the following equation:

$$\text{Dry matter digestibility } (\%) = \left(1 - \frac{\text{Feed marker concentration}}{\text{Fecal marker concentration}}\right) \times 100.$$

If the feed marker concentration is 10% in a given diet, fecal concentrations of 20 and 30% indicate dry-matter digestibilities of 50 and 67%, respectively. Unfortunately, the perfect marker is yet to be found. Many of the so-called nondigestible plant markers are not completely recoverable in the feces. The excretion of added markers, such as chromium, can very diurnally, and the marker may not move with the nutrient being examined (Gasaway et al., 1975, 1976; Ruggiero and Whelan, 1977).

Such problems have stimulated the development of additional digestion procedures, particularly in vitro. In vitro refers to the use of a laboratory digestion procedure that may only remotely be related to animal functioning. In vitro digestion trials are most popular in estimating ruminant digestibilities because they attempt to stimulate the fermentation and acid protease hydrolysis occurring in vivo. The ingredients in the fermentation step are (a) a carefully weighed

5.0–1.0 g food sample; (b) 40 ml of a buffer-nutrient solution; (c) 10 ml of rumen fluid obtained from a fistulated or recently killed animal; (d) a small flask or vial within which the above components are mixed; (e) bottled carbon dioxide, which is used to purge the flasks of all oxygen; and (f) a water bath or oven within which the *in vitro* fermentation can be maintained at 38–40°C. The general philosophy of the *in vitro* fermentation procedure is to maximize digestion opportunity such that the limitation imposed by the plant–microbial interaction establishes the end point and not an unfavorable pH shift or limiting nutrient.

Fermentation generally lasts for 48 hr, since particles are often retained in the rumen for prolonged periods and the *in vitro* digestive process normally has reached an end point because of either toxic end product accumulation or substrate exhaustion (Fig. 13.2). The contents of the fermentation flask can either be acidified with hydrochloric acid and incubated with pepsin for 24–48 hr (Tilley and Terry, 1963) or refluxed in neutral detergent (Goering and Van Soest, 1970). The acid–pepsin hydrolysis attempts to simulate abomasal digestion, while refluxing in neutral detergent removes microbial cells and soluble plant debris. The residues of either procedure are filtered, washed, dried, and weighed. The amount of dry-matter disappearance is an estimate of digestibility.

In vitro dry-matter digestibilities of forages consumed by deer overestimate the observed *in vivo* values (Fig. 13.3). The overestimate is higher when cow inocula is used and essentially confirms the difference even in the inocula of concentrate versus roughage–bulk feeders (Prins and Greelen, 1971; Short *et al.*,

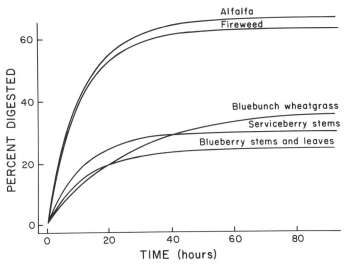

Fig. 13.2. Rates of *in vitro* dry-matter digestion of five forages using mule deer inocula. (From Milchunas *et al.*, 1978, courtesy of the Colorado Division of Wildlife.)

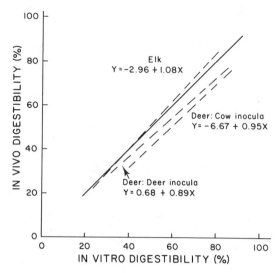

Fig. 13.3. In *vivo–in vitro* comparisons of dry-matter digestibilities using the Tilley–Terry fermentation-acid pepsin hydrolysis procedure. The deer *in vitro* samples were digested with both cow (Palmer and Cowan, 1980) and deer inocula (Palmer *et al.*, 1976, Ruggiero and Whelan, 1976; Milchunas *et al.*, 1978; Mould and Robbins, 1982). The elk *in vitro* samples were digested with elk inocula only (Mould and Robbins, 1982). Solid line is the 1:1 relationship.

1974). *In vitro–in vivo* digestibilities for elk are essentially identical. However, plants containing toxic or inhibitory secondary plant compounds may have lower digestibilities *in vitro* than *in vivo* since the open-flow nature of the rumen, including absorption, dilution, and passage, is not simulated *in vitro* (Person *et al.*, 1980).

A hybrid of both the *in vivo* and *in vitro* procedures that takes advantage of the rumen environment is the nylon bag technique. A 0.5–1.0 g sample is carefully weighed and placed in a small-pore nylon bag that is subsequently suspended in the rumen of a fistulated animal. A major value of the nylon bag technique is that it provides a mechanism to examine the rate of rumen digestion. For example, while forbs and shrubs have a very rapid initial disappearance, with rumen digestion being completed in 48 hr, digestion of mature grasses is much slower and can continue in excess of 96 hr (Short *et al.*, 1974; White and Trudell, 1980). The major problem with the nylon bag technique is correcting for the passage of fine rumen or sample particles through the pores of the bag. Since the bag and its contents are usually washed under a gently flowing stream of water when removed from the rumen, the major error is in the passage of fine, non-digested particles out of the bag. While one can place in the bag sample particles larger than the pore size, microbial digestion has the potential of fragmenting larger particles into smaller ones that could move through the pores.

A third digestion procedure is similar to the intent of crude fiber analysis in chemically identifying the nondigestible fraction of a food sample. While we clearly recognize the difficulties of easily dividing plant samples into fractions on the basis of digestibility, animal tissues may be more amenable to chemically partitioning digestible fractions. For example, Angerman et al., (1974) extracted an homogenized mouse sample with neutral detergent. The residue of mostly hair, bones, teeth, and some collagen and elastin does represent the principal nondigestible components of a mouse.

Digestibilities are termed either apparent or true. The two terms refer not to the accuracy of a digestion measurement but to whether the digestibility coefficient is a gross (apparent) or net (true) function. For example, apparent digestibilities are determined by the following equation:

$$\text{Apparent digestibility (\%)} = \frac{\text{Amount consumed} - \text{Fecal excretion}}{\text{Amount consumed}} \times 100.$$

However, fecal losses include both nondigested feed residues as well as animal metabolic secretions. True digestibilities can be determined by correcting the total fecal loss for the metabolic losses:

$$\text{True digestibility (\%)} = \frac{\text{Amount consumed} - (\text{Total fecal excretion} - \text{Metabolic losses})}{\text{Amount consumed}} \times 100.$$

True and apparent digestibilities differ for only those feed constituents in which the animal adds a metabolic fecal component. For example, fecal protein,

TABLE 13.1
Relationships between the Feed Content and True and Apparent Digestibilities of a Highly Digestible Feed Fraction Whose Ingestion Produces a Constant Metabolic Fecal Loss

Feed content (g/100 g feed)	True digestible amount[a] (g/100 g feed)	Nondigested amount (g/100 g feed)	Metabolic loss (g/100 g feed)	Apparent digestible amount[b] (g/100 g feed)	Apparent digestibility coefficient[c] (%)
0.0	0.00	0.00	5.0	−5.00	−∞
5.0	4.75	0.25	5.0	−0.25	−5.0
10.0	9.50	0.50	5.0	4.50	45.0
20.0	19.00	1.00	5.0	14.00	70.0
30.0	28.50	1.50	5.0	23.50	78.3
100.0	95.00	5.00	5.0	90.00	90.0

[a] True digestibility = 95%.
[b] Apparent digested amount = intake minus total fecal excretion.
[c] Apparent digestibility coefficient = apparent digested amount ÷ feed content.

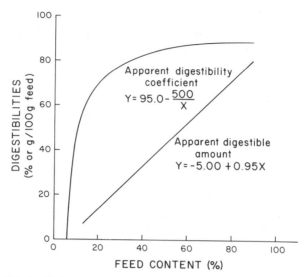

Fig. 13.4. Relationship between apparent digestibility coefficients, apparent digestible amount, and feed content for the example in Table 13.1.

lipids, minerals, or vitamins can be either from the feed or from the animal. When apparent digestibilities include a metabolic loss, true digestibility coefficients will always be higher than the apparent coefficient. Conversely, the digestibility of plant cell wall components are always true, since the animal does not synthesize and excrete cellulose, hemicellulose, lignin, or cutin. For highly digestible feed components in which a constant metabolic loss occurs, the apparent digestibility is a curvilinear function of dietary concentration (Fig. 13.4; Table 13.1). However, the apparent digestible amount (feed content times apparent digestibility) is always a linear function when the same assumptions prevail (Fig. 13.4; Table 13.1). Note that the equations for both the linear and curvilinear lines describing apparent digestive functions are composed of the true digestibility coefficient and the metabolic fecal loss. Negative apparent digestibilities occur when metabolic and nondigested fecal losses exceed the amount of intake.

B. Digestibilities

1. Protein

Apparent digestibility coefficients for dietary protein are curvilinear functions of dietary content (Fig. 13.5). The curvilinearity is caused by the generally high true digestibility of dietary protein and the continuous excretion of metabolic fecal nitrogen in all animals. Because of the lack of understanding of factors determining apparent protein digestibilities, several early wildlife nutritionists were reluctant to report negative coefficients (Forbes *et al.*, 1941; Bissell *et al.*, 1955).

Apparent protein digestibilities are highest in neonates consuming milk and carnivores consuming meat diets, because metabolic losses are minimal and concentrations of feed protein are high. For example, the apparent digestibility of milk protein consumed by elk calves was 97.5% and meat protein consumed by foxes and fishers 94.5±1.3% (W. R. Inman, 1941; W. R. Inman and Smith, 1941; Davison, 1975; C. T. Robbins *et al.*, 1981). When an entire animal, including bones, teeth, and hair or feathers, is consumed, the apparent protein digestibility will be much lower. For example, the apparent protein digestibility of small birds and mammals consumed by fishers and badgers was 82.0±4.2% (Jense, 1968; Davison, 1975).

True digestibilities of dietary protein range from 65 to 100% and average 88.7±7.1% (Table 13.2). The true digestibility of the protein in pelleted diets (85.9±5.1) fed to ruminants is generally lower than that in diets of long hays or browse (96.8±1.2%). This may be due to either (*a*) artifact, protein-bound lignin formed during the pelleting process or (*b*) gastrointestinal transit that is too rapid for complete digestion of the pellets because of fine grinding. The true digestibility of meat diets continues to reflect the amount of nondigestible hair,

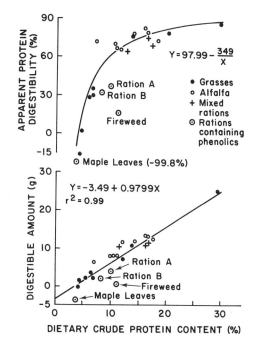

Fig. 13.5. Digestible protein as a function of dietary protein concentration in elk. Note that rations containing soluble phenolics have a lower apparent digestible protein content than grasses and alfalfa. The regression is for only the grasses, alfalfa, and mixed rations that are low or free of soluble phenolics. (From Mould and Robbins, 1982, courtesy of the *Journal of Wildlife Management.*)

TABLE 13.2
True Digestibilities of Dietary Protein Consumed by Wildlife

Species or group	Diet	Digestibility coefficient	Reference
Monogastrics and cecal digestors			
Black-tailed jackrabbit	Alfalfa, forbs, and shrub leaves	98.5	Data of Nagy *et al.* (1976)
Capybara		87.0	Parra (1978)
Domestic rabbit	Fish meal, almond hulls, and oat hay	92.0	Slade and Robinson (1970)
Fisher	Coturnix quail	89.3	Davison (1975)
	Deer meat	98.4	
	Small mammals	86.4	
	Snowshoe hares	93.1	
Guinea pig	Fish meal, almond hulls, and oat hay	94.0	Slade and Robinson (1970)
Howler monkey	Leaf and fruit diets	82.7	Milton *et al.* (1980)
Mink	Animal proteins	84.5	Glem-Hansen and Eggum (1974)
Rhesus monkey	Commercially pelleted diet	87.4	Data of R. C. Robbins and Gavan (1966)
Rock hyrax	Carrots, apples, oats, and beet pulp	65.0	Hume *et al.* (1980)
Small rodents	Grasses and seeds	87.4	Figure 8.3
Snowshoe hare	Browse and browse-concentrate mixtures	91.3	Data of Holter *et al.* (1974), Walski and Mautz (1977)
Macropod marsupials			
Brushtail possum	Concentrate diets with added fiber		Wellard and Hume (1981)
	Low fiber	100.0	
	High fiber	91.0	

(*continued*)

feathers, or other proteinaceous material. For example, the protein of deer meat when fed to fishers is 10–14% more digestible than that of whole small mammals or birds.

The estimation of true digestibility as the regression slope between protein intake and apparent digestible amount assumes that the excretion of metabolic fecal nitrogen remains constant and that other dietary components have either no effect or a constant effect. Several secondary plant compounds can reduce the digestibility of dietary protein. For example, tannins and other soluble phenolics can form nondigestible complexes with protein, and proteinase inhibitors in many seeds can prevent the digestive actions of host enzymes in simple-stom-

TABLE 13.2 *Continued*

Species or group	Diet	Digestibility coefficient	Reference
Euro	Hays and straw	87.9	Data of Hume (1974)
Red kangaroo	Hays and straw	85.3	Data of Foot and Romberg (1965), Hume (1974)
Wallaby	Alfalfa, oat straw, concentrates	83.7	Data of Kennedy and Hume (1978), Hume and Dunning (1979), Hume (1977a)
Wallaby	Oat straw-concentrates	87.0	Barker (1968)
Ruminants			
Bush duiker	Pelleted diet	83.5	Data of Arman and Hopcraft (1975)
Caribou and reindeer	Pellet and lichen diets	78.8	Data of McEwan and Whitehead (1970), Jacobsen and Skjenneberg (1972)
Eland	Pelleted diet	83.2	Data of Arman and Hopcraft (1975)
Hartebeest	Pelleted diet	87.4	Data of Arman and Hopcraft (1975)
Red deer	Pelleted diet	96.0	Data of Maloiy *et al.* (1970)
Thomson's gazelle	Pelleted diet	84.7	Data of Arman and Hopcraft (1975)
White-tailed deer	Pelleted diet	84.4	S. H. Smith *et al.* (1975)
White-tailed deer	Pelleted diet	89.4	Holter *et al.* (1979a)
Bison	Hay	96.8	Data of Richmond *et al.* (1977)
Elk	Hay	98.0	Mould and Robbins (1981)
White-tailed deer and mule deer	Hay and browse	95.7	C. T. Robbins *et al.* (1975)

ached animals (Richardson, 1977). The soluble phenolics of fireweed (*Epilobium angustifolium*) and maple leaves reduced the apparent digestible amount of protein fed to elk (Fig. 13.5). When both high-protein, phenolic-free and low-protein, tannin-containing rations are fed to animals and the apparent protein digestibilities are used to estimate graphically the true digestibility, the slope of the line increases above 1 and the Y intercept or metabolic fecal nitrogen estimate becomes much higher than when only tannin-free rations are fed. For example, when fox squirrels consumed both seeds and tannin-containing nuts and red deer consumed hays and tannin-containing deciduous tree and shrub leaves, implausible true protein digestibilities of 104% occurred (Baumgras, 1944; Van de Veen,

1979). The assumption that the higher fecal nitrogen losses occurring when animals consume tannin-containing rations are simply increased metabolic fecal nitrogen losses (Osbourn *et al.*, 1971) is unfounded, since tannin binding could occur with either plant or animal protein during mastication and digestion. Additionally, since the tannin–protein complex is reversible at appropriate pHs (Berenbaum, 1980), dissociation and recombination may occur several times during gastrointestinal passage.

2. *Plant Neutral Detergent Solubles and Fiber*

The apparent digestibility of plant neutral detergent solubles is a curvilinear function of their dietary concentration. However, as observed with protein, when one compares apparent digestible amount to intake, the relationship is linear (Fig. 13.6). Neutral detergent solubles of grasses and plants with minimal inhibitory secondary plant compounds are uniformly 98–100% digestible in all herbivores (Van Soest, 1967; Short and Reagor, 1970; Parra, 1978). The lignified plant cell wall does not impede their digestion. However, secondary plant compounds, such as the tannin–protein complex, will reduce the apparent digestibility of the neutral detergent solubles.

Digestibilities of neutral detergent fiber are functions of the relative concentration of structural digestion inhibitors, particularly lignin, cutin, and silica; type

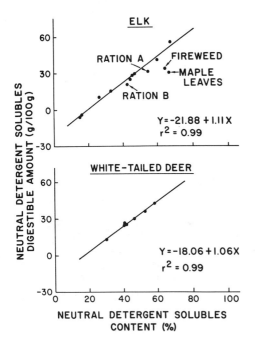

Fig. 13.6. Apparent digestible amount of neutral detergent solubles when consumed by elk and white-tailed deer. Rations containing phenolics are indicated. (From Mould and Robbins, 1982, courtesy of the *Journal of Wildlife Management*.)

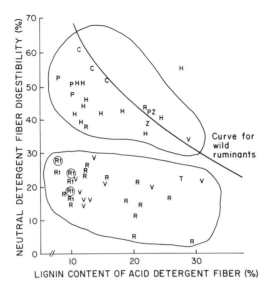

Fig. 13.7. Relationship between the lignin concentration of the acid detergent fiber and the neutral detergent fiber digestibility in several ruminant and nonruminant herbivores. Species are horse (H), zebra (Z), Przewalski's horse (PZ), capybara (C), domestic pig (P), domestic rabbit (R), laboratory rat (R+), laboratory rat whose cecum was inoculated with rumen fluid (circled R+), meadow vole (V), and tortoise (T). Curve for wild ruminants is an average for white-tailed deer and elk. (Adapted from Parra, 1978, by permission of the Smithsonian Institution Press from *The Ecology of Arboreal Folivores*, G. Gene Montgomery, ed., Fig. 3, p. 214. © Smithsonian Institution, Washington, D.C.)

of gastrointestinal tract; and rate of passage. For example, the digestibility of the NDF decreases in ruminants as the lignin–cutin concentration increases (Fig. 13.7). Silica reduces the NDF digestibility of grasses by three units per unit of biogenic silica (Van Soest and Jones, 1968). Ruminants and large nonruminant herbivores digest more of the NDF than do small nonruminant herbivores. Structural inhibitors do not limit NDF digestion in small nonruminant herbivores, because the rate of passage is too fast for complete utilization of even the available cellulose and hemicellulose. Although the small nonruminant herbivores digest approximately 20% of the NDF, the variation in observed digestibilities is probably due to level of intake and other factors that affect the rate of passage within each species (Ambuhl et al., 1979).

It has long been suggested that not all ruminants are equal in their abilities to digest fiber (Hungate et al., 1959; Prins and Geelen, 1971; Short et al., 1974; Van de Veen, 1979). Numerous comparative studies of wild and domestic ruminants have found differences in fiber digestibility (Maloiy et al., 1970; Gebcznska et al., 1974; Arman and Hopcraft, 1975; Hoppe et al., 1977; Richmond *et al.*, 1977; Schaefer *et al.*, 1978). However, an approximately equal number of studies have concluded that the wild or domestic ruminant is the most efficient. The hypothesis that domestication has altered digestive efficiency is simply the wrong hypothesis in most cases and is certainly not testable by comparing deer and sheep or elk, bison, and cattle. Differences in fiber digestibility must ultimately be relatable to rumen capacity and gastrointestinal passage phenomena.

The general hypothesis is that, since small ruminants have a proportionately smaller fermentation capacity ($Y = 0.1050X^{1.05}$ where X is body weight in kilograms and Y is fermentation contents in kilograms) per unit of energy requirement ($W^{0.75}$) than do larger ruminants, the rate of passage must be faster in the smaller species (Parra, 1978; Van Soest, 1981). If the rate of passage becomes fast enough, the digestibility of plant fiber must decrease. While rumen capacity and body size are interrelated, major interspecific differences in animals of similar size do occur. For example, rumen–reticulum capacity in domestic sheep is approximately double that of similar sized deer (Prins and Geelen, 1971). The general hypothesis is difficult to test experimentally, since digestion trials are usually conducted at less than ad libitum intakes, and small seasonal differences in digestibility may confound data interpretation (Westra and Hudson, 1981). However, North American elk weighing five times more than white-tailed deer and having a proportionately larger rumen relative to body metabolism are approximately 10% more efficient in digesting NDF than are the small deer (Mould and Robbins, 1982).

3. Dry Matter and Energy

The understanding of factors determining dry-matter digestibility is basic to predicting energy digestibility. For most herbivore, granivore, and carnivore diets, dry-matter and energy digestibilities are virtually synonymous (Tables 13.3 and 13.4). The digestibility of carnivore diets can be reduced if significant amounts of nondigestible hair, feathers, exoskeletons, claws, beaks, or teeth are ingested (Jense, 1968; Davison, 1975; Moors, 1977). Since excretion of metabolic fecal energy is minimal in carnivores, coefficients of apparent and true digestible energy differed by only 1.93 digestible units in mink (Farrell and Wood, 1968).

The digestible dry-matter content of herbivore diets can be estimated as the sum of the relative digestible amounts of the neutral detergent solubles and fiber. For example, apparent dry-matter digestibilities (Y in percentage) for feeds with minimal inhibitory secondary plant compounds ingested by elk and white-tailed deer can be predicted by the following equations:

Elk:

$$Y = (1.11\ NDS - 21.88) + NDF\left(\frac{176.92 - 40.50\ \log_e X}{100}\right)$$

White-tailed deer:

$$Y = (1.06 \ NDS - 18.06) + NDF \left(\frac{161.39 - 36.95 \ \log_e X}{100} \right),$$

where NDS (neutral detergent solubles) and NDF (neutral detergent fiber) are in percentages and X is the lignin–cutin content of the acid detergent fiber (Mould and Robbins, 1982). For an early season grass containing 40% NDF and a lignin–cutin to ADF ratio of 15%, the apparent dry-matter digestibilities are 71.6% for elk and 70.1% for white-tailed deer. Excretion of metabolic dry matter in white-tailed deer is a relatively constant 14.4 units, whereas it varies as a function of apparent digestible dry matter (X) in elk $(Y = 25.03 - 0.1481X)$. Thus, true digestibilities in the preceding example are 86.0% for elk and 84.5% for white-tailed deer.

One of the underlying, theoretical problems in conducting digestion trials and predicting digestibility of herbivore diets is whether components or characteristics of one ingested food affect the digestibility of other feeds in a mixed diet. The potential lack of independence is termed associative digestion (Table 13.5). Because of the complexity of the problem, few wildlife studies have attempted to determine if associative digestion effects occur. Dietz *et al.* (1962) observed that when mule deer were fed alfalfa pellets and browse (mountain mahogany and bitterbrush) both separately and in combination, the apparent protein and dry-matter digestibilities of mixed diets were below what was predicted from the digestibilities of the feeds when fed separately. However, one of the confounding problems in this study that commonly occurs is that the deer consumed the

TABLE 13.3

Relationships between Apparent Digestible Dry Matter (X) and Energy (Y) for Several Wild Herbivores and Granivores

Species or group	Diet	Equation	R^2	Reference
Elk	Hay	$Y = -0.61 + 0.98X$	0.98	Mould and Robbins (1982)
Mule deer	Hay, browse, pelleted diets	$Y = -1.48 + 1.04X$	0.99	Milchunas *et al.* (1978)
Mule deer and white-tailed deer	Hay, browse, pelleted diets	$Y = -0.71 + 0.99X$	0.94	C. T. Robbins *et al.* (1975)
Roe deer	Browse	$Y = -1.47 + 1.02X$	0.99	Drozdz and Osiecki (1973)
Small rodents	Grains, nuts, forbs, grasses, mosses, pelleted diets	$Y = +0.64 + 1.00X$	0.98	Hanson and Cavender (1973), C. C. Smith and Follmer (1972), Batzli and Cole (1979)
Snowshoe hares	Browse and concentrates	$Y = -9.23 + 1.16X$	0.97	Holter *et al.* (1974), Mautz *et al.* (1976)

TABLE 13.4

Average Apparent Digestible Energy and Dry-Matter Coefficients for Seeds, Nuts, and Animal
Tissue Consumed by Wild Mammals[a]

Feeding type	Diet	Average apparent digestible	
		Energy	Dry matter
Carnivores	Mammals or birds	89.4 ± 4.9	88.8 ± 9.1
	Fish	93.9 ± 2.8	
	Invertebrate larvae	91.2 ± 5.3	90.5 ± 6.7
Granivores	Seeds or nuts (excluding hulls of nuts)	89.6 ± 5.3	89.5 ± 4.7

[a]Data from W. R. Inman, 1941; W. R. Inman and Smith, 1941; Baumgras, 1944; Golley, 1960;
Hawkins and Jewell, 1962; Buckner, 1964; Golley et al., 1965; Brisbin, 1966; Sharp, 1967; Drozdz,
1968; Jense, 1968; Barrett, 1969; Johnson and Groepper, 1970; Gebczynski et al., 1972; C. C. Smith
and Follmer, 1972; Vogtsberger and Barrett, 1973; Davison, 1975; Gorecki and Grygielska, 1975;
Litvaitis and Mautz, 1976; Pernetta, 1976; Moors, 1977; Stueck et al., 1977; Green and Eberhard,
1979; Schreiber, 1979; Ashwell-Erickson and Elsner, 1981; Rau et al., 1981.

browse at approximately one-half to one-third the level at which the alfalfa was
consumed. The incorporation of plants containing soluble phenolics in mixed
diets fed to elk did not affect the digestibility of the other feeds (Mould and
Robbins, 1982). Thus, far more thorough studies of associative digestion are
needed.

II. METABOLISM OF ABSORBED NUTRIENTS

A. Metabolizable Energy

Digestible energy must be further partitioned to determine the amount of food
energy that can be used to meet maintenance and production requirements.
Losses of urinary and gaseous energy must be subtracted from the digestible
energy consumed by mammals to estimate metabolizable energy. Because of the
difficulty in separating urine and feces in birds, metabolizable energy coefficients in birds combine losses of fecal, urinary, and gaseous energy to produce
an assimilation efficiency as a function of gross energy.

Mammals:

$$\text{Metabolizable energy coefficient } (\%) = \frac{\text{Digestible energy} - \text{Urinary energy} - \text{Gaseous energy}}{\text{Digestible energy}} \times 100$$

TABLE 13.5
Potential Effect and Direction of Associative Digestion in Herbivores with Forestomach Fermentation

Item	Mode of action	Potential result	
		Single-component diet	Mixed diet
Secondary plant compounds			
Terpenoids	Toxicity to microorganisms at a threshold level	Reduced NDF and dry-matter digestion	When diluted below the threshold level, increased NDF and dry-matter digestion
Tannins and other soluble phenolics	Precipitation of dietary, microbial, and host protein	Reduced protein, dry-matter, and potentially NDF digestion	No additional effect, further reductions if complete precipitation has not occurred in the single-component diet, or increased digestion if toxicity has occurred in the single-component diet
Grinding and pelleting	Increased rate of passage	Reduced digestibility of all components	When diluted with feeds in a long form capable of slowing the rate of passage of the fine pelleted particles, increased digestibilities of all components
Increasing level of intake	1. Increased rate of passage	Reduced NDF and dry-matter digestion above a threshold level	Reduced NDF and dry-matter digestion above a threshold level
	2. Reduced metabolic excretion per unit of feed	Increased apparent dry-matter digestion	Increased apparent protein and dry-matter digestion
Inadequate intake of protein or other essential nutrient	Reduced secretion of digestive enzymes, reduced microbial populations	Reduced digestion of all components	When mixed with other forages containing adequate quantities of the deficient nutrient, increased digestion of all components
High levels of grain added to fibrous diets	Increased microbial emphasis on digestion of readily available carbohydrates		Reduced digestion of NDF and dry matter

Birds:

$$\text{Metabolizable energy coefficient } (\%) = \frac{\text{Gross energy } - \text{ Urinary and fecal energy } - \text{ Gaseous energy}}{\text{Gross energy}} \times 100$$

1. Methane

Methane is produced in the gastrointestinal tract during fermentation. Methane, propionate, and, to a lesser extent, unsaturated fatty acids are hydrogen sinks, because their production or hydrogenation enable gastrointestinal fermentation to continue without producing significant amounts of free hydrogen. However, methane, unlike propionate and saturated fatty acids, cannot be oxidized by mammals or birds and is therefore an energy loss since it is eructed, exhaled, or passed.

Animals with primarily lower-tract fermentation, such as the rock ptarmigan and snowshoe hare, do not produce significant amounts of methane (Table 13.6) (von Engelhardt, 1978). The chemical pathways and efficiencies of lower-tract fermentation may be far different from those occurring in the forestomach, since only 4–12% of the chemical energy fermented by the ptarmigan ends up as methane as compared to 19% in the ruminant (Gasaway, 1976). Even though the red kangaroo has forestomach fermentation, all methane produced in this animal originates in the lower tract (Kempton et al., 1976). Tammar wallabies produce slightly more methane, but it is still less than 2% of the apparent digestible energy (von Engelhardt, 1978). However, since these animals do liberate free hydrogen (Moir and Hungate, 1968; von Engelhardt, 1978) and may completely oxidize elemental hydrogen to water during fermentation (Kempton et al., 1976), methane measurements underestimate the amount of gaseous energy lost. However, if these animals produce the same amount of gaseous energy per unit of fermentation as do ruminants, losses of gaseous energy would rarely exceed 2% of the apparent digestible energy.

Wild ruminants produce significant amounts of methane (Table 13.6) and only very small amounts of free hydrogen (equal to less than 1% of the methane produced; Van Hoven and Boomker, 1981). The reduction in methane production as concentrates are added to hay diets is indicative of the increasing production of propionate and its role as a hydrogen acceptor. The fermentation of browse produces minimal methane, because the fiber is poorly digestible since it is highly lignified. Thus, browse can be viewed as a concentrate diet in which the readily available neutral detergent solubles are simply diluted with large amounts of poorly digestible fiber. Because of the nutritional significance of methane and other gaseous energy losses, far more extensive and systematic studies of these losses are warranted.

TABLE 13.6
Methane Production as a Percentage of Apparent Digestible Energy in Mammals and Apparent
Metabolizable Energy in Birds

Species or group	Diet	Methane production (%)	Reference
Ruminants			
Domestic livestock	All diets	10–15	Blaxter and Clapperton (1965)
Eland	Hay	12.2 ± 0.1	Rogerson (1968)
	Hay and concentrates	10.8 ± 0.5	Rogerson (1968)
	Concentrate	9.8 ± 0.3	Rogerson (1968)
Moose	Concentrate and sawdust pellets	9.8 ± 1.4	Regelin et al. (1981)
	Browse	5.6 ± 2.3	
Red deer	Hay and concentrates	7.2	Simpson et al. (1978)
White-tailed deer	Browse	5.2 ± 0.7	Mautz et al. (1975)
	Browse and concentrates	2.5 ± 1.0	Colovos et al. (1954), Greeley (1956), Silver (1963)
	Concentrates	7.3	Thompson et al. (1973)
Wildebeest	Hay	11.9 ± 0.4	Rogerson (1968)
	Hay and concentrates	10.5 ± 0.8	Rogerson (1968)
Nonruminants			
Grey kangaroo	Alfalfa	0.4 ± 0.2	Kempton et al. (1976)
Rock ptarmigan	Commercial diet, added cellulose	0.2–0.4	Gasaway (1976)
Snowshoe hare	Browse and concentrates	0.6 ± 0.1	Holter et al. (1974)

2. Urinary Energy

Excretion of urinary energy is largely dependent on diet and increases from grain and nut diets to forage and animal tissue diets (Table 13.7). Forage ingestion increases urinary energy excretion because plant foliage contains many high-energy compounds that although absorbed, cannot be oxidized. For example, the benzene ring contained in many plant compounds is excreted in the urine as a conjugate of the amino acid glycine called hippuric acid. Browse rations produce the greatest loss of urinary energy because of the necessary excretion of absorbed terpenoids, phenols, and many other secondary plant compounds. Urinary energy losses also increase as dietary protein content increases (Holter et al., 1974). The higher losses of urinary energy in carnivores consuming high-protein animal tissues occur because many of the amino acids are deaminated and the carbon skeleton is used to meet energy requirements while the nitrogen is excreted.

TABLE 13.7
Urinary Energy Excretion as a Percentage of Apparent Digestible Energy in Wild Mammals[a]

Animal group	Diet	Urinary energy (% ADE)
Fissiped carnivores	Meat or whole vertebrates	9.7 ± 3.9
Lagomorphs	Browse, both deciduous and evergreen	21.6 ± 7.0
Pinniped carnivores	Fish	7.4 ± 2.8
Primates	Ripe figs (18%) and young tree leaves and buds	8.3
	Young leaves and buds (87%) and immature fruit (13%)	10.5
Rodents	Grain or nuts	2.9 ± 1.2
Rodents and lagomorphs	Herbage (alfalfa, green wheat, and forbs)	9.2 ± 2.0
Ruminants	Hay and concentrates	6.6 ± 1.0
	Deciduous browse	11.7 ± 4.5
	Evergreen browse	18.9 ± 0.7

[a]Data from Colovos et al., 1954; Colovos, 1955; Greeley, 1956; Silver and Colovos, 1957; Golley et al., 1965; Drozdz, 1968; Jense, 1968; Rogerson, 1968; Barrett, 1969; Gebczynski et al., 1972; Holter et al., 1974; Davison, 1975; Gorecki and Grygielska, 1975; Litvaitis and Mautz, 1976; Mautz et al., 1976; Shoemaker et al., 1976; Moors, 1977; Stueck et al., 1977; Simpson et al., 1978; Nagy and Milton, 1979; Mould and Robbins, unpublished; Rau, 1981; Ashwell-Erickson and Elsner, 1981.

3. Metabolizable Energy Coefficients

Metabolizable energy coefficients in mammals reflect the differences in losses of methane and urinary energy (Table 13.8). For example, metabolizable energy coefficients are highest in rodents consuming grains or nuts, intermediate in carnivores and small nonruminant herbivores, and lowest in ruminants. Occasionally, authors correct metabolizable energy coefficients to nitrogen balance because urinary energy is, in part, dependent on total nitrogen metabolism. If an animal consuming a test diet is in negative nitrogen balance due to inadequate energy intake, a portion of the urinary energy is being derived from tissue catabolism and is thus not solely a property of the efficiency of food utilization. The correction of metabolizable energy to nitrogen balance is accomplished by adding 7.45 kcal/g of nitrogen deficit in mammals and 8.73 kcal/g deficit in birds and, conversely, subtracting when in positive nitrogen balance (Sibbald, 1981). The nitrogen balance correction can significantly alter the metabolizable energy coefficient when the animals are consuming feed at a level far from maintenance. For example, nitrogen-corrected metabolizable energy coefficients for white-tailed deer increase approximately 10% to 92.4±2.1% for deciduous browse and 83.9±4.1% for evergreen browse because most of the animals experience significant weight loss.

Because metabolizable energy coefficients in birds include losses of digestible energy, lower metabolizable energy coefficients for birds relative to mammals are expected. The very high apparent metabolizable energy coefficients for nectivorous, carnivorous, and granivorous birds are indicative of the high digestibilities of these diets (Table 13.8). If the apparent digestible and metabolizable energy coefficients for carnivorous ($89.4 \times 0.903 = 80.7\%$) and granivorous ($89.6 \times 0.969 = 86.8\%$) mammals are combined into one coefficient compara-

TABLE 13.8
Apparent Metabolizable Energy Coefficients as a Percentage of Apparent Digestible Energy in Wild Mammals and as a Percentage of Gross Energy in Wild Birds[a]

Species or group	Diet	Metabolizable energy coefficient (%)
Mammals		
Fissiped carnivores	Meat or whole vertebrates	90.3 ± 3.9
Lagomorphs	Browse, both deciduous and evergreen	78.2 ± 12.5
Pinniped carnivores	Fish	92.4 ± 2.5
Rodents	Grains or nuts	96.9 ± 1.2
Rodents and lagomorphs	Herbage (alfalfa, green wheat, and forbs)	90.8 ± 2.0
Ruminants		
Domestic livestock	Hay, concentrates, and silage	82
Moose	Deciduous browse	77.9 ± 4.2
Red deer	Hay and concentrates	87.1
White-tailed deer	Concentrates	86.9
	Deciduous browse	83.2 ± 5.4
	Evergreen browse	76.4 ± 1.9
Wildebeest and eland	Hay and concentrates	82.4 ± 0.7
Birds		
Fish-eaters	Fish	77.6 ± 2.1
Granivores	Corn	85.0 ± 1.1
	Sorghum	86.1
	Hulled sunflower seeds	80.2 ± 5.4
Hawks and owls	Meat or whole vertebrates	77.5 ± 5.8
Insectivores	Larvae or adults	79.1 ± 7.9
Nectarivores	Sugar solutions	98.5 ± 0.8

[a]Data from Colovos et al., 1954; Colovos, 1955; Greeley, 1956; Silver and Colovos, 1957; Gibb, 1957; Graber, 1962; Silver, 1963; Kahl, 1964; Kale, 1965; West and Hart, 1966; Drozdz, 1968; Jense, 1968; Rogerson, 1968; Gebczynski et al., 1972; Duke et al., 1973; Labisky and Anderson, 1973; Thompson et al., 1973; Willson and Harmeson, 1973; Evans and Dietz, 1974; Hainsworth, 1974; Davison, 1975; Gorecki and Grygielska, 1975; Litvaitis and Mautz, 1976; Mautz et al., 1976; Shoemaker et al., 1976; Kendeigh et al., 1977; Kushlan, 1977; Moors, 1977; Stueck et al., 1977; Cooper, 1978; Simpson et al., 1978; Kirkwood, 1979; Robel et al., 1979a, 1979b; Collins et al., 1980; Ashwell-Erickson and Elsner, 1981; Rau, 1981; Regelin et al., 1981.

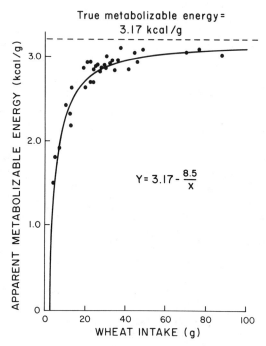

Fig. 13.8. Effect of intake level on the apparent metabolizable energy content of wheat fed to chickens. Gross energy content of wheat as fed was 3.88 kcal/g. (From Sibbald, 1975, courtesy of *Poultry Science.*)

ble to the metabolizable energy coefficient of birds, the total metabolizable efficiencies of birds and mammals are quite similar on these diets.

Avian apparent metabolizable energy coefficients for a single diet are seldom constant, since variation has been attributed to the ingesting species, individual age, ambient temperature, and photoperiod (Kale, 1965; Myrcha *et al.,* 1973; Willson and Harmeson, 1973; Kendeigh *et al.,* 1977). However, it has been recognized that the major variable affecting the observed apparent metabolizable energy coefficient is level of intake (Fig. 13.8). Because the excretory energy loss in birds includes both endogenous and feed fractions, apparent metabolizable energy coefficients are curvilinear functions of intake. Many seeds and nuts which are consumed with their nondigestible hulls or glumes and inhibitory secondary plant compounds will have lower apparent metabolizable energy coefficients than, for example, corn (Robel *et al.,* 1979a, 1979b). Because many of these seeds will also be less preferred items during single-species feeding trials, they are penalized even further because of their low intake.

Consequently, more rapid progress will be made in determining the efficiency of food utilization in birds if apparent metabolizable energy coefficients are corrected for endogenous and metabolic secretions to estimate true metabolizable energy coefficients (Sibbald, 1980). True metabolizable energy coefficients can be determined by feeding each diet at different levels of intake and plotting intake and total excreta energy or metabolizable amount. The endogenous and metabolic losses (i.e., the Y intercept of the various regressions) are a maintenance requirement that can then be balanced against true metabolizable energy intake. True metabolizable energies of agricultural grains have not varied with birds species (Sibbald, 1976), are additive, and apparently have no interaction effect when combined in mixed diets (Sibbald, 1977). For birds ingesting vegetation with significant fiber content, detergent analyses should also be used to provide necessary information to understand and predict fiber digestion.

B. Net Energy Coefficients

Heat is produced during the many digestive and metabolic processes that ultimately convert ingested gross energy to mechanical or tissue energy. The work of digestion, including mastication, peristalsis, secretion, and active transport, and all aspects of fermentation and absorbed nutrient metabolism produce heat because of the inefficiency of metabolic pathways. The net energy coeffi-

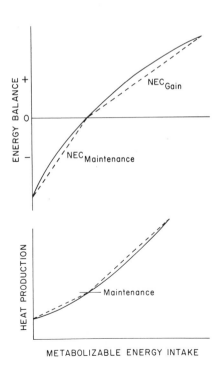

Fig. 13.9. Experimental relationships between metabolizable energy intake, heat production, and energy balance and the linear approximations used to estimate net energy coefficients.

cient is the estimate of the efficiency to which metabolizable energy can be used in maintenance and productive processes.

Of all the efficiency estimates, the net energy coefficient has been the most difficult to estimate and understand since it varies with diet and productive state (Reid et al., 1980). Net energy coefficients for maintenance and production are usually determined by feeding animals housed in thermoneutral environments several different levels of intake. Energy balance or heat production is plotted against metabolizable energy intake (Fig. 13.9). When the regression involves energy balance as the dependent variable, the slope of the regression is the net energy coefficient. However, when heat production is the dependent variable, one minus the slope is the efficiency estimate.

The net energy coefficient measured below energy balance (i.e., $NEC_{maintenance}$) is the efficiency with which metabolizable energy substitutes for tissue energy in meeting maintenance requirements. Because the heat increment even below maintenance is a curvilinear function, an entire range of net energy coefficients are possible. Consequently, the linear expression between basal metabolism and energy balance is often used since it provides an average $NEC_{maintenance}$. Similarly, the NEC_{gain} is curvilinear and decreases with increasing intake (Blaxter, 1974). However, because much of the curvilinearity in the efficiency of gain is restricted to the very high levels of metabolizable energy intake (Holter et al., 1979b), the linear expression can be used to estimate the average NEC_{gain}. Net energy coefficients for feather and hair growth, egg and milk production, and fetal growth are simply the efficiencies to which the metabolizable energy of feed and tissue can be converted into the gross energy of these tissues.

Net energy coefficients for maintenance and gain decrease as the dietary metabolizable energy concentration decreases (Fig. 13.10). The NEC_{gain} decreases far less when pelleted diets are fed to ruminants than when long or chopped forages are fed. These differences in net energy coefficients due to dietary form are partly attributable to a higher percentage of fermentable substrates escaping the rumen when fed finely ground pelleted diets, less energy expenditure in mastication and rumination for pelleted diets, and increased production and more efficient utilization of propionic acid than acetic acid for pelleted diets relative to long hays (Reid et al., 1980). The NEC_{gain} should also reflect the relative percentages of protein and fat being deposited. For example, the NEC_{gain} in simple-stomached animals averages 45% for protein deposition and 75% for fat deposition (Webster, 1980). The NEC for de novo fat deposition in ruminants is virtually identical to that occurring in the simple-stomached animal (Reid et al., 1980). Although the composition of the gain in wild ruminants may be quite different from that occurring in domestic ruminants, the NEC for both gain and maintenance are not significantly different between wild and domestic ruminants (Fig. 13.10). An average NEC_{gain} of 67% has been used in

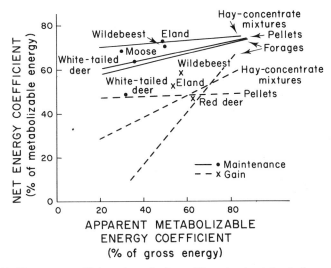

Fig. 13.10. Net energy coefficients determined on wild ruminants relative to the averages for domestic ruminants (lines) from Blaxter (1974) and Blaxter and Boyne (1978). (Sources for wild ruminant data are in Table 13.9.)

bird energetic studies because it bisects the relative efficiencies of protein and fat deposition (Ricklefs *et al.,* 1980; Vleck and Vleck, 1980).

Net energy coefficients also depend on the ratio of specific metabolites required as compared to the end products of digestion. For example, feather synthesis is the least efficient process quantified thus far (Table 13.9). The very low net energy coefficient for molting is in part due to the very low concentration of sulfur-containing amino acids in the ingested foods relative to feathers and therefore the extensive amino acid catabolism necessary for the extraction of the essential amino acids for feather development (Dolnik and Gavrilov, 1979). Similarly, productive net energy coefficients will vary depending on whether the requirement is being met by ingested feed or mobilized tissue. For example, NEC_{milk} in domestic ruminants averages 64% when the milk is produced from feed metabolites, but 82–84% when it is produced from mobilized tissue (Moe *et al.,* 1972; Moe and Tyrrell, 1974).

C. True Versus Apparent Energy Utilization Schemes

Several schematic pathways have been used by nutritionists to visualize the various losses of energy relative to its progression from ingested gross energy to usable net energy (Fig. 13.11). The two systems used have been termed apparent and true relative to the steps where energy losses are subtracted. The most

TABLE 13.9

Net Energy Coefficients as a Percentage of Apparent Metabolizable Energy in Wildlife

Species or group	Diet	Metabolic process	Net energy coefficient	Reference
Mammals				
Monogastrics and cecal digestors				
Fisher	Meat	Maintenance	77.2	Davison (1975)
Seal	Fish	Maintenance	91.2	Ashwell-Erickson and Elsner (1981)
Common vole	Hazelnuts and grains	Gain	88.8	Jagosz et al. (1979)
Dormice	Hazelnuts and carrots	Gain	43.7	Gebczynski et al. (1972)
Gray squirrel	Commercial diet	Gain	51.2	Data of Ludwick et al. (1969)
Mink	Meat and grain	Gain	71.3	Harper et al. (1978)
Snowshoe hare	Browse and concentrates	Gain	78.6	Holter et al. (1974)
Ruminants				
Eland	Hay and concentrates	Maintenance	72.7	Rogerson (1968)
Moose	Browse	Maintenance	68.2	Regelin et al. (1981)

Species	Diet	Function	Value	Reference
White-tailed deer	Browse	Maintenance	63.9	New Hampshire data summarized by C. T. Robbins (1973)
White-tailed deer	Browse	Maintenance	48.3	Mautz et al. (1975)
Wildebeest	Hay and concentrates	Maintenance	70.8	Rogerson (1968)
Pronghorn antelope	Milk and concentrates	Maintenance and gain	73.0	Wesley et al. (1970)
White-tailed deer	Pelleted diet	Maintenance and gain	81.1	Thompson et al. (1973)
Eland	Hay and concentrates	Gain	52.4	Rogerson (1968)
Red deer	Hay and concentrates	Gain	47.0	Simpson et al. (1978)
Wildebeest	Hay and concentrates	Gain	59.3	Rogerson (1968)
Birds				
Bobolink	Commercial diet	Gain, fat	64.0	Gifford and Odum (1965)
Brown honeyeater	Sucrose solutions	Gain, fat	87.5	Collins et al. (1980)
Bobwhite quail	Commercial diet	Egg synthesis	71.2	Case and Robel (1974)
Zebra finch	Commercial diet	Egg synthesis	75.0	El-Wailly (1966)
Granivorous birds	Seeds	Feather molt	4.9	Kendeigh et al. (1977), Dolnik and Gavrilov (1979), King (1980)
Carnivorous and insectivorous birds	Animal tissue	Feather molt	9.1	Kendeigh et al. (1977)

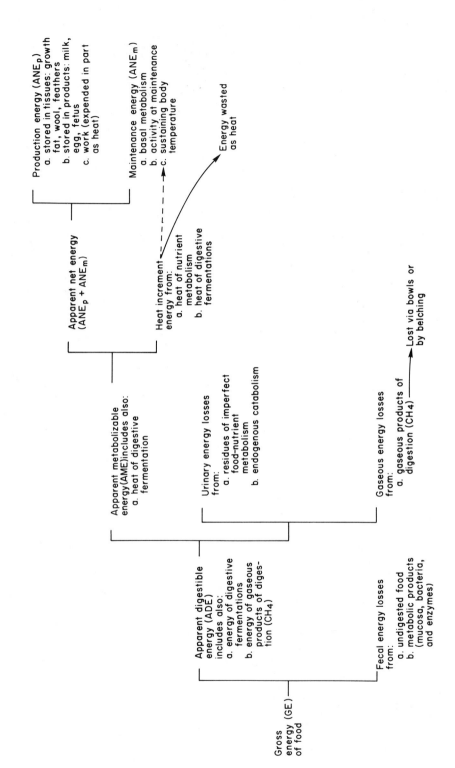

Production energy (ANE$_p$)
a. stored in tissues: growth fat, wool, feathers
b. stored in products: milk, egg, fetus
c. work (expended in part as heat)

Maintenance energy (ANE$_m$)
a. basal metabolism
b. activity at maintenance
c. sustaining body temperature

Energy wasted as heat

Apparent net energy (ANE$_p$ + ANE$_m$)

Heat increment energy from:
a. heat of nutrient metabolism
b. heat of digestive fermentations

Apparent metabolizable energy (AME) includes also:
a. heat of digestive fermentation

Urinary energy losses from:
a. residues of imperfect food-nutrient metabolism
b. endogenous catabolism

Gaseous energy losses from:
a. gaseous products of digestion (CH$_4$)

Lost via bowls or by belching

Apparent digestible energy (ADE) includes also:
a. energy of digestive fermentations
b. energy of gaseous products of diges-tion (CH$_4$)

Fecal energy losses from:
a. undigested food
b. metabolic products (mucosa, bacteria, and enzymes)

Gross energy (GE) of food

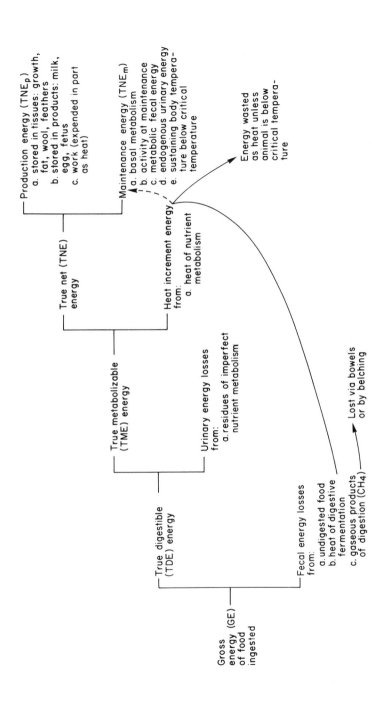

Fig. 13.11. Apparent and true partition of ingested gross energy. (From *Applied Animal Nutrition* by E. W. Crampton and L. E. Harris. W. H. Freeman and Company. Copyright © 1969.)

significant difference between the two systems is in the recognition of metabolic fecal and endogenous urinary energy as simply excretory losses similar to non-digested food residues in the apparent scheme versus a requirement in the true scheme. One cannot interchange the various steps in the schemes, such as to multiply apparent digestible energy by a true metabolizable or net coefficient. Other than the problem of metabolic losses, where a given loss is subtracted in a scheme designed to improve human comprehension of a complex system is certainly inconsequential relative to predicting the efficiency of animal functioning, as long as all energy losses are recognized.

D. Protein Metabolism

1. Biological Value

Protein utilization efficiencies have been studied less by wildlife nutritionists than have similar energy efficiencies. The lack of studies on protein efficiencies is, in part, due to the preoccupation of wildlife nutritionists in understanding energy flow and the fact that protein utilization is not an independent observation but is dependent on many other animal, dietary, and environmental parameters, including energy. However, only one additional efficiency estimate beyond digestible protein is needed to quantify the utilization of absorbed nitrogen, since urine is the primary route of digested nitrogen loss. Absorbed nitrogen not appearing in the urine can be used for maintenance and production.

The efficiency of utilization of absorbed dietary nitrogen is termed biological value (Mitchell, 1924). Several different equations have been used to calculate biological values and include:

$$\text{Biological value (\%)} = \frac{NI - (FN - MFN) - (UN - EUN)}{NI - (FN - MFN)} \times 100,$$

where NI is nitrogen intake, FN is total fecal nitrogen, MFN is metabolic fecal nitrogen, UN is total urinary nitrogen, and EUN is endogenous urinary nitrogen. All components of the equation must be in the same units, such as grams per day or grams per kilogram per day. Since $NI - (FN\text{-}MFN)$ is the true digestible amount of dietary nitrogen, the equation can also be expressed as:

$$\text{Biological value (\%)} = \frac{NI\,(TD) - (UN - EUN)}{NI\,(TD)} \times 100,$$

where TD is the true digestibility of dietary nitrogen. The final form of the equation commonly seen is:

$$\text{Biological value (\%)} = \frac{NB + MFN + EUN}{ADN + MFN} \times 100,$$

where *NB* is the nitrogen balance and *ADN* is apparent digestible nitrogen. The numerator of this equation represents the maintenance (*MFN* and *EUN*) and production (nitrogen balance or retention) requirements, while the denominator is simply another way to express true digestible nitrogen.

Biological values are not constant, but vary with protein source and its amino acid composition, mode of protein digestion, animal age and productive state, digestible energy intake relative to energy requirements, total protein requirement, and internal physiological means of conserving absorbed nitrogen. A deficiency of a specific essential amino acid relative to its requirement would lower the biological value of absorbed protein. Similarly, various productive processes require amino acids in differing concentrations, such as the sulfur-containing amino acid requirement for feather keratin synthesis. While this would tentatively suggest that separate biological values are required to understand each productive process, the total amino acid requirement of the animal will need to be evaluated because excesses of amino acids relative to one process may be useful elsewhere in the body.

Protein metabolism is inseparably related to energy metabolism. Absorbed amino acids can be used as an energy source and the carbon skeleton oxidized when dietary energy is lacking, thus lowering the biological value. The classical determination of biological value always requires that dietary energy be adequate to prevent protein catabolism for energy and that protein concentration be 10% of the diet. The biological value thus determined represents a maximum efficiency that may not describe internal protein metabolism of the free-ranging animal exposed to a wide range of dietary energy to protein ratios.

Biological values decrease with increasing dietary nitrogen intake or protein concentration (Fig. 13.12). The average biological value for the forestomach and cecal fermenters at 10% dietary crude protein (72.9%) is not significantly different from the average for cattle (70.0%) and sheep (65.0%) (ARC, 1965). The variation in the regression (Fig. 13.12) again indicates that, while the relative level of protein intake is an important determinant of biological values, other factors will need to be considered. The biological values of 12 animal proteins fed at 10% dietary crude protein ranged from 87.5 to 35.0% (average of 65.1%) when fed to rats and from 71.3 to 33.8% (average of 55.5%) when fed to mink (Glem-Hansen and Eggum, 1974). Carnivores, in general, may be less efficient in protein utilization than other animals since their diet is normally very high in protein content.

The determination of biological values of ingested protein in birds is generally unrealistic because of the physical impossibility of separating endogenous and exogenous nitrogen in the excrement. Perhaps a more viable approach is to compare nitrogen retention to nitrogen intake (Sibbald, 1975). For example, 22.4% of the nitrogen ingested by Passeriformes and Galliformes on various diets was retained (see Chapter 8, Fig. 8.2). However, if the very low nitrogen intakes are associated with inadequate energy intake and if some of the birds

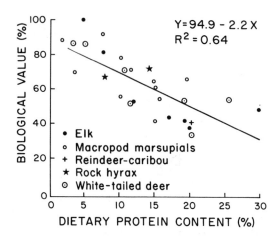

Fig. 13.12. Biological values of dietary protein in marsupials and eutherians with forestomach and/or cecal fermentation. (Data from Ullrey *et al.,* 1967; McEwan and Whitehead, 1970; C. T. Robbins *et al.,* 1974; Hume, 1977; Hume *et al.,* 1980; Mould and Robins, 1981; Wellard and Hume, 1981.)

were at maintenance, the slope of the regression may be displaced as protein is catabolized to meet energy requirements.

2. *Conclusions: Protein Utilization*

The efficiency of ingested protein utilization is described by the product of the true digestibility and the biological value (Fig. 13.13). One of the essential nitrogen conservation mechanisms in mammals with gastrointestinal fermentation is termed urea recycling. Urea recycling refers to the renal retention of urea; its movement via saliva or digestive fluids, or simply by diffusion across the tissue into the gastrointestinal tract; its hydrolysis to ammonia; and its absorption or excretion as ammonia or microbial protein. Urea recycling has been studied in bears (Nelson, 1980), camels (Schmidt-Nielsen *et al.,* 1957), caribou (Wales *et al.,* 1975), elk (Westra, 1978; Mould and Robbins, 1981), golden-mantled and Richardson's ground squirrels (Steffen *et al.,* 1980; Bintz and Torgerson, 1981), llama (Hinderer and Engelhardt, 1975), rabbits (Knutson *et al.,* 1977), rock hyrax (Hume *et al.,* 1980), wallabies (Kennedy and Hume, 1978), and white-tailed deer (C. T. Robbins *et al.,* 1974).

Recycled urea nitrogen entering the gastrointestinal tract often exceeds total dietary nitrogen intake. The relative amount of urea recycled can be as high as 90–95% of the endogenous production and is inversely proportional to protein intake. Urea or uric acid utilization for amino acid synthesis is dependent on gastrointestinal microbial hydrolysis. Efficient utilization of the hydrolyzed urea is dependent on the site of gastrointestinal entry. For example, urea that enters

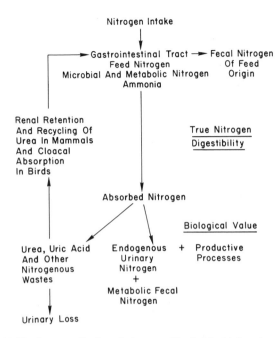

Fig. 13.13. Conceptualization of nitrogen utilization by birds and mammals.

the forestomach fermentation chambers and is converted to microbial protein would be far more useful than that entering the lower tract, because the microbial amino acids produced in the forestomach could be absorbed more efficiently during transit through the small intestine. The reentry of the urea nitrogen from the gastrointestinal tract back into the host has ranged from insignificant levels to 40–50% of the amount recycled (Hume *et al.*, 1980). The biological value of ingested nitrogen will be increased relative to the animal without gastrointestinal fermentation by the extent to which the recycled urea in mammals and the cloacal nitrogen in birds are reabsorbed and used for maintenance and production.

REFERENCES

Agricultural Research Council (ARC). (1965). "The Nutrient Requirements of Farm Livestock. No. 2. Ruminants." Agricultural Research Council, London.

Ambuhl, S., Williams, V. J., and Senior, W. (1979). Effects of caecectomy in the young adult female rat on digestibility of food offered *ad libitum* and in restricted amounts. *Aust. J. Biol. Sci.* **32,** 205–213.

Angerman, J. A., Dapson, R. W., and Smith, M. H. (1974). Effect of age on the indigestible component of a mouse. *J. Mammal.* **55,** 210.

Arman, P., and Field, C. R. (1973). Digestion in the hippopotamus. *E. Afr. Wildl. J.* **11,** 9–17.

Arman, P., and Hopcraft, D. (1975). Nutritional studies on East African herbivores. 1. Digestibilities of dry matter, crude fiber and crude protein in antelope, cattle, and sheep. *Brit. J. Nutr.* **33,** 255–264.

Ashwell-Erickson, S., and Elsner, R. (1981). The energy cost of free existence for Bering Sea harbor and spotted seals. *In* "The Eastern Bering Sea Shelf: Oceanography and Resources" (D. W. Hood and J. A. Calder, eds.), pp. 869–899. Univ. of Washington Press, Seattle.

Barker, S. (1968). Nitrogen balance and water intake in the Kangaroo Island wallaby, *Protemnodon eugenii* (Desmarest). *Aust. J. Exp. Biol. Med. Sci.* **46,** 17–32.

Barrett, G. W. (1969). Bioenergetics of a captive least shrew, *Cryptotis parva. J. Mammal.* **50,** 629–630.

Batzli, G. O., and Cole, F. R. (1979). Nutritional ecology of microtine rodents: Digestibility of forage. *J. Mammal.* **60,** 740–750.

Baumgras, P. (1944). Experimental feeding of captive fox squirrels. *J. Wildl. Manage.* **8,** 296–300.

Berenbaum, M. (1980). Adaptive significance of midgut pH in larval lepidoptera. *Am. Nat.* **115,** 138–146.

Bintz, G. L., and Torgerson, G. E. (1981). The metabolism of [^{14}C] urea by control and starved Richardson's ground squirrels. *Comp. Biochem. Physiol.* **69A,** 551–555.

Bissell, H. D., Harris, B., Strong, H., and James, F. (1955). The digestibility of certain natural and artificial foods eaten by deer in California. *Calif. Fish Game* **41,** 57–78.

Blaxter, K. L. (1974). Metabolisable energy and feeding systems for ruminants. *In* "Seventh Nutrition Conference for Feed Manufacturers, University of Nottingham" (H. Swan and D. Lewis, eds.), pp. 3–25. Avi Publishing Co., Westport, Conn.

Blaxter, K. L., and Boyne, A. W. (1978). The estimation of the nutritive value of feeds as energy sources for ruminants and the derivation of feeding system. *J. Agric. Sci.* **90,** 47–68.

Blaxter, K. L., and Clapperton, J. L. (1965). Prediction of the amount of methane produced by ruminants. *Brit. J. Nutr.* **19,** 511–522.

Brisbin, I. L., Jr. (1966). Energy utilization in a captive hoary bat. *J. Mammal.* **47,** 719–720.

Buckner, C. H. (1964). Metabolism, food capacity and feeding behavior in four species of shrews. *Can. J. Zool.* **42,** 259–279.

Case, R. M., and Robel, R. J. (1974). Bioenergetics of the bobwhite. *J. Wildl. Manage.* **38,** 638–652.

Collins, B. G., Cary, G., and Packard, G. (1980). Energy assimilation, expenditure and storage by the brown honeyeater, *Lichmera indistincta. J. Comp. Physiol.* **137,** 157–163.

Colovos, N. F. (1955). The nutritive evaluation of some forage rations for deer. *New Hampshire Fish Game Dept. Proj. FW-2-R-3* (Mimeo).

Colovos, N. F., Keener, H. A., Davis, H. A., and Terri, A. E. (1954). The nutritive evaluation of some forage rations for deer. *New Hampshire Fish Game Dept. Proj. FW-2-R-2* (Mimeo).

Cooper, J. (1978). Energetic requirements for growth and maintenance of the Cape gannet (Aves; Sulidae). *Zool. Afr.* **13,** 305–317.

Cowan, R. L., Hartsook, E. W., Whelan, J. B., Long, T. A., and Wetzel, R. S. (1969). A cage for metabolism and radioisotope studies with deer. *J. Wildl. Manage.* **33,** 204–208.

Crampton, E. W., and Harris, L. E. (1969). "Applied Animal Nutrition." Freeman, San Francisco.

Davison, R. P. (1975). The efficiency of food utilization and energy requirements of captive female fishers. Master's thesis, Univ. of New Hampshire, Durham.

Dietz, D. R., Udall, R. H., and Yeager, L. E. (1962). Chemical composition and digestibility by mule deer of selected forage species, Cache la Poudre range, Colorado. *Colo. Game Fish Dept. Tech. Publ. No.* **14.**

Dolnik, V. R., and Gavrilov, V. M. (1979). Bioenergetics of molt in the chaffinch (*Fringilla coelebs*). *Auk* **96,** 253–264.

Drozdz, A. (1968). Digestibility and assimilation of natural foods in small rodents. *Acta Theriol.* **13**, 367–389.

Drozdz, A., and Osiecki, A. (1973). Intake and digestibility of natural foods by roe-deer. *Acta Theriol.* **13**, 81–91.

Duke, G. E., Ciganek, J. G., and Evanson, O. A. (1973). Food consumption and energy, water and nitrogen budgets in captive great horned-owls (*Bubo virginianus*). *Comp. Biochem. Physiol.* **44A**, 283–292.

El-Wailly, A. J. (1966). Energy requirements for egg-laying and incubation in the zebra finch, *Taeniopygia castanotis*. *Condor* **68**, 582–594.

Evans, K. E., and Dietz, D. R. (1974). Nutritional energetics of sharp-tailed grouse during winter. *J. Wildl. Manage.* **38**, 622–629.

Farrell, D. J., and Wood, A. J. (1968). The nutrition of the female mink (*Mustela vison*). II. The energy requirements for maintenance. *Can J. Zool.* **46**, 47–52.

Foot, J. Z., and Romberg, B. (1965). The utilization of roughage by sheep and the red kangaroo, *Macropus rufus* (Desmarest). *Aust. J. Agric. Res.* **16**, 429–435.

Forbes, E. B., Marcy, L. F., Voris, A. L., and French, C. E. (1941). The digestive capacities of white-tailed deer. *J. Wildl. Manage.* **5**, 108–114.

Gasaway, W. C. (1976). Methane production in rock ptarmigan (*Lagopus mutus*). *Comp. Biochem. Physiol.* **54A**, 183–185.

Gasaway, W. C., Holleman, D. F., and White, R. G. (1975). Flow of digesta in the intestine and cecum of the rock ptarmigan. *Condor* **77**, 467–474.

Gasaway, W. C., White, R. G., and Holleman, D. F. (1976). Digestion of dry matter and absorption of water in the intestine and cecum of rock ptarmigan. *Condor* **78**, 77–84.

Gebczynska, Z., Kowalczyk, J., Krasinska, M., and Ziolecka, A. (1974). A comparison of the digestibility of nutrients by European bison and cattle. *Acta Theriol.* **19**, 283–289.

Gbeczynski, M., Gorecki, A., and Drozdz, A. (1972). Metabolism, food assimilation and bioenergetics of three species of dormice (*Gliridae*). *Acta Theriol.* **17**, 271–294.

Gibb, J. A. (1957). Food requirements and other observations on captive tits. *Bird Study* **4**, 207–215.

Gifford, C. E., and Odum, E. P. (1965). Bioenergetics of lipid deposition in the bobolink, a trans-equatorial migrant. *Condor* **67**, 383–403.

Glem-Hansen, N., and Eggum, B. O. (1974). A comparison of protein utilization in rats and mink based on nitrogen balance experiments. *Z. Tierphysiol., Tierernahrg. u. Futtermittelkde.* **33**, 29–34.

Goering, H. K., and Van Soest, P. J. (1970). Forage fiber analyses. *USDA Agric. Handbook No.* **379**.

Golley, F. B. (1960). Energy dynamics of a food chain of an old-field community. *Ecol. Monogr.* **30**, 187–206.

Golley, F. B., Petrides, G. A., Rauber, E. L., and Jenkins, J. H. (1965). Food intake and assimilation by bobcats under laboratory conditions. *J. Wildl. Manage.* **29**, 442–447.

Gorecki, A., and Grygielska, M. (1975). Consumption and utilization of natural foods by the common hamster. *Acta Theriol.* **20**, 237–246.

Graber, R. R. (1962). Food and oxygen consumption in three species of owls. *Condor* **72**, 60–65.

Greeley, F. (1956). Nutritive evaluation of some forage rations of deer, *New Hampshire Fish Game Dept. Proj. FW-2-R* (Mimeo).

Green, B., and Eberhard, I. (1979). Energy requirements and sodium and water turnovers in two captive marsupial carnivores: The Tasmanian devil, *Sarcophilus harrisii*, and the native cat, *Dasyurus viverrinus*. *Aust. J. Zool.* **27**, 1–8.

Hainsworth, F. R. (1974). Food quality and foraging efficiency: The efficiency of sugar assimilation by hummingbirds. *J. Comp. Physiol.* **88**, 425–431.

Hanson, R. M., and Cavender, B. R. (1973). Food intake and digestion by black-tailed prairie dogs under laboratory conditions. *Acta Theriol.* **18**, 191–200.

Harper, R. B., Travis, H. F., and Glinsky, M. S. (1978). Metabolizable energy requirement for maintenance and body composition of growing farm-raised male pastel mink *(Mustela vison). J. Nutr.* **108**, 1937–1943.

Hawkins, A. E., and Jewell, P. A. (1962). Food consumption and energy requirements of captive British shrews and the mole. *Proc. Zool. Soc. London* **138**, 137–155.

Hinderer, S., and Engelhardt, W. v. (1975). Urea metabolism in the llama. *Comp. Biochem. Physiol.* **52A**, 619–622.

Holter, J. B., Tyler, G., and Walski, T. (1974). Nutrition of the snowshoe hare *(Lepus americanus). Can. J. Zool.* **52**, 1553–1558.

Holter, J. B., Hayes, H. H., and Smith, S. H. (1979a). Protein requirement of yearling white-tailed deer. *J. Wildl. Manage.* **43**, 872–879.

Holter, J. B., Urban, W. E., Jr., and Hayes, H. H. (1979b). Predicting energy and nitrogen retention in young white-tailed deer. *J. Wildl. Manage.* **43**, 880–888.

Hoppe, P. P., Qvortrup, S. A., and Woodford, M. H. (1977). Rumen fermentation and food selection in East African zebu cattle, wildebeest, Coke's hartebeest and topi. *J. Zool.* **181**, 1–9.

Hume, I. D. (1974). Nitrogen and sulphur retention and fibre digestion by euros, red kangaroos and sheep. *Aust. J. Zool.* **22**, 13–23.

Hume, I. D. (1977a). Maintenance nitrogen requirements of the macropod marsupials *Thylogale thetis,* red-necked pademelon, and *Macropus eugenii,* Tammar wallaby. *Aust. J. Zool.* **25**, 407–417.

Hume, I. D. (1977b). Production of volatile fatty acids in two species of wallaby and in sheep. *Comp. Biochem. Physiol.* **56A**, 299–304.

Hume, I. D., and Dunning, A. (1979). Nitrogen and electrolyte balance in the wallabies *Thylogale thetis* and *Macropus eugenii* when given saline drinking water. *Comp. Biochem. Physiol.* **63A**, 135–139.

Hume, I. D., Rubsamen, K., and Engelhardt, W. v. (1980). Nitrogen metabolism and urea kinetics in the rock hyrax *(Procavia habessinica). J. Comp. Physiol.* **138B**, 307–314.

Hungate, R. E., Phillips, G. D., McGregor, A., Hungate, D. P., and Buechner, H. K. (1959). Microbial fermentation in certain mammals. *Science* **130**, 1192–1194.

Inman, W. R. (1941). Digestibility studies with foxes. II. Digestibility of frozen beef tripe, frozen lip meat, frozen beef hearts and frozen cow udder. *Sci. Agric.* **22**, 33–39.

Inman, W. R., and Smith, G. E. (1941). Digestibility studies with foxes. I. Effect of the plane of nutrition upon the digestibility of meats. *Sci. Agric.* **22**, 18–32.

Jacobsen, E., and Skjenneberg, S. (1972). Some results from feeding experiments with reindeer. *Proc. Int. Reindeer and Caribou Symp.* **1**, 95–107.

Jagosz, J., Gorecki, A., and Pozzi-Cabaj, M. (1979). The bioenergetics of deposit and utilization of stored energy in the common vole. *Acta Theriol.* **24**, 391–397.

Jense, G. K. (1968). Food habits and energy utilization of badgers. Master's thesis. Southa Dakota St. Univ., Brookings.

Johnson, D. R., and Groepper, K. L. (1970). Bioenergetics of North Plains rodents. *Am. Midl. Nat.* **84**, 537–548.

Kalh, M. P. (1964). Food ecology of the wood stork *(Mycteria americanus)* in Florida. *Ecol. Monogr.* **34**, 97–117.

Kale, H. W., II. (1965). Ecology and bioenergetics of the long-billed marsh wren in Georgia salt marshes. *Publ. Nuttall Ornith. Club No.* **5**.

Kempton, T. J., Murray, R. M., and Leng, R. A. (1976). Methane production and digestibility measurements in the grey kangaroo and sheep. *Aust. J. Biol. Sci.* **29**, 209–214.

Kendeigh, S. C., Dolnik, V. R., and Gavrilov, V. M. (1977). Avain energetics. *In* "Granivorous

Birds in Ecosystems'' (J. Pinowski and S. C. Kendeigh, eds.), pp. 127–204. Cambridge Univ. Press, London.

Kennedy, P. M., and Hume, I. D. (1978). Recycling of urea nitrogen to the gut of the tammar wallaby (*Marcropus eugenii*). *Comp. Biochem. Physiol.* **61A**, 117–121.

King, J. R. (1980). Energetics of avian moult. *Proc. Int. Ornithol. Congr.* **17**, 312–317.

Kirkwood, J. K. (1979). The partition of food energy for existence in the kestrel (*Falco tinnunculus*) and the barn owl (*Tyto alba*). *Comp. Biochem. Physiol.* **63A**, 495–498.

Knutson, R. S., Francis, R. S., Hall, J. L., Moore, B. H., and Heisinger, J. F. (1977). Ammonia and urea distribution and urease activity in the gastrointestinal tract of rabbits (*Oryctolagus* and *Sylvilagus*). *Comp. Biochem. Physiol.* **58A**, 151–154.

Kushlan, J. A. (1977). Growth energetics of the white ibis. *Condor* **79**, 31–36.

Labisky, R. F., and Anderson, W. L. (1973). Nutritional responses of pheasants to corn, with special reference to high-lysine corn. *Ill. Nat. Hist. Surv. Bull.* **31**, 87–112.

Litvaitis, J. A., and Mautz, W. W. (1976). Energy utilization of three diets fed to a captive red fox. *J. Wildl. Manage.* **40**, 365–368.

Ludwick, R. L., Fontenot, J. P., and Mosby, H. S. (1969). Energy metabolism of the eastern gray squirrel. *J. Wildl. Manage.* **33**, 569–575.

McEwan, E. H., and Whitehead, P. E. (1970). Seasonal changes in the energy and nitrogen intake in reindeer and caribou. *Can. J. Zool.* **48**, 905–913.

Maloiy, G. M. O., Kay, R. N. B., Goodall, E. D., and Topps, J. H. (1970). Digestion and nitrogen metabolism in sheep and red deer given large or small amounts of water and protein. *Brit. J. Nutr.* **23**, 843–854.

Mautz, W. W. (1971a). Comparison of the $^{51}CrCl_3$ ratio and total collection techniques in digestibility studies of a wild ruminant, the white-tailed deer. *J. Anim. Sci.* **32**, 999–1002.

Mautz, W. W. (1971b). Confinement effects on dry-matter digestibility coefficients displayed by deer. *J. Wildl. Manage.* **35**, 366–368.

Mautz, W. W., Silver, H., and Hayes, H. H. (1975). Estimating methane, urine, and heat increment for deer consuming browse. *J. Wildl. Manage.* **39**, 80–86.

Mautz, W. W., Walski, T. W., and Urban, W. E., Jr. (1976). Digestibility of fresh frozen versus pelleted browse by snowshoe hares. *J. Wildl. Manage.* **40**, 496–499.

Milchunas, D. G., Dyer, M. I., Wallmo, O. C., and Johnson, D. E. (1978). *In vivo/ in vitro* relationships of Colorado mule deer forages. *Colo. Div. Wildl. Spec. Rep. No.* **43.**

Milton, K., Van Soest, P. J., and Robertson, J. B. (1980). Digestive efficiencies of wild howler monkeys. *Physiol. Zool.* **53**, 402–409.

Mitchell, H. H. (1924). A method of determining the biological value of protein. *J. Biol. Chem.* **58**, 873–903.

Moe, P. W., and Tyrrell, H. F. (1974). Observations on the efficiency of utilisation of metabolisable energy for meat and milk production. *In* ''Seventh Nutrition Conference for Feed Manufacturers, University of Nottingham'' (H. Swan and D. Lewis, eds.), pp. 27–35. Avi Publishing Co., Westport, Conn.

Moe, P. W., Flatt, W. P., and Tyrrell, H. F. (1972). Net energy value of feeds for lactation. *J. Dairy Sci.* **55**, 945–958.

Moir, R. J., and Hungate, R. W. (1968). Ruminant digestion and evolution. *In* ''Handbook of Physiology'' (C. F. Code, ed.), pp. 2673–2694. Am. Physiol. Soc., Washington, D.C.

Moors, P. J. (1977). Studies of the metabolism, food consumption and assimilation efficiency of a small carnivore, the weasel (*Mustela nivalis* L.). *Oecologia* **27**, 185–202.

Mothershead, C. L., Cowan, R. L., and Ammann, A. P. (1972). Variations in determinations of digestive capacity of the white-tailed deer. *J. Wildl. Manage.* **36**, 1052–1060.

Mould, E. D., and Robbins, C. T. (1981). Nitrogen metabolism in elk. *J. Wildl. Manage.* **45**, 323–334.

Mould, E. D., and Robbins, C. T. (1982). Digestive capabilities in elk compared to white-tailed deer. *J. Wildl. Manage.* **46**, 22–29.

Murray, R. M., Marsh, H., Heinsohn, G. E., and Spain, A. V. (1977). The role of the midgut caecum and large intestine in the digestion of sea grasses by the dugong (Mammalia: Sirenia). *Comp. Biochem. Physiol.* **56A**, 7–10.

Myrcha, A., Pinowski, J., and Tomek, T. (1973). Energy balance of nestlings of tree sparrows, *Passer m. montanus* (L.), and house sparrows, *Passer d. domesticus* (L.). *In* "Productivity, Population Dynamics and Systematics of Granivorous Birds" (S. C. Kendeigh and J. Pinowski, eds.), pp. 59–83. PWN-Polish Scientific Publishers, Warsaw.

Nagy, K. A., and Milton, K. (1979). Energy metabolism and food consumption by wild howler monkeys (*Alouatta palliata*). *Ecology* **60**, 475–480.

Nagy, K. A., Shoemaker, V. H., and Costa, W. R. (1976). Water, electrolyte, and nitrogen budgets of jackrabbits (*Lepus californicus*) in the Mojave desert. *Physiol. Zool.* **49**, 351–363.

Nelson, R. A. (1980). Protein and fat metabolism in hibernating bears. *Fed. Proc.* **39**, 2955–2958.

Osbourn, D. F., Terry, R. A., Cammell, S. B., and Outen, G. E. (1971). The effect of leuco-anthocyanins in sainfoin (*Onobrychis viciifolia* Scop.) on the availability of protein to sheep and upon the determination of the acid detergent fiber and lignin fractions. *Proc. Nutr. Soc.* **30**, 13A–14A.

Palmer, W. L., and Cowan, R. L. (1980). Estimating digestibility of deer foods by an *in vitro* technique. *J. Wildl. Manage.* **44**, 469–472.

Palmer, W. L., Cowan, R. L., and Ammann, A. P. (1976). Effect of inoculum source on *in vitro* digestion of deer foods. *J. Wildl. Manage.* **40**, 301–307.

Parra, R. (1978). Comparison of foregut and hindgut fermentation in herbivores. *In* "The Ecology of Arboreal Folivores" (G. G. Montgomery, ed.), pp. 205–229. Smithsonian Inst. Press, Washington, D.C.

Pernetta, J. C. (1976). Bioenergetics of British shrews in grassland. *Acta Theriol.* **21**, 481–497.

Person, S. J., Regau, R. E., White, R. G., and Luick, J. R. (1980). *In vitro* and nylon-bag digestibilities of reindeer and caribou forages. *J. Wildl. Manage.* **44**, 613–622.

Prins, R. A., and Geelen, M. J. H. (1971). Rumen characteristics of red deer, fallow deer, and roe deer. *J. Wildl. Manage.* **35**, 673–680.

Rau, J., Murua, R., and Rosenmann, M. (1981). Bioenergetics and food preferences in sympatric southern Chilean rodents. *Oecologia* **50**, 205–209.

Regelin, W. L., Schwartz, C. C., and Franzmann, A. W. (1981). Energy expenditure of moose on the Kenai National Wildlife Refuge. *Kenai Alaska Field Station, Ann. Prog. Rep. Proj. W-17-11.*

Reid, J. T., White, O. D., Anrique, R., and Fortin, A. (1980). Nutritional energetics of livestock: Some present boundaries of knowledge and future research needs. *J. Anim. Sci.* **51**, 1393–1415.

Richardson, M. (1977). The proteinase inhibitors of plants and microorganisms. *Phytochemistry* **16**, 159–169.

Richmond, R. J., Hudson, R. J., and Christopherson, R. J. (1977). Comparison of forage intake and digestibility by American bison, yak, and cattle. *Acta Theriol.* **22**, 225–230.

Ricklefs, R. E., White, S. C., and Cullen, J. (1980). Energetics of postnatal growth in Leach's storm petrel. *Auk* **97**, 566–575.

Robbins, C. T. (1973). The biological basis for the determination of carrying capacity. Doctoral dissertation, Cornell Univ., Ithaca, N.Y.

Robbins, C. T., Prior, R. L., Moen, A. N., and Visek, W. J. (1974). Nitrogen metabolism of white-tailed deer. *J. Anim. Sci.* **38**, 186–191.

Robbins, C. T., Van Soest, P. J., Mautz, W. W., and Moen, A. N. (1975). Feed analyses and digestion with reference to white-tailed deer. *J. Wildl. Manage.* **39**, 67–79.

Robbins, C. T., Podbielancik-Norman, R. S., Wilson, D. L., and Mould, E. D. (1981). Growth and nutrient consumption of elk calves compared to other ungulate species. *J. Wildl. Manage.* **45**, 172–186.

Robbins, R. C., and Gavan, J. A. (1966). Utilization of energy and protein of a commercial diet by rhesus monkeys (*Macaca mulatta*). *Lab. Anim. Care* **16**, 286–291.

Robel, R. J., Bisset, A. R., Clement, T. M., Jr., and Dayton, A. D. (1979a). Metabolizable energy of important foods of bobwhites in Kansas. *J. Wildl. Manage.* **43**, 982–987.

Robel, R. J., Bisset, A. R., Dayton, A. D., and Kemp, K. E. (1979b). Comparative energetics of bobwhites on six different foods. *J. Wildl. Manage.* **43**, 987–992.

Rogerson, A. (1968). Energy utilization by the eland and wildebeest. *Symp. Zool. Soc. London* **21**, 153–161.

Ruggiero, L. F., and Whelan, J. B. (1976). A comparison of *in vitro* and *in vivo* feed digestibility by white-tailed deer. *J. Range Manage.* **29**, 82–83.

Ruggiero, L. F., and Whelan, J. B. (1977). Chromic oxide as an indicator of total fecal output in white-tailed deer. *J. Range Manage.* **30**, 61–63.

Schaefer, A. L., Young, B. A., and Chimwano, A. M. (1978). Ration digestion and retention times of digesta in domestic cattle (*Bos taurus*), American bison (*Bison bison*), and Tibetan yak (*Bos grunniens*). *Can. J. Zool.* **56**, 2355–2358.

Schmidt-Nielsen, B., Schmidt-Nielsen, K., Houpt, T. R., and Jarnum, S. A. (1957). Urea excretion in the camel. *Am. J. Physiol.* **188**, 477–484.

Schreiber, R. K. (1979). Coefficients of digestibility and caloric diet of rodents in the northern Great Basin desert. *J. Mammal.* **60**, 416–420.

Sharp, H. F., Jr. (1967). Food ecology of the rice rat, *Oryzomys palustris* (Harlan), in a Georgia salt marsh. *J. Mammal.* **48**, 557–563.

Shoemaker, V. H., Nagy, K. A., and Costa, W. R. (1976). Energy utilization and temperature regulation by jackrabbits (*Lepus californicus*) in the Mojave Desert. *Physiol. Zool.* **49**, 364–375.

Short, H. L., and Reagor, J. C. (1970). Cell wall digestibility affects forage value of woody twigs. *J. Wildl. Manage.* **34**, 964–967.

Short, H. L., Blair, R. M., and Segelquist, C. A. (1974). Fiber composition and forage digestibility by small ruminants. *J. Wildl. Manage.* **38**, 197–209.

Sibbald, I. R. (1975). The effect of level of feed intake on metabolizable energy values measured with adult roosters. *Poult. Sci.* **54**, 1990–1997.

Sibbald, I. R. (1976). The true metabolizable energy values of several feedingstuffs measured with roosters, laying hens, turkeys and broiler hens. *Poult. Sci.* **55**, 1450–1463.

Sibbald, I. R. (1977). A test of the additivity of true metabolizable energy values of feedingstuffs. *Poult. Sci.* **56**, 363–366.

Sibbald, I. R. (1980). Metabolizable energy in poultry nutrition. *BioScience* **30**, 736–741.

Sibbald, I. R. (1981). Metabolic plus endogenous energy and nitrogen losses of adult cockerels: The correction used in the bioassay for true metabolizable energy. *Poult. Sci.* **60**, 805–811.

Silver, H. (1963). Deer nutrition studies. Laboratory analyses of fish and game and their foods. *New Hampshire Fish Game Dept. Proj. FW-2-R-11* (Mimeo).

Silver, H., and Colovos, N. F. (1957). Nutritive evaluation of some forage rations of deer. *New Hampshire Fish Game Dept. Tech. Circ. No. 15* (Mimeo).

Simpson, A. M., Webster, A. J. F., Smith, J. S., and Simpson, C. A. (1978). The efficiency of utilization of dietary energy for growth in sheep (*Ovis ovis*) and red deer (*Cervus elaphus*). *Comp. Biochem. Physiol.* **59A**, 95–99.

Slade, L. M., and Robinson, D. W. (1970). Nitrogen metabolism in nonruminant herbivores. II. Comparative aspects of protein digestion. *J. Anim. Sci.* **30**, 761–763.

Smith, C. C., and Follmer, D. (1972). Food preferences of squirrels. *Ecology* **53**, 82–91.

Smith, S. H., Holter, J. B., Hayes, H. H., and Silver, H. (1975). Protein requirement of white-tailed deer fawns. *J. Wildl. Manage.* **39**, 582–589.

Steffen, J. M., Rigler, G. L., Moore, A. K., and Riedesel, M. L. (1980). Urea recycling in active golden-mantled ground squirrels (*Spermophilus lateralis*). *Am. J. Physiol.* **239**, R168–R173.

Stueck, K. L., Farrell, M. P., and Barrett, G. W. (1977). Ecological energetics of the golden mouse based on three laboratory diets. *Acta Theriol.* **22,** 309–315.

Tilley, J. M. A., and Terry, R. A. (1963). A two-stage technique for *in vitro* digestion of forage crops. *J. Brit. Grassl. Soc.* **18,** 401–411.

Thompson, C. B., Holter, J. B., Hayes, H. H., Silver, H., and Urban, W. E., Jr. (1973). Nutrition of white-tailed deer. I. Energy requirements of fawns. *J. Wildl. Manage.* **37,** 301–311.

Ullrey, D. E., Youatt, W. G., Johnson, H. E., Fay, L. D., and Bradley, B. L. (1967). Protein requirement of white-tailed deer fawns. *J. Wildl. Manage.* **31,** 679–685.

Van de Veen, H. E. (1979). Food selection and habitat use in the red deer (*Cervus elaphus* L.). Doctoral dissertation, Rijksuniversiteit te Groningen. Groningen, Netherlands.

Van Hoven, W., and Boomker, E. A. (1981). Feed utilization and digestion in the black wildebeest (*Connochaetes gnou,* Zimmerman, 1780) in the Golden Gate Highlands National Park. *S. Afr. J. Wildl. Res.* **11,** 35–40.

Van Soest, P. J. (1967). Development of a comprehensive system of feed analyses and its application to forages. *J. Anim. Sci.* **26,** 119–128.

Van Soest, P. J. (1981). "Nutritional Ecology of the Ruminant." O. B. Books, Corvallis, Oreg.

Van Soest, P. J., and Jones, L. H. P. (1968). Effect of silica in forages upon digestibility. *J. Dairy Sci.* **51,** 1644–1648.

Vleck, C. M., and Vleck, D. (1980). Patterns of metabolism and growth in avian embryos. *Am. Zool.* **20,** 405–416.

Vogtsberger, L. M., and Barrett, G. W. (1973). Bioenergetics of captive red foxes. *J. Wildl. Manage.* **37,** 495–500.

von Engelhardt, W., Wolter, S., Lawrenz, H., and Hemsley, J. A. (1978). Production of methane in two non-ruminant herbivores. *Comp. Biochem. Physiol.* **60A,** 309–311.

Wales, R. A., Milligan, L. P., and McEwan, E. H. (1975). Urea recycling in caribou, cattle, and sheep. *Proc. Int. Reindeer and Caribou Symp.* **1,** 297–307.

Walski, T. W., and Mautz, W. W. (1977). Nutritional evaluation of three winter browse species of snowshoe hares. *J. Wildl. Manage.* **41,** 144–147.

Webster, A. J. F. (1980). The energetic efficiency of growth. *Livestock Prod. Sci.* **7,** 243–252.

Wellard, G. A., and Hume, I. D. (1981). Nitrogen metabolism and nitrogen requirements of the brushtail possum, *Trichosurus vulpecula* (Kerr). *Aust. J. Zool.* **29,** 147–156.

Wesley, D. E., Knox, K. L., and Nagy, J. G. (1970). Energy flux and water kinetics in young pronghorn antelope. *J. Wildl. Manage.* **34,** 908–912.

West, G. C., and Hart, J. S. (1966). Metabolic responses of evening grosbeaks to constant and to fluctuating temperatures. *Physiol. Zool.* **39,** 171–184.

Westra, R. (1978). The effect of temperature and season on digestion and urea kinetics in growing wapiti. Doctoral dissertation, Univ. of Alberta, Edmonton.

Westra, R., and Hudson, R. J. (1981). Digestive function of wapiti calves. *J. Wildl. Manage.* **45,** 148–155.

White, R. G., and Trudell, J. (1980). Habitat preference and forage consumption by reindeer and caribou near Atkasook, Alaska. *Arctic Alpine Res.* **12,** 511–529.

Willson, M. F., and Harmeson, J. C. (1973). Seed preferences and digestive efficiency of cardinals and song sparrows. *Condor* **75,** 225–234.

14

Food Intake Regulation

The efficient exploitation of available food is a vital requirement of all animals.
—Emlen, 1966, p. 611

Feed intake is one of the best regulated of all homeostatic mechanisms. Intake regulation must occur at several levels. For example, the animal must balance nutrient acquisition to meet daily and seasonal metabolic demands through regulating the incidence and intake during individual meals. Sensory inputs necessary to regulate intake can be either physiological, such as a food or animal metabolite whose blood level is an expression of intake and nutrient balances, or physical, such as gastrointestinal capacity. Most animals regulate total feed intake relative to energy balances (Schoener, 1971; Baumgardt, 1974; Panksepp, 1974; Friedman and Stricker, 1976; Krebs and Cowie, 1976; Hainsworth and Wolf, 1979). Correspondingly, all animals must optimize diet selection and intake in order to meet all nutrient requirements within energy, time, and bulk constraints (Westoby 1974, 1978; Pulliam, 1975; Ellis *et al.*, 1976; Kaufman *et al.*, 1980).

I. PHYSIOLOGICAL REGULATION

Although early studies sought an overriding intake control mechanism, a multitude of physiological and physical inputs are monitored and integrated at the hypothalamus–pituitary level to regulate daily and seasonal intake (Bray, 1974; Panksepp, 1974). Physiological control mechanisms have been studied primarily by infusing various metabolites and monitoring changes in the onset, duration, or termination of individual feeding bouts. Blood-borne metabolites that affect total intake include glucose and insulin (glucostatic theory of intake regulation [Mayer, 1955; Booth, 1968]), fatty acids or other lipid metabolites [lipostatic theory (Mayer, 1955; Dowden and Jacobson, 1960; Baile and McLaughlin, 1970)], cholecystokinin [a hormone released by the intestinal mucosa as food enters the small intestine that stimulates gallbladder contraction and pancreatic secretion and reduces intake (G. P. Smith *et al.*, 1974)], and possibly, liver metabolites [hepatostatic theory (Friedman and Stricker, 1976)]. Since food metabolism also liberates heat, changes in body temperature have been proposed as an intake regulation mechanism [thermostatic theory (Strominger and Brobeck, 1953; Andersson and Larsson, 1961; Dinius *et al.*, 1970)]. Metabolites, such as glucose and insulin, that reduce intake in simple-stomached animals are often unimportant in animals with extensive forestomach fermentation because of the differing metabolites of digestion and normal circulation (Holder, 1963; Baile and Martin, 1971).

II. PHYSICAL REGULATION

The physical capacity of the gastrointestinal tract compartments to simply hold and process feed can limit intake when foods of either low nutrient density, low digestibility, or intermittent availability are ingested. Examples of each are the crop storage limitations to the ingestion of dilute nectar solutions by hummingbirds (Hainsworth and Wolf, 1972), forestomach capacity limitations when herbivores ingest poorly digested forages (Montgomery and Baumgardt, 1965), and stomach or crop capacities when predators gorge on large or temporarily abundant prey items (Kaufman *et al.*, 1980). Another meal cannot be ingested until gastrointestinal bulk is reduced by either digestion and absorption or passage. Complete digestion can be as rapid as 30 min for sugar solutions consumed by hummingbirds, or as slow as many days for available cell wall carbohydrates consumed by ruminants.

A. Rate of Passage

Rate of passage becomes of paramount importance when foods containing considerable nondigestible matter are ingested. Passage is defined as the flow of undigested matter within or through the entire gastrointestinal tract per unit of time. Passage has been studied by adding inert, nonabsorbable markers to the feed and collecting ingesta samples at various sites along the gastrointestinal tract, in feces, or in oral pellets in such species as hawks and owls that void most nondigestible constituents orally rather than anally. Of all the decisions to be made in studying passage rate, none is more important than the choice of the marker(s).

Since passage is a timed phenomenon, ingesta and the corresponding excreta must be identifiable. Inert markers, which are added to a specific meal, provide an identifiable tag as the food moves through the gastrointestinal tract. Food markers have included dyes (Sibbald *et al.*, 1962; Kostelecka-Myrcha and Myrcha, 1964a, 1964b, 1965; Cowan *et al.*, 1970; Dean *et al.*, 1980), nail polish

(Walsberg, 1975), glass beads (Bailey, 1968), plastic tubing or strips (Clemens and Phillips, 1980; Clemens and Maloiy, 1981; Milton, 1981), radioactive and stable metallic isotopes (Duke *et al.*, 1968; Mautz and Petrides, 1971; Dawson, 1972; Gasaway *et al.*, 1975), rare earth elements (Hartnell and Satter, 1979), cotton (Balgooyen, 1971), brine shrimp eggs (Malone, 1965a, 1965b), poly-ethylene glycol (Clemens and Phillips, 1980), and charcoal (Sibbald *et al.*, 1962). The most significant marker development in recent years has been the observation that certain rare earth elements, such as samarium, cerium, and lanthanum, bind to plant fiber in a nondissociable complex as long as the fiber at the binding site itself is not digested. By using a combination of markers in mixed and single component rations, the rates of passage of different dietary components can be simultaneously studied.

Passage phenomena would be easy to understand if the gastrointestinal tract were simply a tube through which food moved in discrete pulses (Fig. 14.1). One of the few animal groups in which excretion is a pulsed phenomenon, readily relatable to the time of ingestion, comprises raptors and several other birds in which nondigested matter is orally cast at regular intervals (Duke, 1977; Below, 1979). Meal-to-pellet intervals average 11–12 hr in owls and 22 hr in hawks and falcons (C. R. Smith and Richmond, 1972; Duke *et al.*, 1976; Fuller *et al.*, 1979; Fuller and Duke, 1979; Duke *et al.*, 1980). However, the precise interval is affected by the amount and temporal sequence of food eaten, food composition, level of hunger, occurrence of additional prey, and external cues, such as dawn. When pooling and mixing occur anywhere within the gastrointestinal tract, such as in the crop, gizzard, ceca, and large intestine of birds and in the stomach, cecum, and large intestine of mammals, passage can be prolonged and diffused (Fig. 14.1).

Passage rate terminology includes turnover rate, transit time or minimum retention time, mean retention time, and maximum retention time. Turnover

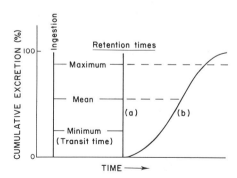

Fig. 14.1. Excretion characteristics of a meal fed to hypothetical animals in which (a) no mixing or pooling occurs as food moves through the gastrointestinal tract and (b) pooling and mixing occur.

usually refers to the rate at which food passes from either an individual gastrointestinal segment or the entire tract by both digestion and passage. Transit time is the time between feeding and first appearance of the marker in the feces. Mean retention time is the integrated, average time between marker ingestion and excretion. Transit time and mean retention time would be identical if mixing did not occur and marker excretion was complete at one defecation. As mixing occurs, mean retention time and transit time become increasingly different. Maximum retention time is the interval between feeding and the last excretion of the marker. Maximum retention time in animals with extensive gastrointestinal mixing is almost meaningless, because it incorporates both the time necessary for physical breakdown and, more importantly, simply the mathematical probability that the last particle has of being diluted and washed out of a specific gastrointestinal pool.

Transit time and mean retention time increase as animal weight and gastrointestinal complexity increase (Figs. 14.2–14.4). By far the most overriding determinants of passage kinetics are gastrointestinal specializations for the normal diet. For example, transit and mean retention times are longest in folivores, providing time for fiber digestion in the stomach or lower tract, and shortest in carnivores, frugivores, and granivores, where stomach and small intestine vertebrate enzymes are the predominant digestive agents.

In the mammalian carnivore studies, many of the authors either have not distinguished between transit and mean retention times or have quantified only transit time. While the difference in these two parameters in carnivores is in part the delay in stomach retention and digestion, some of the difference is undoubtedly the time necessary to accumulate sufficient bulk in the colon to stimulate defecation. The extent to which the latter is predominating minimizes the digestive importance of the time difference, even though some fermentive hydrolysis and volatile fatty acid absorption is occurring in the lower tract (Stevens, 1978; Clemens and Stevens, 1979). Granivore and frugivore passage kinetics can be

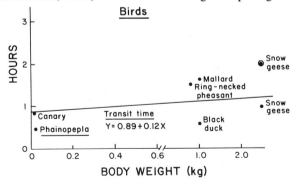

Fig. 14.2. Transit time (points) and mean retention time (circled points) in birds. (Data from Malone, 1965a, 1965b; Duke *et al.,* 1968; Grandy, 1972; Walsberg, 1975; Burton *et al.,* 1979.)

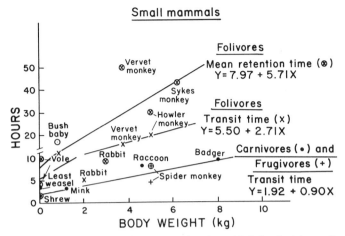

Fig. 14.3. Transit time (points) and mean retention times (circled points) in small mammals. (Data from Wood, 1956; Short, 1961; Sibbald *et al.*, 1962; Kostelecka-Myrcha and Myrcha, 1964a, 1964b, 1965; Clemens and Stevens, 1979; Clemens and Phillips, 1980; Bleavins and Aulerich, 1981; Clemens and Maloiy, 1981; Harlow, 1981; Milton, 1981; Van Soest, 1981.)

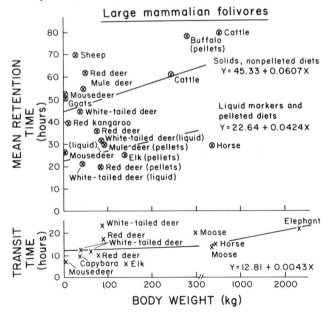

Fig. 14.4. Transit time (points) and mean retention time (circled points) in large herbivorous mammals. (Data from Gill, 1959, 1961; Foot and Romberg, 1965; Cowan *et al.*, 1970; Mautz, 1970; Mautz and Petrides, 1971; Thorne and Dean, 1973; Milchunas *et al.*, 1978; Milne *et al.*, 1978; Morat and Nordin, 1978; Parra, 1978; Schaefer *et al.*, 1978; Sanchez-Hermosillo and Kay, 1979; Van Soest, 1981; Westra, undated.) The buffalo (pellets) and horse are not included in the regressions.

more difficult to interpret if food remnants are retained and digested in the ceca or enlarged large intestine. For example, the ceca of pheasants and ptarmigan retained ingesta three to four times longer than was required for passage through the rest of the tract (Duke *et al.*, 1968; Gasaway *et al.*, 1975). The differences between transit and mean retention times are maximum in ruminants (Fig. 14.4).

Some of the variation in passage rates is due to the choice of markers, their association with different feed components, and different passage kinetics of specific feed components. Rarely can a single marker be used to describe the passage rate phenomena of all undigestible dry matter. For example, when mistletoe berries are fed to *Phainopepla* (a common, small passerine of arid and semiarid areas of Mexico and southwestern United States), the seeds and exocarp are excreted in a mirror-image sequence of each other. As a berry enters the gizzard, the seed and pulp are extruded into the duodenum by gizzard contractions, while the exocarp is retained. This sequence is repeated 8–16 times before the accumulated exocarps are passed into the duodenum. Since the exocarps are sandwiched in the gizzard in the exact order of berry entrance, the exocarp of the last berry to enter the gizzard is the first to move into the duodenum. Thus, the seed and exocarp of the first berry consumed in a sequence are respectively the first and last excreted (Walsberg, 1975). Similarly, markers associated with the slowly digesting fiber fed to ruminants pass much more slowly than soluble ones, which move with the gastrointestinal fluids. For example, soluble markers fed to ruminants move approximately twice as fast as particulate markers on long feed (Fig. 14.4). Lignin passes much more slowly than even a general particulate marker (Faichney, 1980). Thus, many passage rates are simultaneously occurring because of the mixing of different meals in ruminants and the different passage rates of the various chemical components of an individual meal (Poppi *et al.*, 1980).

Pelleted diets fed to elk, deer, and moose have mean retention times approximately equal to the liquid pool (Fig. 14.4) (Regelin *et al.*, 1981). The mechanical grinding of the pelleted diets has apparently reduced the particles to a size capable of passing out of the rumen as soon as the pellet disintegrates in these species. The difference in the regressions between the pelleted diets and the long forages (approximately 23 hr) is conceivably the time necessary to fragment the larger particles through rumination and microbial digestion to a size capable of passing out of the rumen–reticulum. The much longer retention of pellets in the buffalo (Fig. 14.4) and domestic cattle (Schaefer *et al.*, 1978) suggests a very different omasal or rumen filtering mechanism in the grazing animals than in the browsers. Less selective rumen retention by the browser would be advantageous because of the generally higher lignification and lower digestibility of browse cell walls than grass cell walls. Cecal retention is relatively less important in ruminants than in other groups because of the prolonged rumen retention times.

Cecal retention (3.8 hr) averaged 19% of rumen mean retention time (20.0 hr) in red deer (Milne *et al.*, 1978).

B. Digestion Versus Passage

Digestion and passage are interacting, competing processes. Increasing rates of passage will reduce digestibility if insufficient time is available for complete hydrolysis. For example, reduced fiber and dry-matter digestion in red kangaroos and red deer relative to domestic sheep were associated with faster rates of passage in the wild species (Foot and Romberg, 1965; Milne *et al.*, 1978). However, longer retention times in forestomach herbivores are useful only when consuming forage cell walls that are poorly lignified and therefore highly digestible. The prolonged retention of thoroughly lignified plant fiber cannot improve its digestibility (Van Soest, 1981; Mould and Robbins, 1982). Similarly, digestive efficiency in the carnivorous badger increased 10% when passage rate slowed by 18% (Harlow, 1981).

Thus, all animals must contend with an "efficiency–velocity" continuum in both the evolutionary and immediate time frame. For example, passage rates in the nonruminant horse are less than half those occurring in large ruminants (Fig. 14.4). When the two groups are consuming highly digestible forage diets, the reduced digestive efficiency in the horse is balanced by the ability to consume and pass more food per unit of time. However, since the ruminant must invest more time to process each unit of food, the ruminant tends to be more selective, choosing foods of higher digestibility than does the horse (Bell, 1971; Janis, 1976; Milton, 1981; Van Soest, 1981).

Because the ability of an animal to meet its requirements when physical capacity is limiting is a function of the rate of digestion, rate of passage, and volume of the limiting gastrointestinal compartment, very general relationships between these three parameters are expected. The capacity of several limiting gastrointestinal compartments, such as the rumen–reticulum of ruminants, cecum and large intestine of nonruminant herbivores, crops of hummingbirds, and cheek pouches of herbivorous heteromyid rodents, are linear functions of body weight (i.e., body weight to the power of 1) (Hainsworth and Wolf, 1972; Parra, 1978; Morton *et al.*, 1980; Van Soest, 1981). Correspondingly, energy expenditures are generally a function of metabolic body weight ranging from $W^{0.69}$ to $W^{0.75}$. Thus, the larger animal or species will be in a more favorable energetic state than the smaller species when low-quality food is abundant because more food can be gathered, transported, and processed relative to the energy requirements (Parra, 1978; Case, 1979; Morton *et al.*, 1980; Van Soest, 1981). Similarly, the difference between the exponents of capacity and requirements ($W^{1.0}/W^{0.75} = W^{0.25}$) suggests that the larger animal with the relatively

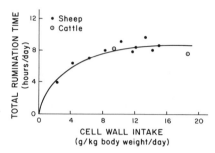

Fig. 14.5. Rumination time as a function of neutral detergent fiber intake. (Data from Welch and Smith, 1969a, 1969b, 1971; Bae *et al.*, 1979.)

slower rate of passage can afford to consume more fibrous, slower digesting forage than can the smaller species (Milton, 1979; Hanley, 1980). However, one must not carry these general relationships too far since, for example, rumen–reticulum capacity also varies markedly with the animal's feeding niche (i.e., browsers or concentrate selectors versus grazers).

Passage rate in ruminants generally increases as intake increases. Passage rate can increase only if rumination time or efficiency increases or the average particle size leaving the rumen increases. Rumination time increases curvilinearly as fiber intake increases (Fig. 14.5). Note that, at high levels of fiber intake, rumination requires 8–9 hr per day. Thus, it is certainly conceivable that the time requirement for rumination is a major determinant of foraging time and therefore of intake in these animals. Because of the curvilinearity in the relationship between total intake and rumination time, the amount of rumination per unit of fiber decreases as cell wall intake increases (Bae *et al.*, 1979). Since changes in the efficiency of rumination may be inadequate to compensate for the reduced rumination time per unit of fiber at higher intakes, fecal particle size often increases with increasing intake (Van Soest, 1981). Consequently, small reductions in fiber or dry-matter digestion can occur as intake increases (Mautz, 1970; Milchunas *et al.*, 1978; Westra and Hudson, 1981; Van Soest, 1981).

III. SYNTHESIS: INTAKE REGULATION

Intake regulation changes from primarily physical to physiological as food nutritive value increases (Fig. 14.6). At very low nutritive values, limited gastrointestinal capacities and passage rates prevent an animal from meeting its energy requirements. As nutritive value increases, the animal is ultimately able to ingest enough dry matter to meet its energy requirements. Once nutritive value is high enough to overcome physical limitations, physiological regulation main-

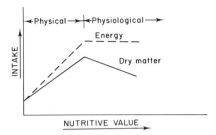

Fig. 14.6. Generalized relationship between nutritive value and intake. (From Montgomery and Baumgardt, 1965, courtesy of the *Journal of Diary Science.*)

tains a constant energy intake at increasing nutritive values by decreasing dry-matter intake.

Several examples supporting such a generalized theory of intake regulation are available. For example, hummingbirds increased meal size from 0.03 g to 0.16 g as nectar concentration decreased from 1.2 molar to 0.4 molar (Collins and Cary, 1981). Similarly, dry-matter intake in deer is a bell-shaped function decreasing on either side of approximately 2.2 kcal apparent digestible energy per gram dry matter (Fig. 14.7). Since many forage diets consumed by deer contain approximately 4.4 kcal/g of dry matter, intake is physically regulated in this example below approximately 50% apparent digestible energy and physiologically regulated above this level (Ammann *et al.,* 1973). However, the critical level of nutritive quality determining the switch from physical to physiological intake regulation should be an inverse function of body size and undoubtedly varies with animal requirements and digestive and metabolic capabilities.

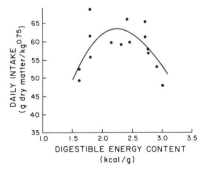

Fig. 14.7. Dry-matter intake as a function of digestible energy content in deer. (From Spalinger, 1980, courtesy of D. E. Spalinger, Department of Zoology, Washington State University, Pullman; data of Ammann *et al.,* 1973; Alldredge *et al.,* 1974; Wallmo *et al.,* 1977.)

IV. FORAGING STRATEGIES

The wildlife nutrition and ecology literature is replete with observations of food habits. However, only in recent years has the focus of these studies shifted from purely descriptive to analytical as attempts to understand the cause–effect basis of foraging are increased. It is increasingly recognized that food habits, foraging movement patterns, and energy and time expenditures are the integrated expressions of the animal's perception of its requirements relative to its nutritional environment (Schoener, 1971).

The optimal foraging strategy of an animal is determined by the simultaneous solution of various cost–benefit functions. For example, foraging costs may include energy and time expenditures for pursuit, handling, and ingestion of food, increased thermoregulatory costs, the risk of predation, reduced reproduction or territorial maintenance activities, or the potential consumption of toxic or inhibitory compounds. Ultimately, the solution of the cost–benefit functions are related to the biological fitness of the forager.

A. Food Selection

Food habits are commonly determined by direct observation of captive or wild animals; analyses of identifiable gastrointestinal or excretory fragments, such as the oral pellets of owls or the cuticular fecal remnants of herbivore diets (Fig. 14.8); or the measurement of utilization plots or transects (Korschgen, 1971). While it is rarely difficult to simply list foods ingested, quantification of the relative amounts ingested can be extremely difficult. For example, the commonly used fecal analysis for herbivore diets requires corrections for differences in digestion and identification (Davitt, 1979; Fitzgerald and Waddington, 1979).

Investigators compiling reported food habits, particularly for generalist herbivores, are often struck by the diversity of foods consumed. Food habits determined for a particular species–habitat interaction are potentially different from all other species–habitat interactions (Nudds, 1980). Thus, if one is simply interested in descriptions of food habits, an infinite number of studies are possible (i.e., N mammal and bird species $\times$ N past nutritional histories $\times$ N environments $\times$ . . .). Thus, we must recognize that, while data on food habits are important, listings of food habits in themselves are only a superficial facade of nutritional understanding. Knowledge and understanding of the animal–food–environment interactions determining the observed food habits must ultimately take precedence over ad infinitum listing.

Early attempts to explain food habits of herbivores revolved around correlations between the preference ranking of a food and its nutrient or chemical content. Significant correlations have been reported between preference and soluble carbohydrates, protein, plant fiber, ether extract, many minerals and vitamins, secondary plant compounds, and organic acids (Westoby, 1974). Such

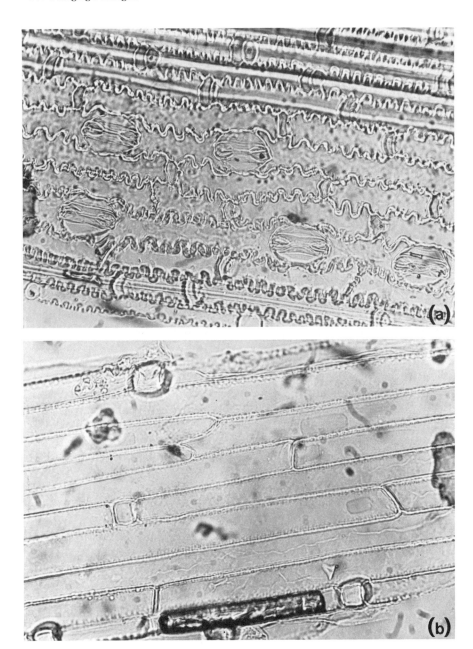

Fig. 14.8. Epidermal cells of (a) *Festuca idahoensis* and (b) *Koelaria cristata* as examples of the way in which microscopic fecal analyses can be used to determine herbivore food habits. (Courtesy of B. B. Davitt, Department of Forestry and Range Management, Washington State University, Pullman.)

correlations relative to herbivore food habits are almost impossible to evaluate since many plant chemical components are reflections of each other. For example, plant parts high in protein are often low in fiber and high in digestible dry matter and energy. Because of the molecular complexity of plants, it is not surprising that significant correlations between animal preference and different plant constituents are virtually as numerous as the number of authors attempting such analyses. Similarly, the statistical significance of correlations between preference and feed components are often area and time dependent. Thus, understanding the cause–effect basis of food habits is far more complex than significant, or even insignificant, correlations.

In further evaluations of food habits as a component of the overall foraging strategy, most investigations have used energy as the currency of the various cost–benefit functions (Schoener, 1971; Krebs and Cowie, 1976; Pyke et al., 1977; Belovsky, 1978; Hamilton and Busse, 1978; Hainsworth and Wolf, 1979; Powell, 1979; Stenseth and Hansson, 1979; Rau et al., 1981; Tinbergen, 1981). The preoccupation with energy occurs because it is a measurable requirement of all animals, it is often in limited supply, and the body is replete with sensory and regulatory systems for energy intake monitoring. Energy optimization models in which energy intake per unit of time is maximized have often predicted food habits that agreed quite closely with field observations. However, models based solely on energy considerations will be inadequate for many species unless such other nutrient constraints as protein and minerals, plant cell walls, toxins, and digestive inhibitors are included (Pulliam, 1975; Belovsky, 1978; Milton, 1979; Hanley, 1980; Oates et al., 1980; Farentinos et al., 1981; Jung and Batzli, 1981). When nutrient constraints and secondary plant compounds are included in foraging models, the complexity mushrooms because the value of each food item is no longer independent of all others, but rather each food item subsequently has an absolute value as well as a relative value that is dependent on the composition of all other dietary components (Ellis et al., 1976).

The selection of foods on the basis of their nutrient content requires a perception of one's requirements, an internal recognition of a food's value in meeting rather specific requirements, but not necessarily a taste or smell for each nutrient. For example, even though sodium is one of the few nutrients that is tasted, animals given free access in a cafeteria-style experiment to the 30 or 40 purified, required nutrients are able to select a diet that enables growth about equal to that of animals consuming balanced laboratory diets (Richter, 1955). Rozin (1976) has suggested that food selection is based on a categorization of foods as (a) novel and of unknown nutritional value; (b) familiar–dangerous; (c) familiar–safe; and (d) familiar–beneficial. For animals exposed to many different foods, the avoidance of toxins while meeting the necessary requirements could be accomplished by:

1. Treating new foods with extreme caution (i.e., neophobic).

2. Being able to learn quickly to eat or reject particular foods and needing to ingest only minute quantities to do so.
3. Having the capacity to seek out and eat plants containing highly specific classes of nutrients and to balance nutrient ingestion relative to the spectra of nutrient requirements.
4. Having to ingest a number of different staple foods over a short period of time and to indulge simultaneously in a continuous food sampling program.
5. Preferentially feeding on the foods with which they are familiar and continuing to feed on them for as long as possible.
6. Preferring to feed on foods that contain only minor amounts of toxic secondary plant compounds (Freeland and Janzen, 1974).

Thus, generalist plant consumers must continuously sample a broad spectrum of available foods in order to monitor changes in essential nutrients, fiber content, secondary plant compounds, and availability. It is not uncommon to see herbivore diets in which 90% of the ingesta is composed of 3 or 4 species capable of meeting all immediate nutrient requirements while in the remaining 10% are 10–15 species that presumably are being sampled.

B. Foraging Costs and Efficiencies

If feeding and the search for food are the predominant activities of most animals (Rozin, 1976), then activity patterns, daily energy expenditures, and overall fitness sbould reflect the costs and efficiencies with which food can be gathered. Foraging costs are most easily measured in units of time and energy. The efficient animal will minimize time and energy expenditures for food gathering while maximizing digestible energy intake. Since necessary minimum foraging times and energy expenditures are inverse curvilinear functions of food availability, foraging efficiency when defined as energy intake per unit of energy expenditure will increase as food availability increases (Norberg, 1977; F. B. Gill and Wolf, 1978; Homewood, 1978; Hainsworth and Wolf, 1979). At high levels of food availability, foraging efficiency becomes asymptotic, since food availability no longer limits intake per unit of foraging effort.

As food density decreases, foraging effort must increase. Although the animal can alter foraging strategies by selecting larger prey (Schoener, 1979; Kaufman et al., 1980; Kaufman, 1980), becoming less selective (Nudds, 1980), or choosing more favorable food patches in the total environment (MacArthur and Pianka, 1966; Charnov, 1976; Tinbergen, 1981), foraging efficiency must ultimately decrease as the animal is forced to expend more time and energy in acquiring the necessary food. The critical threshold of decreasing food density at which foraging effort must increase is a function of body weight. For example, limited food availability to a moose could easily be a superabundance to a herbivorous vole. Ultimately, food availability becomes so low that requirements cannot be met, because continued foraging would simply increase requirements faster than intake. At this point, animals can either hibernate, emigrate, or reduce foraging effort to conserve body reserves. The general patterns of intake and foraging effort are apparent in studies of elk and caribou (Fig. 14.9).

An inadequate understanding of foraging strategies has been a major weakness of many wildlife biologists. It has often been assumed that if a moderate amount of food is available, almost irrespective of its density or distribution, the animal will effectively consume it. A classical example of the error in this assumption can be seen each winter in Yellowstone National Park. The northern winter range in the park is about 100,000 ha and contains $120,918 \times 10^3$ kg of forage (Houston, 1976). Using daily consumption rates of captive elk to predict the requirements of free-ranging elk, Barmore (1972) suggested that a wintering population of 10,000 elk would consume only 5% of the forage and thus were not food limited. However, 47% of the winter range is composed of vegetation types having standing crops of 337 kg or less per hectare, and 54% had 673 kg or less per hectare. These forage measurements did not include the effect of snow (which at times can be several feet deep), very poor palatability and digestibility of many of the forages, and animal social interactions that can reduce the useful forage density to a fraction of that measured. To the contrary of Barmore's conclusion, forage availability relative to the energy requirements of elk living in an exceedingly harsh environment is the major determinant of elk survival in Yellowstone during winter.

A seemingly contradictory observation to this conclusion is that wintering elk spend less time feeding in the park during winter than during any other season (Craighead et al., 1973). While this observation might be interpreted as suggesting that elk are having no problem in meeting their food requirements, the more

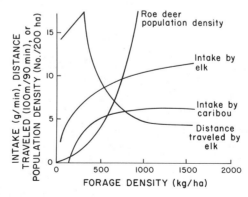

Fig. 14.9. Forage consumption and foraging effort in elk and caribou and population levels in roe deer. (Data from White et al., 1975; Bobek, 1977; Collins et al., 1978.)

appropriate conclusion is that they are simply in that portion of the foraging efficiency curve where nutrient availability is so low compared to foraging costs that the only feasible approach is to reduce costs and conserve limited body reserves (Gates and Hudson, 1979). As a further manifestation of the interaction between food availability and foraging strategies, roe deer population densities are a mirror image of the general forage density–forage intake curves (Fig. 14.9). Since foraging strategies are the animal's integration of its requirements relative to its perception of the food resources, additional studies in this area will have broad implications throughout the entire field of wildlife nutrition and management.

REFERENCES

Alldredge, A. W., Lipscomb, J. F., and Whicker, F. W. (1974). Forage intake rates of mule deer estimated with fallout cesium-137. *J. Wildl. Manage.* **38,** 508–516.

Ammann, A. P., Cowan, R. L., Mothershead, C. L., and Baumgardt, B. R. (1973). Dry matter and energy intake in relation to digestibility in white-tailed deer. *J. Wildl. Manage.* **37,** 195–201.

Andersson, B., and Larsson, S. (1961). Influence of local temperature changes in the preoptic area and the rostral hypothalamus on the regulation of food and water intake. *Acta Physiol. Scandinavica* **52,** 75–89.

Bae, D. H., Welch, J. G., and Smith, A. M. (1979). Forage intake and rumination by sheep. *J. Anim. Sci.* **49,** 1292–1299.

Baile, C. A., and McLaughlin, C. L. (1970). Feed intake of goats during volatile fatty acid injections into four gastric areas. *J. Dairy Sci.* **53,** 1058–1063.

Baile, C. A., and Martin, F. H. (1971). Hormones and amino acids as possible factors in the control of hunger and satiety in sheep. *J. Dairy Sci.* **54,** 897–906.

Bailey, J. A. (1968). Rate of food passage by caged cottontails. *J. Mammal.* **49,** 340–342.

Balgooyen, T. G. (1971). Pellet regurgitation by captive sparrow hawks (*Falco sparverius*). *Condor* **73,** 382–385.

Barmore, W. J. (1972). A computer simulation model of elk distribution, habitat use, and forage consumption on winter range in Yellowstone National Park. Yellowstone National Park, Mammoth.

Baumgardt, B. R. (1974). Control of feeding and the regulation of energy balance. *Fed. Proc.* **33,** 1139.

Bell, R. H. V. (1971). A grazing ecosystem in the Serengeti. *Sci. Am.* **225,** 86–93.

Belovsky, G. E. (1978). Diet optimization in a generalist herbivore: The moose. *Theor. Pop. Biol.* **14,** 105–134.

Below, T. H. (1979). First reports of pellet ejection in 11 species. *Wilson Bull.* **91,** 626–628.

Bleavins, M. R., and Aulerich, R. J. (1981). Feed consumption and food passage time in mink (*Mustela vison*) and European ferrets (*Mustela putorius furo*). *Lab. Anim. Sci.* **31,** 268–269.

Bobek, B. (1977). Summer food as the factor limiting roe deer population size. *Nature* **268,** 47–49.

Booth, D. A. (1968). Effects of intrahypothalamic glucose injection on eating and drinking elicited by insulin. *J. Comp. Physiol. Psychol.* **65,** 13.

Bray, G. A. (1974). Endocrine factors in the control of food intake. *Fed. Proc.* **33,** 1140–1145.

Burton, B. A., Hudson, R. J., and Bragg, D. D. (1979). Efficiency of utilization of bulbrush rhizomes by lesser snow geese. *J. Wildl. Manage.* **43,** 728–735.

Case, T. J. (1979). Optimal body size and an animal's diet. *Acta Biotheor.* **28**, 54–69.

Charnov, E. L. (1976). Optimal foraging: The marginal value theorem. *Theor. Pop. Biol.* **9**, 129–136.

Clemens, E. T., and Maloiy, G. M. O. (1981). Organic acid concentrations and digesta movement in the gastrointestinal tract of the bushbaby (*Galago crassicaudatus*) and Vervet monkey (*Cercopithecidae pygerythrus*). *J. Zool.* **193**, 487–497.

Clemens, E., and Phillips, B. (1980). Organic acid production and digesta movement in the gastrointestinal tract of the baboon and Sykes monkey. *Comp. Biochem. Physiol.* **66A**, 529–532.

Clemens, E. T., and Stevens, C. E. (1979). Sites of organic acid production and patterns of digesta movement in the gastrointestinal tract of the raccoon. *J. Nutr.* **109**, 1110–1117.

Collins, B., and Cary, G. (1981). Short-term regulation of food intake by the brown honeyeater, *Lichmera indistincta*. *Comp. Biochem. Physiol.* **68A**, 635–640.

Collins, W. B., Urness, P. J., and Austin, D. D. (1978). Elk diets and activities on different lodgepole pine habitat segments. *J. Wildl. Manage.* **42**, 799–810.

Cowan, R. L., Jordan, J. S., Grimes, J. L., and Gill, J. D. (1970). Comparative nutritive values of forage species. *In* "Range and Wildlife Habitat Evaluation-A Research Symposium," pp. 48–56. USDA Misc. Publ. 1147.

Craighead, J. J., Craighead, F. C., Jr., Ruff, R. L., and O'Gara, B. W. (1973). Home ranges and activity patterns of nonmigratory elk of the Madison drainage herd as determined by biotelemetry. *Wildl. Monogr.* **33.**

Davitt, B. B. (1979). Elk summer diet composition and quality on the Colockum multiple use research unit, central Washington. Master's thesis, Washington St. Univ., Pullman.

Dawson, N. J. (1972). Rate of passage of a non-absorbable marker through the gastrointestinal tract of the mouse (*Mus musculus*). *Comp. Biochem. Physiol.* **41A**, 877–881.

Dean, R. E., Thorne, E. T., and Moore, T. D. (1980). Passage rate of alfalfa through the digestive tract of elk. *J. Wildl. Manage.* **44**, 272–273.

Dinius, D. A., Kavanaugh, J. F., and Baumgardt, B. R. (1970). Regulation of food intake in ruminants. 7. Interrelations between food intake and body temperature. *J. Dairy Sci.* **53**, 438–445.

Dowden, D. R., and Jacobson, D. R. (1960). Inhibition of appetite in dairy cattle by certain intermediate metabolites. *Nature* **188**, 148–149.

Duke, G. E. (1977). Pellet egestion by a captive chimney swift (*Chaetura pelagica*). *Auk* **94**, 385.

Duke, G. E., Petrides, G. A., and Ringer, R. K. (1968). Chromium-51 in food metabolizability and passage rate studies with the ring-necked pheasant. *Poult. Sci.* **47**, 1356–1364.

Duke, G. E., Evanson, O. A., and Jegers, A. (1976). Meal to pellet intervals in 14 species of captive raptors. *Comp. Biochem. Physiol.* **53A**, 1–6.

Duke, G. E., Fuller, M. R., and Huberty, B. J. (1980). The influence of hunger on meal to pellet intervals in barred owls. *Comp. Biochem. Physiol.* **66A**, 203–207.

Ellis, J. E., Wiens, J. A., Rodell, C. F., and Anway, J. C. (1976). A conceptual model of diet selection as an ecosystem process. *J. Theor. Biol.* **60**, 93–108.

Emlen, J. M. (1966). The role of time and energy in food preference. *Am. Nat.* **100**, 611–617.

Faichney, G. J. (1980). Measurement in sheep of the quantity and composition of rumen digesta and of the fractional outflow rates of digesta constituents. *Aust. J. Agric. Res.* **31**, 1129–1137.

Farentinos, R. C., Capretta, P. J., Kepner, R. E., and Littlefield, V. M. (1981). Selective herbivory in tassel-eared squirrels: Role of monoterpenes in Ponderosa pines chosen as feeding trees. *Science* **213**, 1273–1275.

Fitzgerald, A. E., and Waddington, D. C. (1979). Comparison of two methods of fecal analysis of herbivore diet. *J. Wildl. Manage.* **43**, 468–473.

Foot, J. Z., and Romberg, B. (1965). The utilization of roughage by sheep and the red kangaroo, *Macropus rufus* (Desmarest). *Aust. J. Agric. Res.* **16**, 429–435.

Freeland, W. J., and Janzen, D. H. (1974). Strategies in herbivory by mammals: The role of plant secondary compounds. *Am. Nat.* **108**, 269–289.

Friedman, M. I., and Stricker, E. M. (1976). The physiological psychology of hunger: A physiological perspective. *Psychol. Rev.* **83**, 409–431.

Fuller, M. R., and Duke, G. E. (1979). Regulation of pellet egestion: The effects of multiple feedings on meal to pellet intervals in great horned owls. *Comp. Biochem. Physiol.* **62A**, 439–444.

Fuller, M. R., Duke, G. E., and Eskedahl, D. L. (1979). Regulation of pellet egestion: The influence of feeding time and soundproof conditions on meal to pellet intervals of red-tailed hawks. *Comp. Biochem. Physiol.* **62A**, 433–438.

Gasaway, W. C., Holleman, D. F., and White, R. G. (1975). Flow of digesta in the intestine and cecum of the rock ptarmigan. *Condor* **77**, 467–474.

Gates, C. C., and Hudson, R. J. (1979). Effects of posture and activity on metabolic responses of wapiti to cold. *J. Wildl. Manage.* **43**, 564–567.

Gill, F. B., and Wolf, L. L. (1978). Comparative foraging efficiencies of some montane sunbirds in Kenya. *Condor* **80**, 391–400.

Gill, J. (1959). The rate of passage of foodstuffs through the alimentary tract of the elk. *Int. Union Game Biol.* **4**, 155–164.

Gill, J. (1961). The rate of passage of foodstuffs through the alimentary tract of red deer (*Cervus elaphus* L.) and some physical characteristics of the feces. *Int. Union Game Biol.* **5**, 161–174.

Grandy, J. W., IV. (1972). Digestion and passage of blue mussels eaten by black ducks. *Auk* **89**, 189–190.

Hainsworth, F. R., and Wolf, L. L. (1972). Crop volume, nectar concentration, and hummingbird energetics. *Comp. Biochem. Physiol.* **42**, 359–366.

Hainsworth, F. R., and Wolf, L. L. (1979). Feeding: An ecological approach. *Adv. Study Behav.* **9**, 53–96.

Hamilton, W. J., III., and Busse, C. D. (1978). Primate carnivory and its significance to human diets. *BioScience* **28**, 761–766.

Hanley, T. A. (1980). Nutritional constraints on food and habitat selection by sympatric ungulates. Doctoral dissertation, Univ. of Washington, Seattle.

Harlow, H. J. (1981). Effect of fasting on rate of food passage and assimilation efficiency in badgers. *J. Mammal.* **62**, 173–177.

Hartnell, G. F., and Satter, L. D. (1979). Determination of rumen fill, retention time and ruminal turnover rates of ingesta at different stages of lactation in dairy cows. *J. Anim. Sci.* **48**, 381–392.

Holder, J. M. (1963). Chemostatic regulation of appetite in sheep. *Nature* **200**, 1074–1075.

Homewood, K. M. (1978). Feeding strategy of the Tana mangabey (*Cercocebus galeritus galeritus*) (Mammalia: Primates). *J. Zool.* **186**, 375–391.

Houston, D. B. (1976). The Northern Yellowstone Elk. Parts III and IV. Vegetation and Habitat Relations. Yellowstone Nat. Park, Wyoming.

Janis, C. (1976). The evolutionary strategy of the Equidae and the origins of rumen and cecal digestion. *Evolution* **30**, 757–774.

Jung, H. G., and Batzli, G. O. (1981). Nutritional ecology of microtine rodents: Effects of plant extracts on the growth of Arctic microtines. *J. Mammal.* **62**, 286–292.

Kaufman, L. W. (1980). Foraging cost and meal patterns in ferrets. *Physiol. Behav.* **25**, 139–141.

Kaufman, L. W., Collier, G., Hill, W. L., and Collins, K. (1980). Meal cost and meal patterns in an uncaged domestic cat. *Physiol. Behav.* **25**, 135–137.

Korschgen, L. J. (1971). Procedures for food-habits analyses. *In* "Wildlife Management Techniques," 3rd ed. (R. H. Giles, Jr., ed.), pp. 233–250. Wildl. Soc., Washington, D.C.

Kostelecka-Myrcha, A., and Myrcha, A. (1964a). The rate of passage of foodstuffs through the alimentary tracts of certain *Microtidae* under laboratory conditions. *Acta Theriol.* **9**, 37–53.

Kostelecka-Myrcha, A., and Myrcha, A. (1964b). Rate of passage of foodstuffs through the alimentary tract of *Neomys fodiens* (Pennant, 1771) under laboratory conditions. *Acta Theriol.* **9,** 371–373.

Kostelecka-Myrcha, A., and Myrcha, A. (1965). Effect of the kind of indicator on the results of investigations of the rate of passage of foodstuffs through the alimentary tract. *Acta Theriol.* **10,** 229–232.

Krebs, J. R., and Cowie, R. J. (1976). Foraging strategies in birds. *Ardea* **64,** 98–116.

MacArthur, R. H., and Pianka, E. R. (1966). On optimal use of a patchy environment. *Am. Nat.* **100,** 603–609.

Malone, C. R. (1965a). Dispersal of plankton: Rate of food passage in mallard ducks. *J. Wildl. Manage.* **29,** 529–533.

Malone, C. R. (1965b). Marking food for studying its rate of passage in poultry. *Poult. Sci.* **44,** 298–299.

Mautz, W. W. (1970). Digestibility and rate of passage of fiber in winter feeds by deer. *New Hampshire Fish Game Dept., Proj. W-51-R-4.*

Mautz, W. W., and Petrides. G. A. (1971). Food passage rate in the white-tailed deer. *J. Wildl. Manage.* **35,** 723–731.

Mayer, J. (1955). Regulation of energy intake and body weight: The glucostatic theory and the lipostatic hypothesis. *Ann. N.Y. Acad. Sci.* **63,** 15–43.

Milchunas, D. G., Dyer, M. I., Wallmo, O. C., and Johnson, D. E. (1978). *In vivo/in vitro* relationships of Colorado mule deer forages. *Colo. Div. Wildl. Spec. Rep. No.* **43.**

Milne, J. A., Macrae, J. C., Spence, A. M., and Wilson, S. (1978). A comparison of voluntary intake and digestion of a range of forages at different times of the year by the sheep and the red deer (*Cervus elaphus*). *Brit. J. Nutr.* **40,** 347–357.

Milton, K. (1979). Factors influencing leaf choice by howler monkeys: A test of some hypotheses of food selection by generalist herbivores. *Am. Nat.* **114,** 362–378.

Milton, K. (1981). Food choice and digestive strategies of two sympatric primate species. *Am. Nat.* **117,** 496–505.

Montgomery, M. J., and Baumgardt, B. R. (1965). Regulation of food intake in ruminants. 1. Pelleted rations varying in energy concentration. *J. Dairy Sci.* **48,** 569–574.

Morat, P., and Nordin, M. (1978). Maximum food intake and passage of markers in the alimentary tract of the lesser mousedeer. *Malaysia Appl. Biol.* **7,** 11–17.

Morton, S. R., Hinds, D. S., and MacMillen, R. E. (1980). Cheek pouch capacity in heteromyid rodents. *Oecologia* **46,** 143–146.

Mould, E. D., and Robbins, C. T. (1982). Digestive capabilities in elk compared to white-tailed deer. *J. Wildl. Manage.* **46,** 22–29.

Norberg, R. A. (1977). An ecological theory on foraging time and energetics and choice of optimal food-searching method. *J. Anim. Ecol.* **46,** 511–529.

Nudds, T. D. (1980). Forage preference: Theoretical considerations of diet selection by deer. *J. Wildl. Manage.* **44,** 735–740.

Oates, J. F., Waterman, P. G., and Choo, G. M. (1980). Food selection by the South Indian leaf-monkey, *Presbytis johnii,* in relation to leaf chemistry. *Oecologia* **45,** 45–46.

Panksepp, J. (1974). Hypothalamic regulation of energy balance and feeding behavior. *Fed. Proc.* **33,** 1150–1165.

Parra, R. (1978). Comparison of foregut and hindgut fermentation in herbivores. *In* "The Ecology of Arboreal Folivores" (G. G. Montgomery, ed.), pp. 205–229. Smithsonian Inst. Press, Washington, D.C.

Poppi, D. P., Norton, B. W., Minson, D. J., and Hendricksen, R. E. (1980). The validity of the critical size theory for particles leaving the rumen. *J. Agric. Sci.* **94,** 275–280.

Powell, R. A. (1979). Ecological energetics and foraging strategies of the fisher (*Martes pennanti:*). *J. Anim. Ecol.* **48,** 195–212.

Pulliam, H. R. (1975). Diet optimization with nutrient constraints. *Am. Nat.* **109**, 765–768.

Pyke, G. H., Pulliam, H. R., and Charnov, E. L. (1977). Optimal foraging: A selective review of theory and tests. *Quart. Rev. Biol.* **52**, 137–154.

Rau, J., Murua, R., and Rosenmann, M. (1981). Bioenergetics and food preferences in sympatric southern Chilean rodents. *Oecologia* **50**, 205–209.

Regelin, W. L., Schwartz, C. C., and Franzmann, A. W. (1981). Energy expenditure of moose on the Kenai National Wildlife Refuge. *Kenai Field St. Ann. Rep., Kenai, Alaska.*

Richter, C. P. (1955). Self-regulatory functions during gestation and lactation. *Trans. Conf. Gestation* **2**, 11–93.

Rozin, P. (1976). The selection of foods by rats, humans, and other animals. *Adv. Study Behav.* **6**, 21–76.

Sanchez-Hermosillo, M., and Kay, R. N. B. (1979). Retention time and digestibility of milled hay in sheep and red deer (*Cervus elaphus*). *Proc. Nutr. Soc.* **38**, 123A.

Schaefer, A. L., Young, B. A., and Chimwano, A. M. (1978). Ration digestion and retention times of digesta in domestic cattle (*Bos taurus*), American bison (*Bison bison*), and Tibetan yak (*Bos grunniens*). *Can. J. Zool.* **56**, 2355–2358.

Schoener, T. W. (1971). Theory of feeding strategies. *Ann. Rev. Ecol. Syst.* **2**, 369–404.

Schoener, T. W. (1979). Generality of the size-distance relation in models of optimal feeding. *Am. Nat.* **114**, 902–914.

Short, H. L. (1961). Food habits of a captive least weasel. *J. Mammal.* **42**, 273–274.

Sibbald, I. R., Sinclair, D. G., Evans, E. V., and Smith, D. L. T. (1962). The rate of passage of feed through the digestive tract of the mink. *Can. J. Biochem. Physiol.* **40**, 1391–1394.

Smith, C. R., and Richmond, M. E. (1972). Factors influencing pellet egestion and gastric pH in the barn owl. *Wilson Bull.* **84**, 179–186.

Smith, G. P., Gibbs, J., and Young, R. C. (1974). Cholecystokinin and intestinal satiety in the rat. *Fed. Proc.* **33**, 1146–1149.

Spalinger, D. E. (1980). Mule deer habitat evaluation based upon nutritional modeling. Master's thesis, Univ. of Nevada, Reno.

Stenseth, N. C., and Hansson, L. (1979). Optimal food selection: A graphic model. *Am. Nat.* **113**, 373–389.

Stevens, C. E. (1978). Physiological implications of microbial digestion in the large intestine of mammals: Relation to dietary factors. *Am. J. Clin. Nutr.* **31**, S161–S168.

Strominger, J. L., and Brobeck, J. R. (1953). A mechanism of regulation of food intake. *Yale J. Biol. Med.* **25**, 383–390.

Thorne, T., and Dean, R. (1973). Evaluation of pelleted alfalfa hay as winter feed for elk. *Wyoming Game Fish Dept., Proj. FW-3-R20.*

Tinbergen, J. M. (1981). Foraging decisions in starlings (*Sturnus vulgaris* L.). *Ardea* **69**, 1–67.

Van Soest, P. (1981). Nutritional ecology of the ruminant. O and B Books, Corvallis, Oreg.

Wallmo, O. C., Carpenter, L. H., Regelin, W. L., Gill, R. B., and Baker, D. L. (1977). Evaluation of deer habitat on a nutritional basis. *J. Range Manage.* **30**, 122–127.

Walsberg, G. E. (1975). Digestive adaptations of *Phainopepla nitens* associated with the eating of mistletoe berries. *Condor* **77**, 169–174.

Welch, J. G., and Smith, A. M. (1969a). Influence of forage quality on rumination time in sheep. *J. Anim. Sci.* **28**, 813–818.

Welch, J. G., and Smith, A. M. (1969b). Effects of varying amount of forage intake on rumination. *J. Anim. Sci.* **28**, 827–830.

Welch, J. G., and Smith, A. M. (1971). Effect of beet pulp and citrus pulp on rumination activity. *J. Anim. Sci.* **33**, 472–475.

Westoby, M. (1974). An analysis of diet selection by large generalist herbivores. *Am. Nat.* **108**, 290–304.

Westoby, M. (1978). What are the biological bases for varied diets? *Am. Nat.* **112**, 627–631.

Westra, R. (undated). Management of captive mule deer. Department of Animal Science, Univ. of Alberta, Edmonton.

Westra, R., and Hudson, R. J. (1981). Digestive function in wapiti calves. *J. Wildl. Manage.* **45,** 148–155.

White, R. G., Thomson, B. R., Skogland, T., Person, S. J., Russell, D. E., Holleman, D. F., and Luick, J. R. (1975). Ecology of caribou at Prudhoe Bay, Alaska. *Univ. Alaska Biol. Pap. Special Rep. No.* **2.**

Wood, A. J. (1956). Time of passage of food in mink. *Black Fox Magazine and Modern Mink Breeder* **39,** 12–13.

15

Computer Models of the Nutritional Interaction

Because of the multiplicity of factors involved, the possible variations in each factor and the interactions among the parameters, a synthesis of the whole process to show energy [or matter] flow through a population . . . , the consequent impact on the ecosystem, (and the possible effects of various management strategies) can best be accomplished through use of computer simulation models.

—Kendeigh et al., 1977, p. 343

The applications of a knowledge of wildlife nutrition are very broad and quite basic to understanding many facets of animal ecology. Many field studies of the interaction between wildlife populations and their food resources have been conducted. Only through understanding of the animal and its interaction with the ecosystem is manipulation and control of the system possible. Although a comprehensive review of these studies is beyond the role of this text, several general findings indicating the importance of the nutritional interaction in ecosystem functioning are warranted.

Although small mammals and birds generally consume less than 5% of the primary productivity of temperate and tropical ecosystems, selective feeding on seeds and seedlings can alter plant distribution, abundance, and form (Fleming, 1975; Golley et al., 1975; French et al., 1976; Wiens and Dyer, 1977; Wiens, 1977; Schitoskey and Woodmansee, 1978; Reichman, 1979). Very significant consumption rates and impacts by small mammals can occur in less complex ecosystems, such as arctic ecosystems, agricultural fields, and forestry plantations (Batzli, 1975; Golley et al., 1975; Grodzinski et al., 1977; Wiens and Dyer, 1977). For example, lemmings protected by snow cover consumed or destroyed 30–100% of the mosses and grasses in several arctic ecosystems (Batzli, 1975). Large herbivorous mammals also can be significant ecosystem components through both consumption and destruction, such as by trampling, of plant biomass (Buechner, 1952; Pegau, 1970; Constan, 1972; Sinclair, 1974; Caughley, 1976). For example, wildebeest consumed 90% of all grass biomass in the Serengeti (Sinclair, 1974).

Predation by insectivores and carnivores can also have significant impacts (Wiens, 1977; Wiens and Dyer, 1977; Caughley *et al.*, 1980). For example, long-billed marsh wrens and red-winged blackbirds may consume 30–50% of the insect standing crop (Kale, 1965; Brenner, 1968). Similarly, sea bird communities were estimated to have consumed 22–29% of the annual fish production in the neritic zone (Wiens and Scott, 1975; Furness, 1978). Thus, the ecosystem effect of animal food consumption can range from minor or subtle to a major determinant of energy and matter flow.

Paradoxically, it is often very difficult to demonstrate clearly food limitations in many animal populations (Newton, 1980). It certainly would be incorrect to suggest that food is the only factor controlling animal populations. While food is important and regulates many populations, predation, social interactions, disease, pesticides and pollutants, and competition for other resources, such as nest sites, can be equally important.

An intimate understanding of ecosystem functioning relative to the nutritional interaction, even in the most simple system, often must utilize computer modeling as a tool because of the very complex, dynamic, and subtle nature of living systems. Computers provide an essential tool for physically processing the data and equations necessary to evaluate a complex set of nutritional interactions. The development of a computer model follows a logical sequence in which the conceptual idea is transferred from the investigator's thoughts to a mathematical model appropriate for computer interfacing (Fig. 15.1). The mechanistic model describing the efficiencies and magnitude of energy or matter flow within each nutritional process is developed from the types of data discussed in preceding chapters. Although the individual bits of information are often viewed as being generated from basic research, their compilation and use in applied analyses is relevant to such diverse topics as diet formulation for captive animals and the evaluation of various field management programs. The computer model must be continuously refined as goals are redefined and critical experiments are conducted to provide more specific information or to test the model's biological validity.

Computer models, even if in a very simplified form relative to our recognition of the complexity of living systems, can be extremely helpful via component or

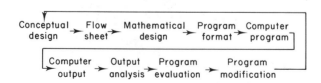

Fig. 15.1. Steps in system modeling. (From *Wildlife Ecology: An Analytical Approach*, by Aaron N. Moen. W. H. Freeman and Company. Copyright © 1973.)

error analyses in indicating the net effect of variability in any single parameter (Furness, 1978). Such analyses can indicate where more precision in understanding is needed or where existing information, irrespective of the error or variability associated with the measurement, is adequate. Modeling efforts and sensitivity analyses are essential if we are to avoid the innumerable studies whose goal is merely to "add another brick to the temple of science [Platt, 1964, p. 351]" irrespective of their value to increasing our collective understanding of wildlife nutrition and ecology.

Several computer models have been developed that incorporate current nutritional knowledge of specific wildlife (Moen, 1973; Wiens and Scott, 1975; French *et al.*, 1976; Grodzinski *et al.*, 1977; Wiens, 1977; Furness, 1978; Innis, 1978; Swift *et al.*, 1980; Barkley *et al.*, 1980). In most of these models, the various nutritional requirements and digestive and metabolic efficiencies interact to estimate food intake. Rather than offering definitive answers, many of the models have reemphasized the poorly understood, immense complexity of the natural world. While this may be frustrating to the individual investigator, continued progress in understanding wildlife nutrition will be immensely rewarding relative to understanding the productivity of wildlife populations and implementing meaningful management programs.

REFERENCES

Barkley, S. A., Batzli, G. O., and Collier, B. D. (1980). Nutritional ecology of microtine rodents: A simulation model of mineral nutrition for brown lemmings. *Oikos* **34**, 103–114.

Batzli, G. O. (1975). The role of small animals in arctic ecosystems. *In* "Small Mammals: Their Productivity and Population Dynamics" (F. B. Golley, K. Petrusewicz, and L. Ryszkowski, eds.), pp. 243–268. Cambridge Univ. Press, London.

Brenner, F. J. (1968). Energy flow in two breeding populations of red-winged blackbirds. *Am. Midl. Nat.* **79**, 289–310.

Buechner, H. K. (1952). Winter-range utilization by elk and mule deer in southeastern Washington. *J. Range Manage.* **5**, 76–80.

Caughley, G. (1976). Wildlife management and the dynamics of ungulate populations. *Appl. Biol.* **1**, 183–246.

Caughley, G., Grigg, G. C., Caughley, J., and Hill, G. J. E. (1980). Does dingo predation control densities of kangaroos and emus? *Aust. Wildl. Res.* **7**, 1–12.

Constan, K. J. (1972). Winter foods and range use of three species of ungulates. *J. Wildl. Manage.* **36**, 1068–1076.

Fleming, T. H. (1975). The role of small mammals in tropical ecosystems. *In* "Small Mammals: Their Productivity and Population Dynamics" (F. B. Golley, K. Petrusewicz, and L. Ryszkowski, eds.), pp. 269–298. Cambridge Univ. Press, London.

French, N. R., Grant, W. E., Grodzinski, W., and Swift, D. M. (1976). Small mammal energetics in grassland ecosystems. *Ecol. Monogr.* **46**, 201–220.

Furness, R. W. (1978). Energy requirements of seabird communities: A bioenergetics model. *J. Anim. Ecol.* **47**, 39–53.

Golley, F. B., Ryszkowski, L., and Sokur, J. T. (1975). The role of small mammals in temperate forests, grasslands and cultivated fields. *In* "Small Mammals: Their Productivity and Population Dynamics" (F. B. Golley, K. Petrusewicz, and L. Ryszkowski, eds.), pp. 223–241. Cambridge Univ. Press, London.

Grodzinski, W., Makomaska, M., Tertil, R., and Weiner, J. (1977). Bioenergetics and total impact of vole populations. *Oikos* **29**, 494–510.

Innis, G. S. (Ed.). (1978). "Grassland Simulation Model." Springer-Verlag, New York.

Kale, H. W., II. (1965). Ecology and bioenergetics of the long-billed marsh wren in Georgia salt marshes. *Publ. Nuttall Ornith. Club No.* **5.**

Kendeigh, S. C., Wiens, J. A., and Pinowski, J. (1977). Epilogue. *In* "Granivorous Birds in Ecosystems" (J. Pinowski and S. C. Kendeigh, eds.), pp. 341–344. Cambridge Univ. Press, London.

Moen, A. N. (1973). "Wildlife Ecology: An Analytical Approach." Freeman, San Francisco.

Newton, I. (1980). The role of food in limiting bird numbers. *Ardea* **68**, 11–30.

Pegau, R. E. (1970). Effect of reindeer trampling and grazing on lichens. *J. Range Manage.* **23**, 95–97.

Platt, J. R. (1964). Strong inference. *Science* **146**, 347–353.

Reichman, O. J. (1979). Desert granivore foraging and its impact on seed densities and distributions. *Ecology* **60**, 1085–1092.

Schitoskey, F., Jr., and Woodmansee, S. R. (1978). Energy requirements and diet of the California ground squirrel. *J. Wildl. Manage.* **42**, 373–382.

Sinclair, A. R. E. (1974). The natural regulation of buffalo populations in East Africa. IV. The food supply as a regulating factor, and competition. *E. Afr. Wildl. J.* **12**, 291–311.

Swift, D. M., Ellis, J. E., and Hobbs, N. T. (1980). Nitrogen and energy requirements of North American cervids in winter—A simulation study. *Proc. Int. Reindeer/ Caribou Symp.* **2**, 244–251.

Wiens, J. A. (1977). Model estimation of energy flow in North American grassland bird communities. *Oecologia* **31**, 135–151.

Wiens, J. A., and Scott, J. M. (1975). Model estimation of energy flow in Oregon coastal seabird populations. *Condor* **77**, 439–452.

Wiens, J. A., and Dyer, M. I. (1977). Assessing the potential impact of granivorous birds in ecosystems. *In* "Granivorous Birds in Ecosystems" (J. Pinowski and S. C. Kendeigh, eds.), pp. 205–266. Cambridge Univ. Press, London.

Index